FRICTION STIR WELDING AND PROCESSING VII

TMS2013

142nd Annual Meeting & Exhibition

FRICTION STIR WELDING AND PROCESSING VII

Proceedings of a symposium sponsored by
the Shaping and Forming Committee of
the Materials Processing & Manufacturing Division of
TMS (The Minerals, Metals & Materials Society)

Held during the
TMS 2013 Annual Meeting & Exhibition
San Antonio, Texas, USA
March 3-7, 2013

Edited by

Rajiv Mishra
Murray W. Mahoney
Yutaka Sato
Yuri Hovanski
Ravi Verma

Editors
Rajiv Mishra
Murray W. Mahoney
Yutaka Sato
Yuri Hovanski
Ravi Verma

ISBN 978-3-319-48582-9 ISBN 978-3-319-48108-1 (eBook)
DOI 10.1007/978-3-319-48108-1

Chemistry and Materials Science: Professional

Published by Springer International Publishers, Switzerland, 2016
Reprint of the original edition published by John Wiley & Sons, Inc., 2013, 978-1-11860-578-3

Printed on acid-free paper

This Springer imprint is published by Springer Nature
The registered company is Springer International Publishing AG
The registered company address is: Gewerbestrasse 11, 6330 Cham, Switzerland

TABLE OF CONTENTS
Friction Stir Welding and Processing VII

Friction Stir Processing

Friction Stir Welding: High Temperature Materials - I

Friction Stir Welding: High Temperature Materials - II

Friction Stir Welding: Light Materials - I

Friction Stir Welding: Light Materials - II

Friction Stir Welding and Processing: Modeling and Controls

Editors

Rajiv Mishra is a Professor of Materials Science and Engineering in the Department of Materials Science and Engineering at the University of North Texas. He is also the UNT Site Director of the NSF I/UCRC for Friction Stir Processing and a Fellow of ASM International. His highest degree is Ph.D. in Metallurgy from the University of Sheffield, United Kingdom (1988). He has received a number of awards which include the Firth Pre-doctoral Fellowship from the University of Sheffield, the Brunton Medal for the best Ph.D. dissertation in the School of Materials from the University of Sheffield in 1988, the Young Metallurgist Award from the Indian Institute of Metals in 1993, Associate of the Indian Academy of Sciences in 1993, and the Faculty Excellence Awards from the University of Missouri-Rolla in 2001, 2002, 2003, 2004, 2005, 2006, and 2007. He has authored or co-authored 230 papers in peer-reviewed journals and proceedings and is principal inventor of four U.S. patents. His current publication-based h-index is 38 and his papers have been cited more than 5,400 times. He has co-edited a book on friction stir welding and processing, and edited or co-edited 13 TMS conference proceedings. He is the in-coming chair of the SMD and will serve on the TMS Board of Directors beginning March 2013 as the TMS Structural Materials Director. He serves on the editorial board of *Materials Science and Engineering A*, *Science and Technology of Welding and Joining*, and *Advances in Materials Science and Engineering*.

Murray W. Mahoney: Prior, Manager/Senior Scientist, Structural Metals Department, Rockwell Scientific; B.S., physical metallurgy, UC Berkeley; M.S., physical metallurgy, UCLA. Mr. Mahoney has over 44 years experience in physical metallurgy and related disciplines. Most recently his work has centered on the development of friction stir welding (FSW) and friction stir processing (FSP), a thermomechanical metal working process to control the microstructure in structural alloys to enhance specific properties for aerospace, marine, automotive, and oil and gas applications. This work has led to the introduction of FSW to join materials considered unweldable, and FSP to improve superplasticity in structural alloys, enhance mechanical properties, improve room temperature formability, and reduce corrosion sensitivity. These studies have resulted in a more complete understanding of microstructural evolution during friction stir joining and processing. The primary research emphasis has been to improve manufacturing efficiency while both reducing fabrication costs and producing lighter weight structures that perform better in their environment. Previously, Mr. Mahoney has worked in superplasticity, solid state joining, nondestructive evaluation, corrosion, fatigue, and creep in structural metals. Mr. Mahoney has authored or co-authored more than 110 published papers, many on FSW and FSP, and has been awarded 21 United States patents. Further, Mr. Mahoney has organized and hosted a number of FSW/FSP symposia including the first International FSW symposium, and participated as co-editor and author on the first reference book on friction stir welding and processing. Mr. Mahoney was honored as the Rockwell Scientific Technologist of the year in 2001.

Yutaka Sato is currently a Associate Professor in the Department of Materials Processing at Tohoku University, Japan. He earned a Ph.D. in Materials Processing at Tohoku University (2001). His Ph.D. thesis was titled "Microstructural Study on Friction Stir Welds of Aluminum Alloys." He participated in friction stir research of steels at Brigham Young University for a year in 2003. He is a member of Sub-commission III-B WG-B4 at IIW, which is a working group to build international standardization of friction stir spot-based processes. His work has focused on metallurgical studies of friction stir welding and processing for more than a decade. He has obtained fundamental knowledge on development of grain structure, texture evolution, joining mechanism, behavior of oxide-layer on surface, properties-microstructure relationship, and so on. Recently, he has centered on developing friction stir welding of steels and titanium alloys, and new tool materials. He has received a number of awards including District Contribution of Welding Technology Award from Japan Welding Society in 2005, Kihara Award from Association for Weld Joining Technology Promotion in 2008, Prof. Koichi Masubuchi Award from AWS in 2009, Murakami Young Researcher Award from the Japan Institute of Metals in 2010, Aoba Foundation Award in 2010, and Honda Memorial Young Researcher Award in 2011. He has authored or co-authored more than 190 papers in peer-reviewed journals and proceedings.

Yuri Hovanski is a Research Engineer at Pacific Northwest National Laboratory. He earned a B.S. degree in Mechanical Engineering at Brigham Young University, and then completed his M.S. degree in Mechanical Engineering at Washington State University. As a member of TMS, he serves as the secretary for the Shaping and Forming Committee. He also participates on the ASM Joining Committee, and various other societies including SAE, AWS, and others. He has participated in friction stir-related research for more than a decade investigating weld formability, abnormal grain growth, and the influence of post-weld microstructure and texture on mechanical properties. More recently, he has focused on the development of low-cost solutions for friction stir welding, in- troducing cost efficient solutions for thermal telemetry, new tool materials and tooling for high volume production techniques in the automotive community. His ground-breaking introduction of scribe friction stir welding has further enabled solid-state joining of vastly dissimilar materials, including combinations of dissimilar metals, metal-polymer, and metal-composite joint combinations. He continues this effort today working with various industrial groups to further transition friction stir technologies into mainstream manufacturing. He actively reviews friction stir related literature for several publications and has documented his work in more than 25 publications.

Ravi Verma is a Staff Researcher in the Manufacturing Systems Research Laboratory at the General Motors Global Research and Development Center in Warren, Michigan. He has over 20 years of industrial and academic R&D experience in structural metallic alloys, processing, structure, properties, forming and simulation, with over 40 publications in archival journals and conference proceedings. More recently, he has focused on joining and durability of lightweight metallic structures, with particular emphasis on friction stir welding. He has interest in the influence of post-weld microstructure and texture on mechanical properties. Before joining GM, he worked as a research scientist at the University of Michigan, Ann Arbor, working in the area of deformation processing of high temperature metallic alloys and fiber-reinforced MMCs. He is a recipient of GM's Charles L. McCuen, Chainman's Honors, and "Boss" Kettering Awards. He holds a Ph.D. in Materials Science from the University of Bombay, India.

FRICTION STIR WELDING AND PROCESSING VII

Friction Stir Processing

Friction Stir Welding and Proceesing VII
Edited by: Rajiv Mishra, Murray W. Mahoney, Yutaka Sato, Yuri Hovanski, and Ravi Verma
TMS (The Minerals, Metals & Materials Society), 2013

MICROSTRUCTURE REFINEMENT AND HOMOGENIZATION OF NON-DEFORMING CONSTITUENT DISTRIBUTIONS DURING FSW/P

Jeffrey Woertz[1], Sarath Menon[2], Terry McNelley[2]

[1]United States Naval Academy, Department of Mechanical Engineering, 590 Holloway Rd., Annapolis, MD, 21402, USA;
[2]Naval Postgraduate School, Department of Mechanical and Aerospace Engineering, 700 Dyer Rd., Monterey, CA 93940 USA;

Keywords: Friction Stir, Homogenization, Redistribution, AA356

Abstract

Weld nugget / stir zone microstructures often exhibit highly homogeneous distributions of non-deforming constituent phases as well as refinement of such phases and grain structures. The role of recrystallization in grain refinement during FSW/P is well understood. However, the processes involved in refinement of particle size and homogenization of particle distributions during FSW/P are less well understood. FS of an as-cast AA356 alloy has been conducted over a wide range of RPM values using both threaded and smooth tools. The conversion of the primary - eutectic microstructure to a highly refined and homogeneous dispersion of Si particles in a fine-grained Al matrix will be documented. The roles of advective mixing, diffusional processes, and discontinuous shearing of material will be delineated and quantitative measures of homogeneity will be discussed. The potential roles of tool and process conditions in homogenization are also considered.

Homogenization via Advective Transport

FSP is highly effective for intermixing of constituents in many alloys [1]-[3]. Given that observation, mass transport, either advective or diffusive, must occur to account for the observed constituent redistribution [4]. Here, model eutectic systems, Al-7%Si (Na modified) and AA356 are considered, negating the need to account for the effects of phase transformations in the FSP temperature range. Experimental and analytical approaches showed that deformation-assisted, high-temperature diffusive processes do not appear to be sufficient, either in magnitude or duration, to account for the particle motion required to achieve the observed levels of redistribution [5]. Calculations can be carried out for temperatures approaching the melting point of the Al matrix to show that the induced hydrostatic pressure gradients would transport particles at particle drift velocities (~1 to 0.1 diameters/sec) that are insufficient to account for observed particle redistribution [5], [6]. Since FSP constituent mixing occurs in the very short time of a single tool rotation, i.e. less than 0.1 seconds, it is reasonable to conclude that an advective component of transport must be significant.

Here, the term ‘advective’ refers to the transport of entrained material resulting from the velocity of a surrounding viscous-like flow field. In the case of a particle embedded in a highly plastic solid near its melting temperature, such behavior can occur. The advection equation in one dimension is given in Equation 1.

$$\frac{\partial u}{\partial t} + w(x,t)\frac{\partial u}{\partial x} = 0 \qquad \text{Equation 1}$$

Where: $u(x,t)$ represents some unknown particle concentration

$w(x,t)$ represents some local velocity field

Requiring the initial condition: $u(x,t=0) = u_o(x)$

The internal structure of the solid material cannot be observed in real time during FS. However, by imposing a simple shear and rotation on a digitally captured volume of base material in binary format, insight can be gained as to how constituent redistribution might be occurring. As a basis for this approach, microstructural examination as illustrated in Figure 1 for AA356 FSP samples containing FS-induced voids showed evidence of discrete "extruded" layers that are repeatedly stripped off at the tool's pin-material interface [5]. These 'extruded' layers have been previously noted and correspond to the selected IPM / RPM combination [7], [8]. Further inspection (shown in Figure 1b), however, provides evidence of significantly thinner sub-layers (ranging from 5 to 15 μm in thickness) that appear to slide over one another as the material is sheared [5], and these processes are accompanied by prior particle fracture (shown in Figure 1c and discussed later).

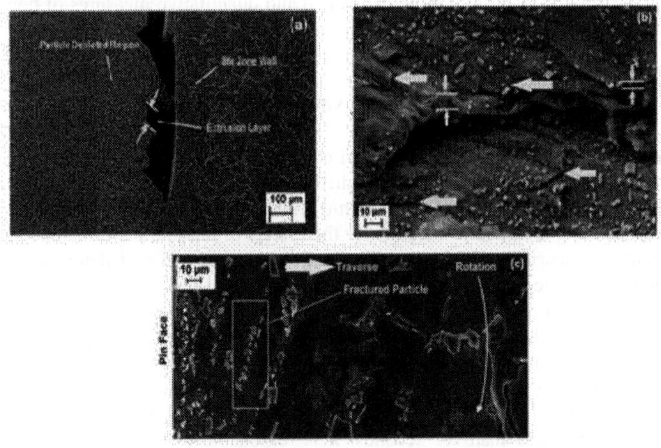

Figure 1 a – c (a) Secondary electron plan view image of 1200 RPM threaded tool extrusion layers approaching the advancing side in a lower section of the SZ where particles completing travel around the tool pin are the same size as particles remaining on the distorted advancing side SZ wall. (b) Secondary electron image of typical sheet-like layer formations inside a 400 RPM void with arrows denote some of the formations and their thicknesses. (c) Secondary electron image of the plan view plane on the front side of the tool for smooth-pin FSP AA356 at 400 RPM. The image shows an acicular particle (entering the SZ) which has fractured but the fragments have not yet separated.

The AA356 shown was friction-stirred over a range of 400 to 4000 RPM at a 3 IPM traversing rate using 3 mm smooth-sided and threaded frustum shaped tool pins. Transverse and mid-depth plan view SZ sections were obtained and mechanically polished using P1200 aluminum oxide paper followed by 6μm, 3μm, and 1μm alumina powders on a microfiber pad. The samples were then electro-polished using a perchloric acid/ethanol/glycerol solution in accordance with [[8]. Secondary electron imaging was conducted with a Zeiss Neon 40 Field Emission Scanning Electron Microscope (SEM) using a 20 keV accelerating voltage and a 5 mm working distance.

The microstructural observations are incorporated into a model illustrated schematically in Figure 2 to illustrate a potential mechanism of redistribution. Velocity and strain gradients may develop as a result of the sticking condition at the pin / matrix interface, resulting in a localized monotonic shear for the volume of material in that region. Instead of bulk deformation, the sliding sub-layers form as a result of the material's plasticity at high temperature and the strain gradient developed across the width of the extrusion layer. In this way, redistribution occurring during FSP can be viewed as a discrete process. Consider a Representative Volume Element (RVE) of material that is sheared at some location. At the next discrete angular location, $\delta\theta$, the same volume of material is again subject to the pin's shearing action, but at a new orientation. The sub-layers, in turn, are sheared across their previous boundaries. By simulating this procedure on a digitized, binary image of base material volume element having periodic boundary conditions [10], synthetic microstructures can be obtained which are remarkably similar in morphology to microstructures of the same material following FSP.

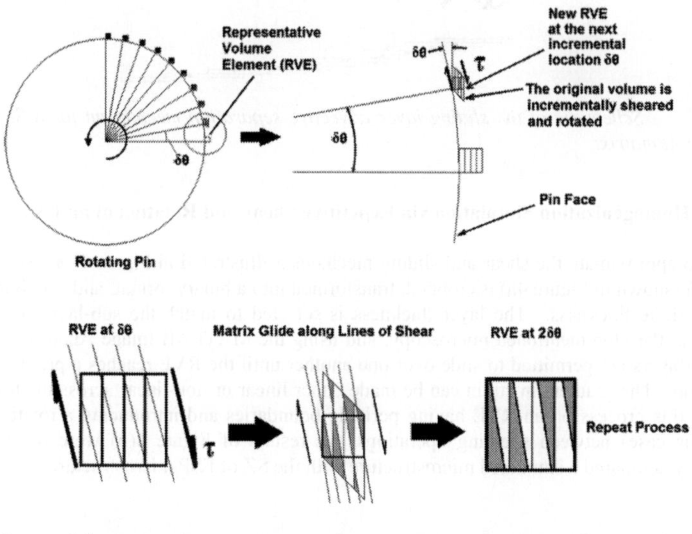

Figure 2 Schematic of an extrusion layer volume element with periodic boundary conditions at the pin-material interface experiencing an incremental rotation while subject to continuous shear.

Particle Fracture

The implication of this discrete layer shearing approach is that constituent particles are fractured prior to transport and that the particle fragments are redistributed as they travel within the segmented regions of the repeatedly sheared RVE. In fact, microstructural evidence (shown in Figure 1b and c) suggests that parent particle fracture is largely complete before traveling around the tool. Without contribution from a diffusive process, a matrix advection mechanism is the only means for particle segments to gain separation. This is shown schematically in Figure 3. If two adjacent particle pieces move apart in this manner, it is necessary to conclude that the pieces are traveling in volumes of matrix material that are moving with different velocities and / or direction, otherwise the final structure would merely resemble a monotonically strained material. It can also be reasoned that a sub-layer or RVE sub-volume has a similar characteristic dimension as the particle fragment it contains.

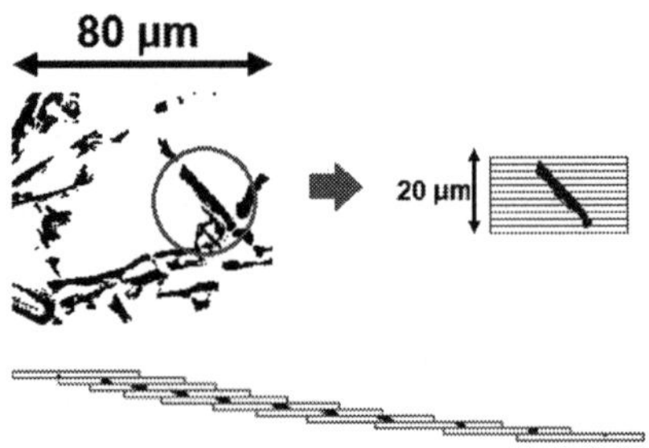

Figure 3 *Schematic of the sliding layer advective separation mechanism for a Si particle within the Al matrix.*

Homogenization Simulation via Repetitive Shear and Rotation of an RVE

To approximate the shear and sliding mechanism illustrated in Figure 2, an SEM image of AA356 (shown in Figure 4a) is cropped, transformed into a binary format, and subdivided into layers of finite thickness. The layer thickness is selected to match the sub-layer thicknesses observed in the aforementioned microscopy, and using the MATLAB Image Analysis Toolbox, these sub-layers are permitted to slide over one another until the RVE reaches a pre-determined shear strain. The gradient in strain can be made either linear or non-linear across the RVE. By repeating this process on an RVE having periodic boundaries and introducing a rotation ($\delta\theta = 90^{\circ}$ in this case) between shearing operations, the results of Figure 4(b) were obtained and statistically compared a simulated microstructure with the SZ of FSP microstructures.

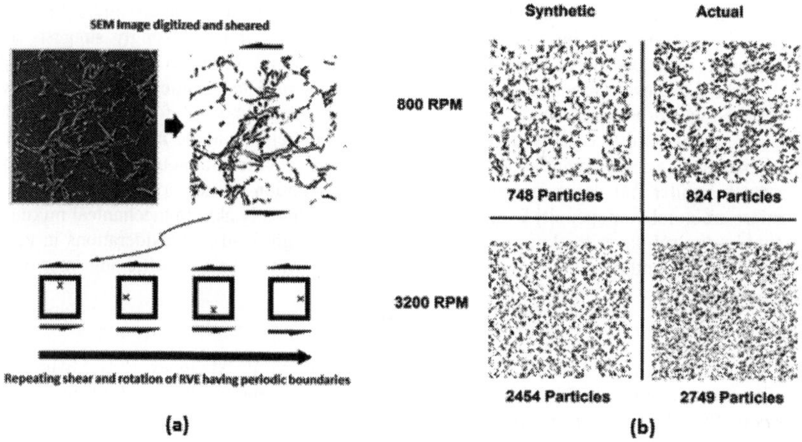

Figure 4 a-b (a) SEM image of AA356 converted into a binary RVE which was then sheared and rotated during repeated operations. (b) Visual comparison of synthetic microstructures (generated using sub-layer layer thicknesses corresponding to expected particle size) and actual FSP microstructures at 800 and 3200 RPM.

Using the MATLAB Image Analysis Toolbox, particle size, aspect ratio, orientation, and spacing characteristics were measured for both the synthetic and FSP samples. The measured data for the two sets of samples were in good agreement (within 10%), supporting the similarity in the microstructures' physical appearance. [5] A homogeneity index was also developed based on the variance of several morphological parameters over multiple length scales. This index, to be discussed in greater detail in a later publication, showed that the synthetically sheared / rotated RVE's and actual FSP samples had particle distributions similar to within measurement error. It was also observed that the degree of homogenization saturates, no longer increasing after some finite number of shearing operations (e.g., after 8 to 12 operations using a repeated shear strain of 1). The results also showed that homogeneity rose linearly with RPM (over the range measured), indicating that a more desirable microstructure might be obtained with higher tool rotation speeds.

Conclusions

In general, the preliminary results suggest that a relatively simple discontinuous geometric shearing approach might be able to explain the redistribution of constituent material in AA356 after FSP. This idea is compatible with the notion that material traveling around the tool returns to a location very near to where it was stripped from the base material. The RVE model implies that not all volumes of material at the pin interface are sheared to the same degree, depending on their original location. As a result, variations in the degree of redistribution should be expected across the SZ. As RPM increases, these variations should diminish, as each volume will be sheared more frequently with most volumes reaching mixing saturation, regardless of starting location.

The geometric similarities observed among the synthetically sheared microstructure image RVE's with periodic boundaries and experimental FSP samples strongly suggests an advective mechanism consistent with those seen in mechanical mixing process [11]. If so, the shearing layer thicknesses required in FSP are on the order of 5 to 15 microns, considerably smaller than the extrusion layer on the pin's retreating side. Layered formations of such thickness were observed to exist when examining void-prone regions in the microscopy of FSP samples taken across a wide range of RPM values. Diffusive processes, although present, appear to be a much smaller and slower contributor to the overall motion of the particles. The results imply microstructural homogeneity is driven by a mechanism mostly akin to mechanical mixing, dominated by the sub-layer thickness. These observations might lead to considerations in FSP parameter selection and possible refinements in tool design with respect to achieving improved homogeneity.

REFERENCES

[1] A. Askari, S. Silling, B. London, M. Mahoney, K. V. Jata, M. W. Mahoney, R. S. Mishra, S. L. Semiatin, and D. P. Field, Eds., "Temperature distribution and resulting metal flow," in *Friction Stir Welding and Processing IV*, 2001, pp. 43–54.

[2] K. Oishi and T. McNelley, "Microstructural modification of as-cast NiAl bronze by friction stir processing," *Metallurgical and Materials Transactions A,* vol. 35, pp. 2951–2961, 9, 2004.

[3] M. L. Santella, T. Engstrom, D. Storjohann and T. Pan, "Effects of friction stir processing on mechanical properties of the cast aluminum alloys A319 and A356," *Scripta Materialia,* vol. 53, pp. 201–206, 7, 2005

[4] J. Philibert, *Atom Movements Diffusion and Mass Transport in Solids,* Les editions de physique. Les Ulis, France, 1991.

[5] Woertz, Jeffrey C., "Redistribution Mechanisms and Quantification of Homogeneity in Friction Stir Welding and Processing of an Aluminum Silicon Alloy," Ph.D. Dissertation, Dept. of Mechanical and Aerospace Eng., Naval Postgraduate School, Monterey, CA, 2012.

[6] X. Dong and Z. Li, "Analytical solution for motion of an elliptical void under a gradient stress field," *Applied Physics Letters,* vol. 94, p. 071909 ,7, 2009.

[7] K. N. Krishnan, "On the formation of onion rings in friction stir welds," *Materials Science and Engineering: A,* vol. 327, pp. 246–251, 4/30, 2002.

[8] K. Kumar and S. V. Kailas, "The role of friction stir welding tool on material flow and weld formation," *Materials Science and Engineering: A,* vol. 485, pp. 367–374, 6/25, 2008.

[9] C. J. Smithells, E. A. Brandes, "Metallography," in *Smithells Metals Reference Book* (5th Edition), Butterworth and Co. Ltd, New York, 1978, pp. 300–302.

[10] Y. Beygelzimer, "Vortices and mixing in metal during sever plastic deformation," *Material Science Forum,* vol. 683, pp. 213–224, 2011.

[11] N. Harnby, M.F. Edwards, A. W. Nienow, *Mixing in the Process Industries*, 2nd ed. London: Butterworth-Heinemann, 1997.

Friction Stir Welding and Proceesing VII
Edited by: Rajiv Mishra, Murray W. Mahoney, Yutaka Sato, Yuri Hovanski, and Ravi Verma
TMS (The Minerals, Metals & Materials Society), 2013

NANO-SIZED GRAIN REFINEMENT USING FRICTION STIR PROCESSING

Brian Thompson[1], Kevin Doherty[2], Jianqing Su[3], Rajiv Mishra[3]

[1]EWI; 1250 Arthur E. Adams Drive; Columbus, Ohio, 43221, USA
[2]U.S. ARL; Aberdeen Proving Grounds; Aberdeen, Maryland, 21005, USA
[3]UNT; 1155 Union Circle #305310; Denton, Texas, 76203, USA

Keywords: Friction Stir Processing, Magnesium, Nano-size grains

Abstract

A key characteristic of a friction stir weld is a very fine grain microstructure produced as a result of dynamic recrystallization. The friction stir processing (FSP) technique was applied to modify the through thickness microstructure of a monolithic plate of a magnesium alloy. Grain structure refinement in these alloys could have a significant impact on their strength and ductility opening up their use for high performance defense applications. EWI has been investigating the use of the FSP technique to achieve nano-sized grains in a magnesium alloy, AZ31B. Heat input estimations have enabled the prediction of welding parameters and tool geometries that could achieve significant grain refinement. This presentation will summarize the experimental procedures using active cooling and theoretical efforts undertaken in order to achieve an average stir zone grain size of 500 nm. This work was performed under a cooperative agreement between EWI and the U.S. Army Research Laboratory.

Introduction

Magnesium (Mg) alloys have been widely used for structural components in the automotive, aerospace and electronics industry due to their low density, high strength to stiffness ratio, good damping capacity, diecastability and recycling. However Mg alloys also exhibit low formability because of their hexagonal close-packed crystal (HCP) structure with only two independent operative basal slip systems at room temperature. It is well accepted that grain refinement may increase strength and ductility of Mg alloys, potentially leading to wider use in a variety of industries including the defense industry.

In order to obtain a microstructure with improved mechanical behavior, several methods based on plastic deformation have been developed for grain refinement in Mg alloys such as rolling, Equal Channel Angular Processing (ECAP) and Equal Channel Angular Extrusion (ECAE). Recently, friction stir processing (FSP) has emerged as an effective tool for grain refinement [1]. FSP is a solid-state process developed on the basis of the friction stir welding (FSW) technique invented by The Welding Institute (TWI) in 1991 [2]. During FSP, heat is generated by the friction between the tool and the work piece and by the plastic deformation occurring around the tool. The material within the processed zone undergoes intense plastic deformation resulting in a dynamically recrystallized fine grain structure. Grain refinement by FSP in Mg alloys has been reported by several researchers [3-9]. It has been demonstrated that the processed grain structures strongly depended on the processing conditions including processing parameters, tool geometry and cooling rate.

In this work, a systematic investigation was performed to study the effects of processing parameters on the resulting microstructure in AZ31B Mg alloy. Ranges of processing conditions were tested using two different FSP tools in ambient conditions and cooled conditions using a liquid nitrogen cooled anvil. The tool designs and processing conditions selected were aimed at reducing the grain size of the AZ31B Mg alloy to create an Ultra-Fine Grain (UFG) structure averaging 500-nm. Using optical microscopy and the electron back-scatter diffraction (EBSD) technique, grain size, grain structure, grain boundary characteristics and micro-texture were examined for the processed samples. On the basis of measured grain size, combined with the *Zener-Holloman* parameter, the peak temperatures in the nugget zone were calculated and compared with experimentally measured temperatures. The relationships between processing parameters, tool geometry, grain size, and temperature were also established in this work.

Experimental Procedure

Friction Stir Processing Trials

The friction stir processing trials were conducted at EWI on a NOVA-TECH Engineering Friction Stir Welding machine furnished by the U.S. Government. Two tools were designed and fabricated from 350M steel and heat treated to an approximate hardness of 60 HRC. Tool 52283-T02 incorporated a left hand, stepped spiral thread on a truncated cone type pin with negative profile scroll features on the shoulder. Tool 52283-T03 also included negative profile scroll features on the shoulder, but incorporated three blades in place of a single pin. Both tools were designed to process down to a thickness of 2.3-mm. Tool 52283-T03 was the larger of the two tools with a shoulder diameter of 17.8-mm, while tool 52283-T02 had a shoulder diameter of 11.1-mm. The aggressive features on each tool were designed in an effort to promote as much grain refinement as possible.

In order to manage the cooling rate of the AZ31B Mg alloy during processing, a copper anvil capable of being cooled by liquid nitrogen was designed and fabricated. Two inlets allowed liquid nitrogen to flow into four channels underneath the top surface of a copper block. Six outlets directly opposite the two inlets allowed the nitrogen to flow out of the copper block and be captured in an insulated bucket. A single groove machined into the surface of the copper block accommodated a single Type K thermocouple. This thermocouple was used to monitor the temperature at the root of each friction stir processing trial. A small hole was drilled into the bottom of each plate to a depth of approximately 4-mm, placing the thermocouple tip just underneath the processing trial. Surface cooling was added through the use of two copper bars lying adjacent and parallel to the processing direction. These copper bars contained a hole through which liquid nitrogen flowed and then emptied into an insulated bucket. An image of the experimental setup without the top surface copper bars is displayed in Figure 1.

The 6.3-mm thick wrought AZ31B Mg alloy plate used in this work was cut into 305-mm square plates for welding. All cooled processing trials were conducted in a single linear pass positioned on the center of the plate. Temperature data was recorded four times a second using an Omega eight channel USB thermocouple data acquisition module. Prior to conducting processing trials, test cooling runs were conducted to establish a base line temperature to which the anvil could be reduced (-180°C). This temperature was then used as the target temperature at which to cool the anvil prior to and after each processing pass.

Figure 1. FSP Experimental Setup

The processing parameters for each tool and condition are summarized in Table I. After establishing initial parameters through scaling trails, the travel speed was varied while the RPM remained constant for each tool. This resulted in a selection of seven welding conditions between both tools. The variation in travel speeds was designed to provide a selection of heat inputs ranging from low to high. Each of the processing parameters for each tool were run in an ambient condition (no liquid nitrogen cooling) and a cooled condition (with liquid nitrogen cooling) on the same copper anvil.

Table I. Selected Processing Parameters for Each Tool Design

Tool No.	Spindle Speed (RPM)	Travel Speed (mm/min)	Heat Input	Ambient Cross-Section No.	Chilled Cross-Section No.
52283-T02	1200	762	Low	M16	M123
	1200	508	Medium	M15	M121
	1200	356	High	M14	M122
52283-T03	1000	762	Low	M12	M128
	1000	635	Moderate	M13	M127
	1000	508	Medium	M11	M129
	1000	381	High	M17	M130

Microstructural Analysis

Post-processing, each weld was cross-sectioned once for microstructural analysis resulting in 14 samples (7 ambient and 7 cooled). Each cross-section was labeled with a unique identifier as listed in Table I. The samples were polished with SiC papers using alcohol as a lubricant and were swab etched with Kroll's solution. Optical microscopy was conducted at EWI on all samples using an Olympus BX51 microscope. Average grain size was measured over a sampling of 10 grains located in the center of the nugget zone following ASTM E1382 (Region 2, Figure 2). After completion of the optical microscopy, all seven cross-sections from the cooled processing trials were sent to the University of North Texas (UNT) for EBSD analysis.

In preparation for EBSD analysis, the seven samples were re-polished once received at UNT. The cross-sections were ground on progressively finer SiC papers to a 2400 grit finish and then polished with 0.1 μm alumina solution. To obtain high quality Kikuchi diffraction patterns, electro polishing was carried out with a solution of 40 pct phosphoric acid and 60 pct ethanol at ~0 °C and using a voltage of 2 to 3 V. EBSD analysis was conducted at UNT with a FEI Nova NanoSEM 230 equipped with the TSL OIM™ EBSD system. A scanning step size of 0.5 μm was used for all the scans. The error in the determination of the crystallographic orientation of each grain was less than 2°. In order to obtain a highly reliable picture of the microstructure, the resulting data was subjected to the following clean-up procedure: (1) grain dilatation with a grain tolerance angle of 5°, and (2) each grain contains at least two scan points. The different locations in the nugget zone examined using OIM are illustrated in Figure 2.

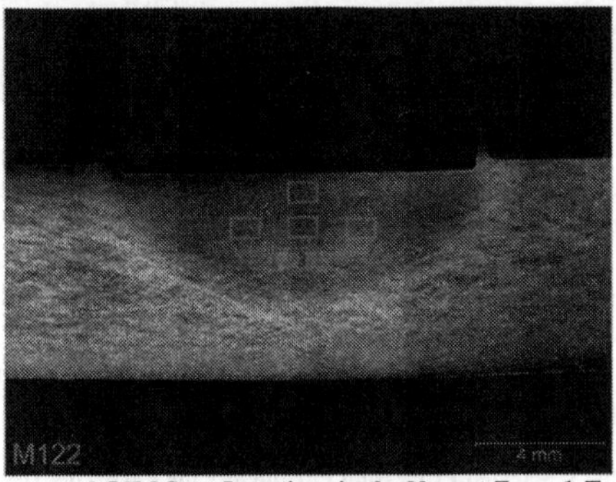

Figure 2. Optical Microscopy and OIM Scan Locations in the Nugget Zone. 1-Top, 2-Center, 3-Bottom, 4-Advancing Side and 5-Retreating Side.

Results and Discussion

Friction Stir Processing Trials

During the processing trials, temperature measurements were only recorded on the trials cooled by liquid nitrogen. Due to the proximity of the thermocouple tip to the processed zone, damage to the thermocouples resulted during two processing trials: M123 and M130. For tool 52283-T02, the peak temperature data was inconclusive for the low heat input trial (M123), reached ~200°C for the medium heat input trial (M121), and reached ~225°C for the high heat input trial (M122). Cooling back down to 0°C from the peak temperature reached during each of these trails took approximately four seconds. For tool 52283-T03 the peak temperature reached ~250°C for the low heat input trial (M128), reached ~270°C for the moderate heat input trial (M127), reached ~300°C for the medium heat input trials (M129), and was inconclusive for the high heat input trial due to the destruction of the thermocouple. The low heat input and moderate heat input trials took approximately four seconds to cool back down to 0°C from peak temperature while the medium heat input trial took approximately seven seconds.

Optical Microscopy

Optical analysis of the weld cross-sections revealed a reduction in grain size when comparing ambient processing trials to those performed with liquid nitrogen cooling. Ambient trials with tool 52283-T02 resulted in an increasing average grain size when increasing the heat input of the welding conditions. The low heat input trial averaged a grain size of approximately 7,600-nm while the high heat input trial averaged a grain size of approximately 9,300-nm (Figure 3). When these same welding conditions were cooled using liquid nitrogen, the average grain size dropped to approximately 6,300-nm. A similar trend was observed in the welding conditions selected for tool 52283-T03. Average grain size for the ambient trails ranged from 10,700-nm to 14,000-nm, for the low to high heat inputs respectively (Figure 3). When liquid nitrogen cooling was applied to these same welding conditions, the average measured grain size dropped to approximately 7,500-nm.

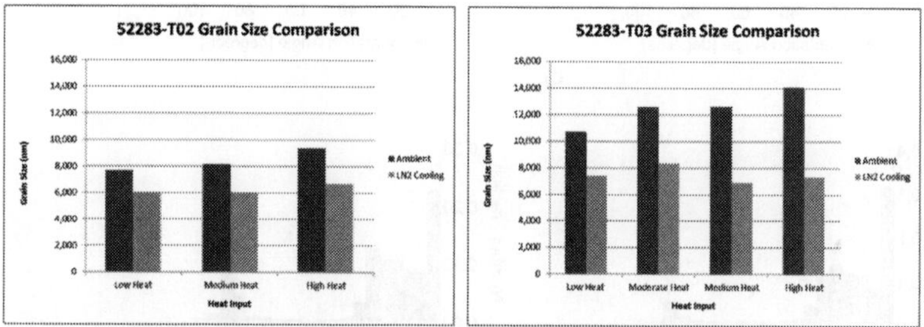

Figure 3. Grain Size Comparisons for each Tool Based Upon Optical Measurements

Orientation Imaging Microscopy

As measured by the TSL OIM™ EBSD system, the average grain sizes in samples M122, M121 and M123 processed by tool 52283-T02 are 2,900-nm, 2,400-nm, and 2,200-nm. Larger grains of 3,200-nm, 3,200-nm, 3,200-nm and 3,300-nm in size were obtained in samples M130, M129, M127 and M128 processed by tool 52283-T03. The present results show that with an increase in tool traverse speed, the processed grain size decreases in the samples processed by tool 52283-T02. However, tool traverse speed has no effect on grain size in the samples processed by tool 52283-T03. The disparity in the grain size measurements between those conducted optically following ASTM E1382 and those conducted using OIM, can be attributed to the inherent resolution and accuracy of the TSL OIM™ EBSD system. As a result, the optical grain size measurements have been used as reference only in preference of those grain sizes observed through OIM.

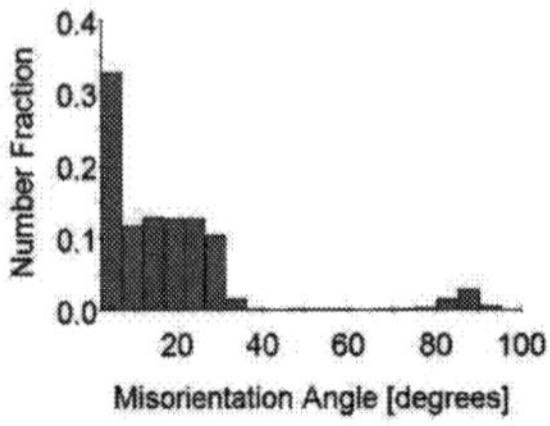

a)

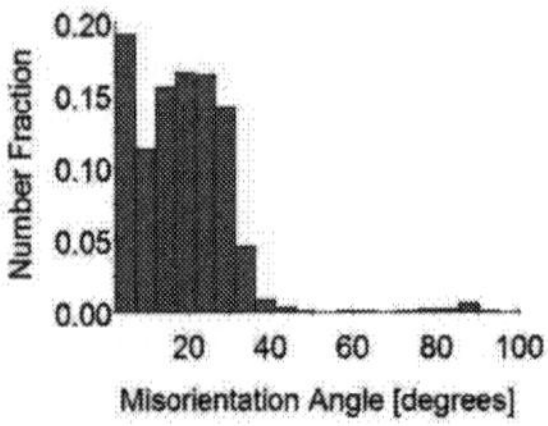

b)

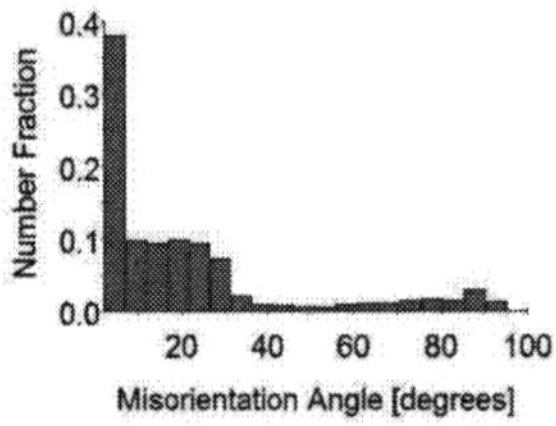

c)

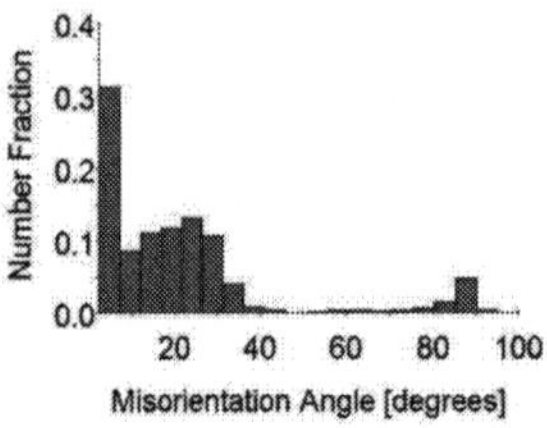

d)

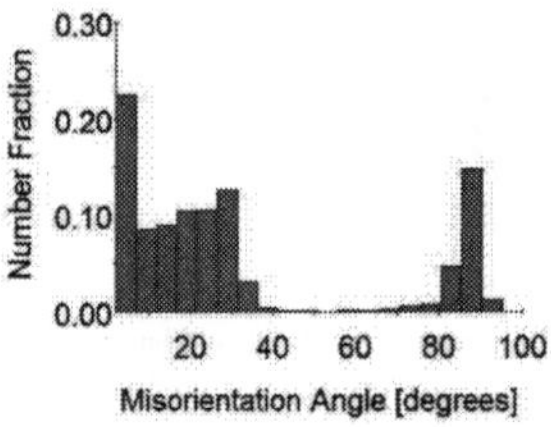

d)

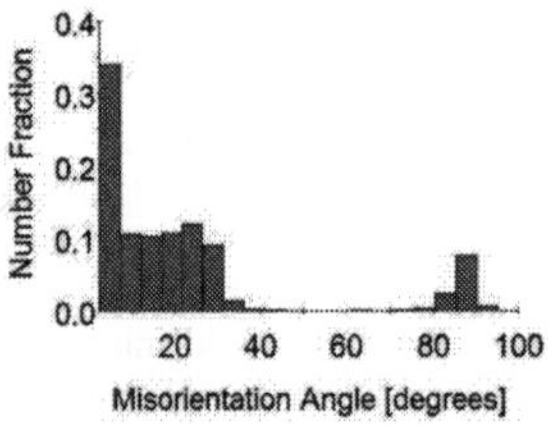

f)

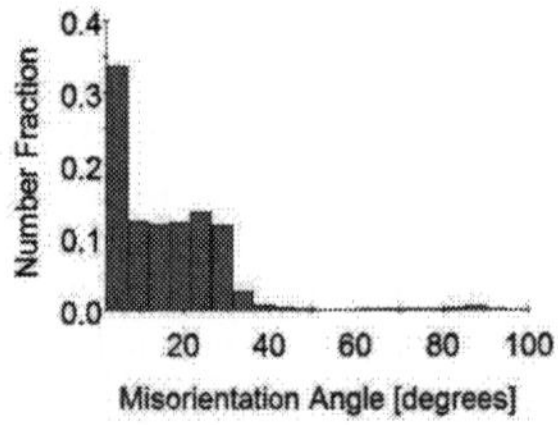

g)

Figure 4. Grain Boundary Misorientation Distributions in Samples a) M122, b) M121, c) M123, d) M130, e) M129, f) M127 and g) M128

In all the cases, the liquid nitrogen cooled samples show similar grain boundary character (Figure 4). The grain boundary misorientation distributions range mainly from 2-35°, followed by a wide gap in the interval of ~35-80°, and then show a peak in ~80-90° which may correspond to the formation of twins with misorientation close to the ideal 86° <11-20> relationship. All of the processed samples have a strong fiber texture with the c-axes of the grains about 35-55° away from processing direction.

Temperature Calculation The Zener–Holloman parameter, combining working temperature and strain rate, has been shown to impose influence on the resulting grain size in extruded Mg based alloys [10]. This parameter appears to be useful in predicting the resulting grain size. Similarly, using the measured grain size, working temperature can be calculated using the following equation.

$$T = \frac{Q}{R(\ln Z - \ln \dot{\varepsilon})} \tag{1}$$

The measured grain size and calculated peak temperature in the nugget for the samples processed by different tools under various processing parameters are summarized in Table II. The influence of the processing parameters on resulting grain size and temperature is displayed in Figure 5.

Table II. Summary of Grain Size and Peak Temperature at Nugget Center

Tool No.	Chilled Cross-Section No.	Spindle Speed (RPM)	Travel Speed (mm/min)	Grain Size (nm)	Calculated Peak Temperature (°C)	Measured Root Temperature (°C)
52283-T02	M123	1200	762	2,100	340	N/A
	M121	1200	508	2,400	352	200
	M122	1200	356	2,900	368	225
52283-T03	M128	1000	762	3,300	383	250
	M127	1000	635	3,200	380	270
	M129	1000	508	3,200	379	300
	M130	1000	381	3,200	380	N/A

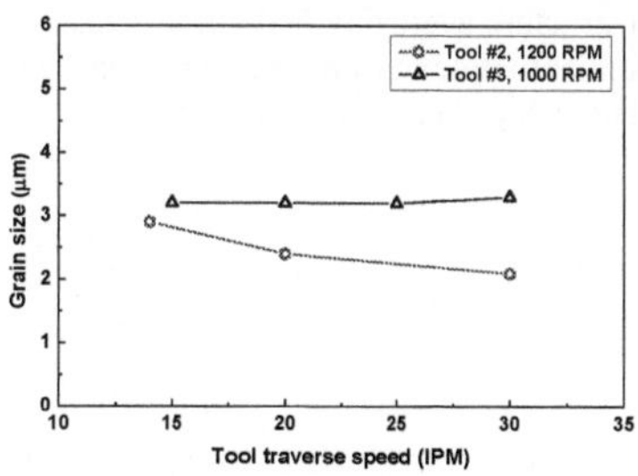

a)

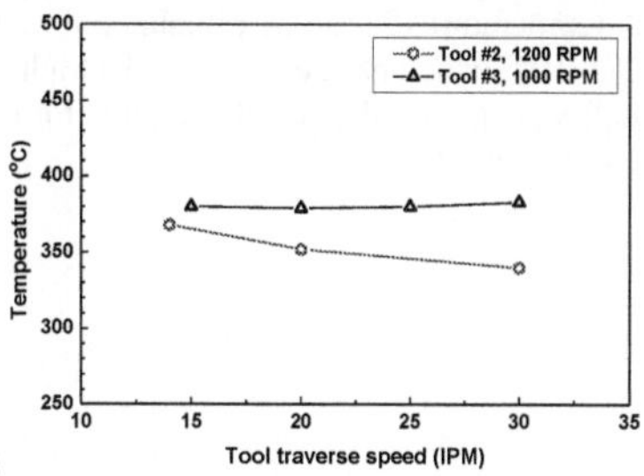

b)

Figure 5. Effect of Tool Traverse Speed (IPM) on a) grain size and b) temperature

For tool 52283-T02, an increase in the tool traverse speed causes a decrease in the peak temperature resulting in a finer grain structure in the nugget zone. This result is consistent with those reported by many researchers in different metals and alloys. However, it is noted that there is almost no difference in grain size in the samples processed by tool 52283-T03 under different tool traverse speeds. This result suggests that to discuss the effect of processing parameters on the resulting microstructure, the tool geometry must be considered. It is also believed that a lower tool rotation rate with tool 52283-T02 would produce a finer grain structure, especially using liquid nitrogen cooling.

Tool Geometry Effects on Heat Input During FSP During FSW or FSP, peak temperature in the nugget zone plays a vital role in determining the processed grain structure. The most important factor in controlling the peak temperature is heat input. Considering power consumed by the machine during the welding process, Khandkar et al. [12] measured the torque and the rotation rate during welding, from which the radial heat flux $\dot{q}(r)$ was calculated as:

$$\dot{q}(r)=\frac{P_{av}r}{\frac{2}{3}\pi R_{Shoulder}^{3}+2\pi R_{Pin}^{2}H_{Pin}} \tag{2}$$

where P_{av} is the average power input calculated from the measured torque and rotation rate, $R_{Shoulder}$ is the shoulder radius, R_{Pin} is the pin radius and H_{Pin} is the pin height. Thus, heat generation is closely related to tool geometry. As demonstrated through experimental trials, the larger average grain sizes for the 52283-T03 tool (7,500-nm) are a result of its larger geometry and therefore higher heat input even though the RPM was less than that of the 52283-T02 tool (6,300-nm). Therefore, it is believed that a tool with a smaller shoulder and pin can decrease heat input caused by friction and plastic deformation. This combined with low heat input generating parameters, could result in finer grain structures created during FSP, especially when using liquid nitrogen cooling.

Fabrication of UFG Structures The purpose of this work was to produce UFG structures averaging 500-nm in size in AZ31B Mg alloy using FSP. On the bases of the relationships between grain size (*d*)-the Zener-Hollomon parameter (*Z*) and the Zener-Hollomon parameter (*Z*)-peak temperature (*T*) in the FSP AZ31B Mg alloy [11], the Zener-Hollomon parameter (*Z*) and peak temperature (*T*) can be calculated for a grain structure averaging 500-nm in size. For tool 52283-T02, which has demonstrated a higher potential to produce a finer grain structure due to its smaller geometry, the calculated plots for the relationships between *d-Z* and *d-T* are displayed in Figure 6.

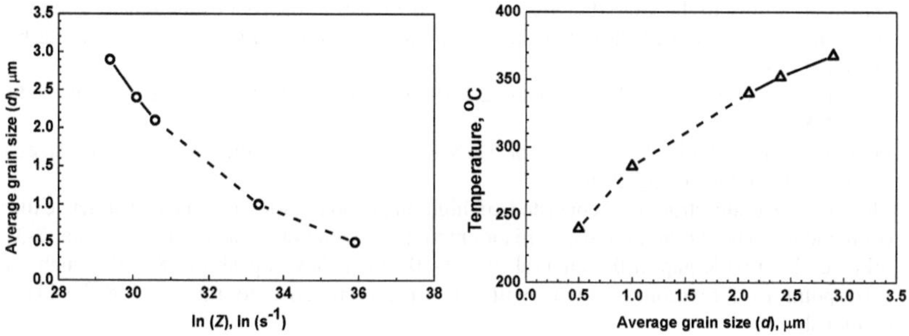

(a) (b)

Figure 6. Plots for the Relationships between a) d-Z and b) d-T for tool 52283-T02. The experimental range already explored is depicted by continuous lines and the dash line represents extrapolation using the constitutive relationships

Future Work

To create an UFG structure averaging 500-nm in size, a lower peak temperature of about 240°C is required. In the present results, average grain size as small as 2,100-nm was obtained by using tool 52283-T02, which corresponds to a peak temperature of ~340°C in the nugget zone. Decreasing the peak temperature during FSP could be accomplished by increasing the cooling rate and/or decreasing the heat input. Since liquid nitrogen cooling has been used to control the cooling rate in this present work, it would be very difficult to further increase the cooling rate while using the tools as currently designed. Therefore, to obtain a finer grain structure, a reduction in heat input must be achieved through tool design and processing parameter modification. From the relationships of *d-Z* and *Z-T* [11], the temperature can be calculated as;

$$T = \frac{Q}{R[\frac{9.0 - \ln d}{0.27} - \ln(2\pi R_m \frac{r_e}{L_e})]} \quad (3)$$

For a small tool with a fast cooling rate, the r_e/L_e can be regarded as the ratio of tool pin radius to tool pin length. Using the current dimensions of tool 52283-T02, a 500-nm grain size corresponds to a temperature of ~240°C. In the current trials, this tool was able to produce a 2,100-nm average grain size at a peak temperature of ~340°C. This means that in order to decrease the peak temperature from 340 to 240°C using the current processing conditions, the dimensions of tool 52283-T02 would need to be decreased by 1/2 to 2/3. However, this could have practical implications in terms of volume of processed material per unit production time.

Conclusions

1) Liquid nitrogen cooling of FSP trials in AZ31B Mg alloy resulted in lower average grain sizes as compared to those produced in ambient temperatures.
2) OIM resulted in a more accurate measurement of average grain size than optical methods.

3) Due to proximity to the weld, the measured temperatures in the root of the nugget zone were less than those calculated; however they both exhibit an upward trend with increasing heat input and tool geometry.
4) Fine grain structures of 2,900, 2,400 and 2,100-nm were observed in specimens processed by tool 52283-T02.
5) In comparison to tool 52283-T02, tool 52283-T03 resulted in larger average grain sizes of 3,300, 3,200, 3,200 and 3,200-nm.
6) Most of the grain structures consisted of high-angle boundaries but have distinct textural components. The grain boundary misorientation distributions range mainly from 2-35°, followed by a wide gap in the interval of ~35-80°, then show a peak in ~80-90° which may correspond to formation of twins with misorientation close to the ideal 86° <11-20> relationship.
7) The FSP samples showed strong fiber textures with the c-axes of the grains about 35°-55° away from the processing direction.
8) The calculated temperatures ranged from 340 to 368°C for the samples processed by tool 52283-T02, and are about 380°C for the samples processed by tool 52283-T03.
9) With an increase in tool traverse speed, grain size decreases in the nugget zone in samples processed by tool 52283-T02. However, there is almost no change in grain size in the samples processed by tool 52283-T03, suggesting a complex effect of strain and temperature caused by tool geometry.
10) To produce UFG structures averaging 500-nm in size, smaller tools with dimensions 2/3~1/2 those of tool 52283-T02, as well as a lower tool rotation rate are suggested.

References

[1] R.S. Mishra, M.W. Mahoney, Mater. Sci. Forum 507 (2001) 357-359.

[2] W.M. Thomas, E.D. Nicholas, J.C. Needham, M.G. Murch, P. Templesmith, C.J. Dawes, Friction Stir Butt Welding, G.B. Patent Application No. 9125978.8, Dec. 1991; U.S. Patent No. 5460317, Oct. 1995.

[3] D. Zhang, M. Suzuki, K. Maruyama, Scripta Mater. 52 (2005) 899-903.

[4] A.H. Feng, Z.Y. Ma, Scripta Mater. 56 (2007) 397-400.

[5] W. Woo, H. Choo, D.W. Brown, P.K. Liaw, Z. Feng, Scripta Mater. 54 (2006) 1859-1864.

[6] C.I. Chang, X.H. Du, J.C. Huang, Scripta Mater. 57 (2007) 209-212.

[7] G. Bhargava, W. Yuan, S.S. Webb, R.S. Mishra, Metall. Mater. Trans. 41 (2010) 13-17.

[8] C.I. Chang, X.H. Du, J.C. Huang, Scripta Mater. 59 (2008) 356–359.

[9] W. Yuan, S.K. Panigrahi, J.-Q. Su and R.S. Mishra Scripta Mater. 65 (2011) 994-997.

[10] Y.N. Wang, C.J. Lee, H.K. Lin, T.T. Huang, J.C. Huang, Mater Sci Forum 2003;426–432:2655.

[11] C.I. Chang, C.J. Lee, J.C. Huang, Scripta Materialia 51 (2004) 509–514.

[12] M.Z.H. Khandkar, J.A. Khan, A.P. Reynolds, Sci. Tech. Weld. Joining, 8, 2003, 165-174.

Friction Stir Welding and Proceesing VII
Edited by: Rajiv Mishra, Murray W. Mahoney, Yutaka Sato, Yuri Hovanski, and Ravi Verma
TMS (The Minerals, Metals & Materials Society), 2013

Fabrication of Carbon Nanotube Reinforced Aluminum Matrix Composites via Friction Stir Processing

Z.Y. Ma*, Z.Y. Liu, B.L. Xiao, W.G. Wang
Shenyang National Laboratory for Materials Science, Institute of Metal Research, Chinese Academy of Sciences, 72 Wenhua Road, Shenyang, Liaoning, 110016, China

***Keywords*:** Carbon nanotubes; Metal-matrix composite; Friction stir processing

Abstract

In this study, CNTs reinforced pure Al, 6061Al and 2009Al composites with uniformly dispersed CNTs were successfully fabricated by a combination of powder metallurgy and subsequent multi-pass friction stir processing (FSP). As the FSP pass increased, the CNT clusters were broken down due to the shear flow of the Al matrix during FSP, however, the CNTs were cut short. After 4-pass FSP, the CNTs were individually dispersed in the aluminum matrix along grain boundaries resulting in a much finer grain size. Although the CNTs were shortened and some Al_4C_3 formed in the matrix, the layer structures of the CNTs were well retained and the CNT-Al interface was good bonded. Compared to the Al matrix, the strength, especially the yield strength of the composites increased significantly.

Introduction

Carbon nanotubes (CNTs) have attracted much attention as the ideal reinforcements for composites because of their extremely high elastic modulus (~ 1 TPa) and strength (30-100 GPa) as well as good thermal and electrical properties [1-4]. By incorporating the CNTs into appropriate matrixes, it is possible to produce the composites with enhanced properties, e.g. increased strength and decreased coefficient of thermal expansion. Although the main research efforts in the past decade were focused on the CNT reinforced polymer or ceramic matrix composites [5,6], a few groups have dedicated themselves to the fabrication of the CNT/metal composites, such as the aluminum matrix composites [7, 8].

The dispersion of the CNTs in the metal matrix is a key challenge for fabricating the CNT/metal composites, because the entangled or bundled CNT clusters are easily induced as a result of their large aspect ratio and the strong Vander force. The entangled and bundled CNTs would reduce either mechanical or physical properties of the CNT/metal composites. Many methods have been developed to incorporate the CNTs into the metal matrix, among these, the two most important methods are ball milling and CNT pre-treatment. However, the CNTs would be severely damaged during milling, and with CNT pre-treatment, the CNT concentration is hardly increased, commonly less than 1 vol.% [7, 9].

Friction stir processing (FSP) is a new metal-working technique. A non-consumable rotating tool with a specially designed pin and shoulder is inserted into a work-piece and moved along the desired path to cover the region of interest. The combination of tool rotation and translation results in severe deformation and thorough mixing of the material in the

* Corresponding author. Tel./Fax: +86-24-83978908; E-mail address: zyma@imr.ac.cn (Z.Y.Ma).

processed zone, achieving localised microstructural modification for specific property enhancement.

FSP has been demonstrated to be an effective method of incorporating the reinforcing particles, e.g., nano-sized particles [10-12], into the metal matrix and homogenising the microstructure of heterogeneous materials, such as cast alloys and composites. Johannes et al. [13] fabricated CNT/aluminium alloy composites using FSP. They inserted the CNTs into the holes or grooves which were pre-machined on the aluminium plates and then subjected the plates to FSP. Their results indicated that the HV hardness of the aluminium alloys was increased after incorporating the CNTs by FSP. Morisada et al. [14] used a similar method to fabricate CNT/magnesium alloy composites. However, the CNT concentration could not be accurately controlled by pre-setting the CNTs into the holes or grooves. Furthermore, no reports of the tensile properties of the composites were included in their studies.

In this work, a modified fabrication processing of the CNT/aluminum composites was used. Firstly, CNT/aluminum composites were fabricated using a conventional powder metallurgy process to form a bulk billet. The billet was then subjected to FSP to improve the distribution of the CNTs. Three types of CNT reinforced aluminum matrix composites, i.e. CNT/Al, CNT/6061Al and CNT/2009Al composites were fabricated, respectively, to verify the wide applicability of this fabrication route. Specially, the microstructure and the tensile properties of the FSP CNT/2009Al composites were investigated in detail.

Experimental

Raw Materials And Composite Fabrication

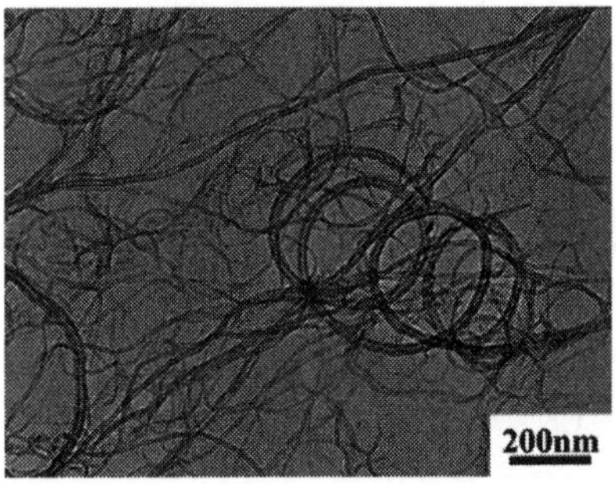

Fig. 1 Morphology of the as-received CNTs.

As-received CNTs (95-98% purity) provided by Tsinghua University, had entangled morphologies with an outer diameter of 10-20 nm and a length of several microns (Fig. 1). No extra pre-treatments were conducted on the CNTs. 1.5 vol.% CNTs were mixed with Al powders, with an average diameter of 10 μm, in a bi-axis rotary mixer running at 50 rpm for 8 h with a 1:1 ball to powder ratio. The as-mixed powders were cold-compacted in a cylinder die, degassed and hot-pressed into cylindrical billets with a diameter of 55 mm and a height of 50 mm. The billets were then hot forged with steel canning at 723 K into disc plates with a thickness of about 10 mm. Then the plates were subjected to in-situ 4-pass FSP at a tool rotation rate of 1200 rpm and a travel speed of 100 mm/min, using a tool with a concave shoulder 20 mm in diameter, a threaded cylindrical pin 6 mm in diameter and 4.2 mm in length. The pin was made of tool steel H13 with HRC of 45. 1.5 vol.% CNT/pure Al and CNT/6061Al (Al-1.0 wt.% Mg-0.6 wt.% Si) composites were fabricated in a same route, respectively. The as-FSP CNT/6061Al composite was solutionized at 803 K for 2 h, water quenched and then aged at 473 K for 12 h.

In order to obtain a better tensile strength for structural application, CNT/2009Al (Al-4.5 wt.% Cu-1.2 wt.% Mg) composites reinforced with 1.5 vol.% and 4.5 vol.% CNTs were fabricated in a similar route but subjected by 1-pass and in-situ 4-pass FSP. For comparison, unreinforced 2009Al was also fabricated under the same conditions. The as-FSP composites and the 2009Al alloy were solutionized at 768 K for 2 h, water quenched and then naturally aged for 4 days.

Characterization Of The Composites

The macroscopic CNT distribution in the matrix, under various fabrication conditions were examined using scanning electron microscopy (SEM, Quanta 600). The microscopic CNT distribution and the grain size of the composites were observed using transmission electron microscopy (TEM, Tecnai G2 20). The CNTs were extracted from 4.5 vol.% CNT/2009Al composite by using hydrochloric acid, then were examined under TEM.

Tensile specimens with a gauge length of 2.5 mm, a width of 1.5 mm and a thickness of 0.8 mm were machined from the FSP composites perpendicular to the FSP direction. Tensile tests were conducted at a strain rate of 1×10^{-3} s^{-1} at room temperature on an Instron 5848 tester.

Results and discussion

Dispersion Of CNTs

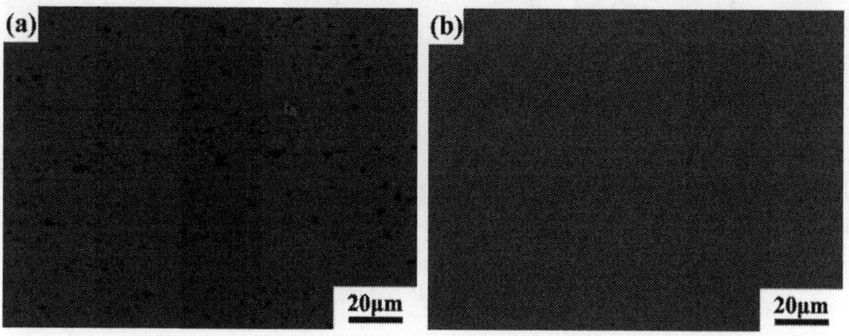

Fig. 2 CNT distribution of 1.5 vol.% CNT/Al composites: (a) before FSP, (b) after 4-pass FSP.

Fig. 2 shows the CNT distribution of 1.5 vol.% CNT/Al composites before and after 4-pass FSP. For the forged composite (Fig. 2(a)), obvious CNT clusters were observed. This is understandable, as the as-received CNTs were entangled with each other and the mixing and forging processes were not effective enough to disperse the CNTs into the aluminum matrix. Also, nearly no aluminum could be observed in the clusters because the CNTs could hardly be soaked by aluminum. After 4-pass FSP, the CNT clusters were completely eliminated, forming CNT/Al composite without CNT clustering. A more obvious contrast in the CNT distribution could be observed between the FSP zone and the un-FSP zone in FSP CNT/6061Al composite (Fig. 3). While obvious CNT clusters were observed in the un-FSP zone, these clusters disappeared completely in the FSP zone.

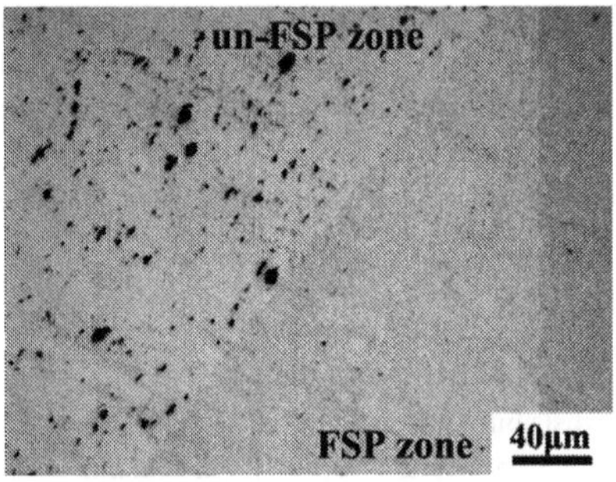

Fig. 3 CNT distribution comparison between FSP zone and un-FSP zone in 1.5 vol.% CNT/6061Al composite.

Fig. 4 shows the CNT distributions in 4.5 vol.% CNT/2009Al composites under different conditions. For the forged composite (Fig. 4(a)), obvious CNT clusters were observed. After 1-pass FSP, the large CNT clusters were significantly reduced, leaving only much smaller clusters (Fig. 4(b)). After 4-pass FSP, even the smallest clusters could no be observed (Fig. 4(c)). This indicates that the entangled CNTs were completely broken up and dispersed into the aluminum matrix due to intense stirring effect of the rotating threaded pin during FSP.

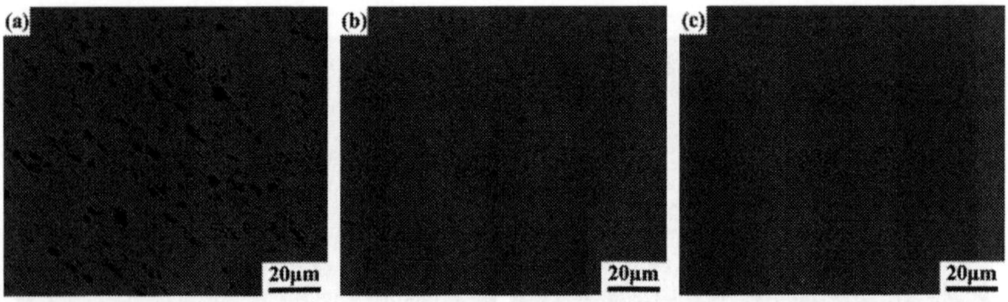

Fig. 4 SEM images showing CNT distribution in 4.5 vol.% CNT/2009Al composites: (a) forged, (b) 1-pass FSP, and (c) 4-pass FSP.

SEM fractograph of the tensile sample shown in Fig. 5(a) revealed that the CNTs were singly dispersed in the aluminum matrix of the 4.5 vol.%CNT/2009Al composite. However, the CNTs were found randomly arranged. Fig. 5(b) shows the CNT morphology extracted from the composites after 4-pass FSP. The length of CNTs could still remain about 400 nm. It was indicated that the CNTs in the composite were shorter than the as-received ones due to significant stirring and breakup during FSP. The shear effect during FSP first cut off the entangled CNTs and the fragments were subsequently dispersed into the aluminum matrix due to the plastic flow of the aluminum.

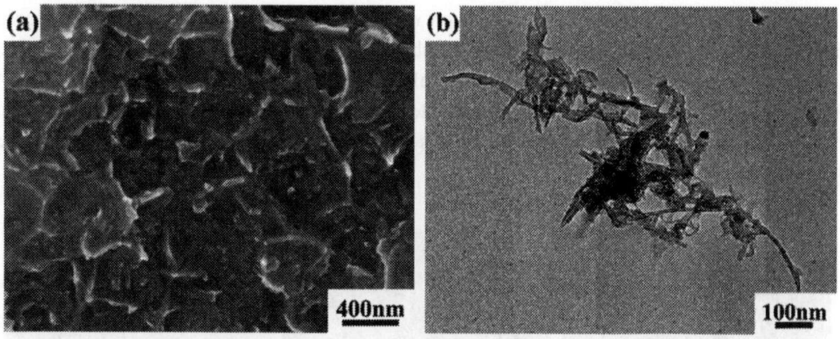

Fig. 5 (a) SEM fractograph of 4.5 vol.% CNT/2009Al composite, (b) CNTs extracted from 4.5 vol.% CNT/2009Al composite.

Microstructure Of The CNT/2009Al Composites

Fig. 6 shows the grain microstructure of the 4-pass FSP 2009Al alloy and CNT/2009Al composites. Significantly refined grain sizes were commonly observed in the FSP materials, such as aluminum alloys and magnesium alloys [15,16], due to the occurrence of dynamic recrystallization. The 4-pass FSP 2009Al had an average grain size of about 4 μm (Figs. 6(a)). By comparison, the composites had a much finer grain size. The average grain size of the 1.5 vol.% and 4.5 vol.% CNT/2009Al composites was ~1.8 and 0.8 μm, respectively (Figs. 6(b) and (c)). This indicates that the grain size decreased as the CNT concentration increased. Some of the grains, on the boundaries of which CNTs were observed as shown by the inset in Fig. 6(c), were even finer. This is attributed to effective pinning effect of the nano-sized CNTs on the grain boundaries, which hindered the growth of the recrystallized grains during FSP. Thus, the grains in the composites showed a much finer size compared with those in the 2009Al alloy, and the grain size of the composites became finer as the FSP pass increased because more CNTs were dispersed in the aluminum matrix.

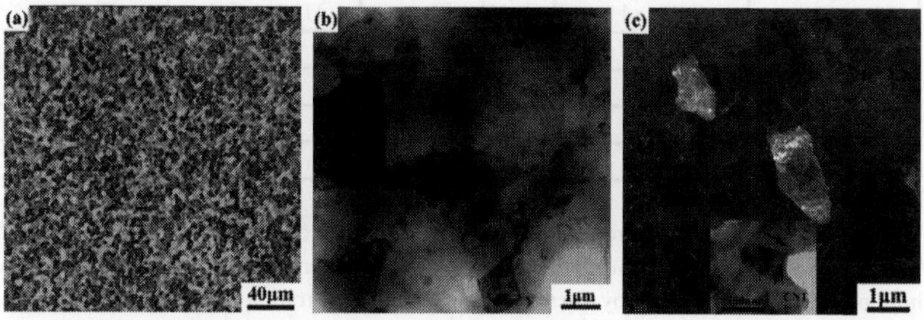

Fig. 6 Grain structures of 4-pass FSP samples: (a) 2009Al (OM), (b) 1.5 vol.% CNT/2009Al (TEM, bright field image), (c) 4.5 vol.% CNT/2009Al (TEM, dark field image).

Fig. 7 shows the fine microstructure of the interfacial regions in the 4.5 vol.% CNT/2009Al composite after 4-pass FSP. It was noted that most of the interfaces between the CNT wall and the Al matrix were clean without reaction (Fig. 7(a)). However, the Al_4C_3 could be found in the composite (Fig. 7(b)). The sizes of the Al_4C_3 were within 50 nm. The temperature during the composite fabrication was relatively low, with the maximum value of 560°C in hot pressing. It is thought that a number of the defect sites existed in the as-received CNTs and FSP further increased the number of defect sites because the CNTs were cut short. Thus, the Al_4C_3 was formed in the defect sites during the composite fabrication. However, most of the layer structures of the CNT walls were well retained and the CNT-Al interfaces were of good bonding (Fig. 7(c)).

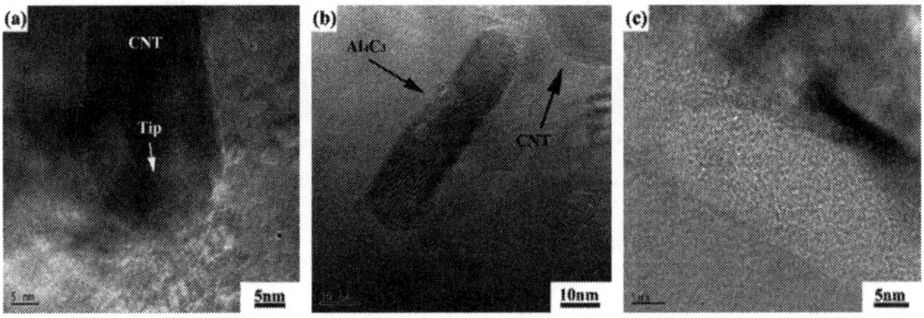

Fig. 7 HREM images showing (a) interface between Al and CNT, (b) Al_4C_3 in Al matrix near CNT, (c) layered wall structure of CNTs in 4-pass FSP 4.5 vol.% CNT/2009 Al composite.

Mechanical Properties Of CNT/Al Composites

Table 1 shows the tensile properties of CNT/Al and CNT/6061Al composites. As a result of the disappearance of the CNT clustering, even 1.5 vol.% CNT incorporation led to an increase in both yield strength (YS) and ultimate tensile strength (UTS). It could also be found that, the 6061Al matrix with a higher strength resulted in a much higher composite strength. Thus, for actual structural application a matrix with higher strength, e.g. 2009Al alloy should be used in order to obtain a higher strength.

Table 1. Tensile properties of Al and CNT/Al composites

Matrix alloy	CNT vol.%	YS (MPa)	UTS (MPa)	El (%)
Pure Al	0	94	145	25
	1.5	133	181	20
6061Al	0	260	337	20
	1.5	348	394	15

Table 2 shows the tensile properties of the CNT/2009Al composites. The YS and UTS of forged 1.5 vol.% and 4.5 vol.% CNT/2009Al composites were much lower, even lower than those of the forged 2009Al alloy. Increasing the CNT concentration from 1.5 to 4.5 vol.% led to further decrease in both strength and elongation. In particular, the forged 4.5 vol.% CNT/2009Al composite exhibited a very low elongation of 1%, which is mainly attributed to a great number of large CNT clusters and voids in the clusters. After 1-pass FSP, the strengths of both 1.5 vol and 4.5 vol.% CNT/2009Al composites were significantly improved compared with those of the forged composites. Increasing the number of FSP pass from 1 to 4 resulted in a further strength increase for the CNT/2009Al composites, which was attributed to the further improved distribution of the CNTs in the aluminum matrix, but there was no obvious strength change for the 2009Al. The elongation of the 1.5 vol.% CNT/2009Al composite changed little with increasing the FSP pass, however, the 4.5 vol.% CNT/2009Al composite exhibited a gradually increasing elongation as the number of FSP pass increased.

Table 2. Tensile properties of 2009Al and CNT/2009Al composites

Specimen		*YS (MPa)*	*UTS (MPa)*	**El. (%)**
2009Al alloy	Forged	299	411	12
	1-pass FSP	297	421	13
	4-pass FSP	305	417	15
1.5 vol.% CNT/2009Al	Forged	307	392	8
	1-pass FSP	371	451	7
	4-pass FSP	385	477	8
4.5 vol.% CNT/2009Al	Forged	273	298	1
	1-pass FSP	414	437	2
	4-pass FSP	435	466	4

It is noted that under the forged condition, increasing the CNT volume fraction from 1.5 to 4.5% resulted in a decrease in the YS. However, under the FSP conditions (both 1-pass and 4-pass FSP), increasing the CNT volume fraction from 1.5 to 4.5% resulted in an increase in the YS. Compared with that of the 4-pass FSP 2009Al alloy, the YS of the 4-pass FSP 1.5 vol.% and 4.5 vol.% CNT/2009Al composites increased by 23.9% and 45%, respectively. Many references reported that the YS began to decrease as the CNT concentrations increased to a critical value (commonly smaller than 2 vol.%) [7]. However, in this study, the YS could still increase as the CNT concentration increased to as high as 4.5 vol.%. This is mainly attributed to the homogeneous dispersion of the CNTs in the aluminum matrix, the proper aspect ratio of the CNTs, and significant matrix grain refinement.

It is noted that the UTS of the composites did not exhibit a similar variation tendency. Increasing the CNT volume fraction from 0 to 1.5 % led to an increase in the UTS; however, the UTS began to decrease when the CNT volume fraction was increased further from 1.5 to 4.5 vol%. Another noteworthy phenomenon is that the increase of the UTS for the FSP composites was not as obvious as that of the YS. Compared with the FSP 2009Al, the maximum UTS increase for the 1-pass and 4-pass FSP 1.5 vol.% CNT/2009Al composites was only 7.4% and 13%, respectively. These need a further investigation.

Although the CNTs were uniformly dispersed into the 2009Al matrix after 4-pass FSP, the elongations of the 1.5 vol.% and 4.5 vol.% CNT/2009Al composites were still lower than that of the 2009Al alloy. This is attributed to the fracture mode of the composite. The composite fracture is directly related to the reinforcement cracking and interfacial failure. For the composites with a high fraction of reinforcements, voids tend to form at lower strains during tension due to the stress concentration at the reinforcement-matrix interfaces, resulting in a low level of elongation.

Conclusions

CNT/Al, CNT/6061Al and CNT/2009Al composites with uniformly dispersed CNTs were successfully fabricated by a combination of powder metallurgy and FSP.

The distribution homogenization of the CNTs increased as the number of FSP pass increased. As a result of the dispersed CNTs, the grain size of the FSP composites was significantly refined. The CNTs were shortened compared with the as-received ones, but still remained about 400 nm. The layer structures of the CNTs were well retained though some Al_4C_3 were formed.

The tensile strengths of the FSP composite, especially the YS, showed significant increase compared with the FSP matrix. The YS could be increased even at a relatively high CNT content of 4.5 vol.%. However, the maximum UTS was achieved at 1.5 vol.% CNT.

References

1. A. Krishnan, E. Dujardin, T.W. Ebbesen, P.N. Yianilos, M.M.J. Treacy, "Young's modulus of single-walled nanotubes", Phys Rev B, 58 (20) (1998), 14013-14109.
2. E.W. Wong, P.E. Sheehan, C.M. Lieber, "Nanobeam Mechanics: Elasticity, Strength, and Toughness of Nanorods and Nanotubes", Science, 277 (5334) (1997), 1971-1975.
3. J.P. Salvetat-Delmotte, A. Rubio, "Mechanical properties of carbon nanotubes: a fiber digest for beginners", Carbon, 40 (10) (2002), 1729-1734.
4. P. Poncharal, Z.L. Wang, D. Ugarte, W.A. De, "Electrostatic Deflections and Electromechanical Resonances of Carbon Nanotubes", Science, 283 (5407) (1999), 1513-1516.
5. Y.A. Kim, T. Hayashi, M. Endo, Y. Gotoh, N. Wada, J. Seiyama, "Fabrication of aligned carbon nanotube-filled rubber composite", Scripta Mater, 54 (1) (2006), 31-35.
6. I. Ahmad I, M. Unwin, H. Cao, Y.Q. Zhu, "Multi-walled carbon nanotubes reinforced Al_2O_3 nanocomposites: Mechanical properties and interfacial investigations", Compos Sci Technol, 70 (8) (2010), 1199-1206.
7. C.F. Deng, D.Z. Wang, "Processing and properties of carbon nanotubes reinforced aluminum composites", Mater Sci Eng A, 444 (1-2) (2007),: 138-145.
8. H.J. Choi, G.B. Kwon, G.Y. Lee, D.H. Bae, "Reinforcement with carbon nanotubes in aluminum matrix composites", Scripta Mater, 59 (3) (2008), 360-363.
9. K. Kondoh, T. Threrujirapapong, H. Imai, J. Umeda, B. Fugetsu, "Characteristics of powder metallurgy pure titanium matrix composite reinforced with multi-wall carbon nanotubes", Compo Sci Technol, 69 (7-8) (2009), 1077-1081.
10. P.B. Berbon, W.H. Bingel, R.S. Mishra, C.C. Bampton, M.W. Mahoney, "Friction stir processing: a tool to homogenize nanocomposite aluminum alloys", Scripta Mater, 44 (1) (2001), 61-66.
11. R.S. Mishra, Z.Y. Ma, I. Charit, "Friction stir processing: a novel technique for fabrication of surface composite", Mater Sci Eng A, 341 (1-2) (2003), 307-310.
12. Y. Morisada, H. Fujii, T. Nagaoka, M. Fukusum, "Nanocrystallized magnesium alloy-uniform dispersion of C_{60} molecules" Scripta Mater, 55 (11) (2006), 1067-1070.
13. L.B. Johannes, L.L. Yowell, E. Sosa, S. Arepalli and R.S. Mishra, "Survivability of single-walled carbon nanotubes during friction stir processing", Nanotechnology, 17 (12) (2006), 3081-3084.
14. Y. Morisada, H. Fujii, T. Nagaoka, M. Fukusumi, "MWCNTs/AZ31 surface composites fabricated by friction stir processing", Mater Sci Eng A, 419 (1-2) (2006), 344-348.
15. A.H. Feng, Z.Y. Ma, "Enhanced mechanical properties of Mg-Al-Zn cast alloy via friction stir processing", Scripta Mater, 56 (5) (2007), 397-400.
16. R.S. Mishra, Z.Y. Ma, "Friction stir welding and processing", Mater Sci Eng R 50 (1-2) (2005), 1-78.

Friction Stir Welding and Proceeing VII
Edited by: Rajiv Mishra, Murray W. Mahoney, Yutaka Sato, Yuri Hovanski, and Ravi Verma
TMS (The Minerals, Metals & Materials Society), 2013

Flow behavior of SiC particles as tracer material during the fabrication of MMCs by friction stir processing

Qing-Yu Shi[1], Kai Sun[1], Wei Wang[1], Gao-qiang Chen[1]

[1]Key laboratory for Advanced Materials Processing Technology, Department of Mechanical Engineering, Tsinghua University, Beijing China. 100084

Keywords: Friction stir processing, Material flow, Metal matrix composites

Abstract

Friction stir processing is a novel technique for metal matrix composites fabrication. During the friction stir process, the flow behavior of reinforced particles and base metal are very important. Very little is known about particle flow during fabrication of MMCs by friction stir processing, which is quite different from friction stir welding or friction stir processing without reinforced particles. The reinforced particles blocked the continuous flow of the base metal, significantly affecting the flow behavior. In this study, SiC particles were placed in different positions of the base metal before friction stir processing. The relationship between the initial positions and the final distribution of SiC particles was obtained. When the particles were put in middle layer of the plate and the rotation direction of the tool is opposite to the screw direction of the pin (counter-clockwise), the final distribution area is fairly large and uniform. This is the optimized position for making MMC's by FSP. Bulk metal matrix composites were achieved by placing the reinforced material in this position. The MMC showed a uniform microstructure and excellent tensile properties.

Introduction

Friction stir welding(FSW) is a solid state joining technique invented by The Welding Institute in 1991 [1]. FSW has a number of attributes that can be used to develop a generic tool for microstructural modification and manufacturing. Friction stir processing (FSP) was developed based on the basic concept of FSW [2]. During FSP, a rotating tool with a shoulder and a pin is plunged into the surface of the work piece. As the tool rotates, it feeds forward to cover the region of interest. Fine grain size and superplasticity can be achieved by the process. Several applications have been achieved in the past few years. Ma [3] achieved ultrafine-grained microstructure in friction stir processed Al-Mg-Zr. Mishra[4] studied high strain rate superplasticity in friction stir processed AA7075. Berbon[5] achieved exceptionally high superplasticity at high strain rates in a friction stir processed Al-Mg-Sc alloy. FSP was also used to modify the microstructure of as-cast Al-Si-Mg alloy[6]. Fabrication of MMCs is another important application for FSP. MMC plates are difficult to fabricate by traditional methods, such as particle metallurgy or stir casting. But they could be fabricated by FSP. Several researches have attempted to fabricate MMCs by FSP [7-12].

One of the most important aspects for fabrication of particle reinforced MMC's by FSP is to disperse the reinforced particles uniformly in the base metal. The distribution of reinforced particles is affected by the material flow during FSP. The material flow during FSP is quite complex, depending on the tool geometry, process parameters, and property of base metal and reinforced particles. Some research has been conducted in order to understand the material flow during FSW and FSP without reinforced particles. Liechty[13] used relatively soft plasticine and small steel particles with 1.0 mm in diameter to simulate FSP without reinforced particles. Collligan[14] used steel particles 0.38 mm in diameter embedded into a machined groove as the tracer material. Ke[15] inserted copper foil into the machined grooves of aluminum alloys as tracer material. London[16] put composite materials, which were Al-30%SiC and Al-20%W respectively, at the mid-plane of plate to study the material flow during FSW. These studies helped to understand the material flow during FSW. But the flow situation during the fabrication of MMCs by FSP is quite different from all the studies mentioned above. During FSW, the material is continuous and uniform, and the material flow is continuous. But the continuous flow of base metal is blocked by the reinforced particles during FSP and there is relative slippage among the reinforced particles in fabrication of MMC's. The relative slippage changed the flow behavior in this process. Even when the small steel balls were used in experiments [13, 14], the steel balls were still too large to simulate the distribution of the reinforced particles, which have an average diameter <10 μm. The steel balls can be easily dispersed, which is very different from the reinforced particles.

In order to reveal the flow behavior of reinforced particles during the fabrication of MMCs, SiC particles were used as the tracer material in this study. The objective of this paper is to study the flow behavior during fabrication of metal matrix composites by FSP, and find the optimized position for adding reinforced particles. SiC-AZ63 MMC was fabricated to validate this preferable position.

Experimental Procedures

Two base metal plates were prepared with dimension of 150 mm×60 mm×3 mm each. A slot along the longitudinal direction with a 0.2mm×0.5mm transverse section was cut in one plate. Commercially available SiC particles were selected as the tracer material. The SiC particles were filled into the groove and the plate with the slot was put upon the other plate as one FSP sample. For different samples, the groove was machined at different positions on the plate as shown in figure 1. The samples with different slot positions were defined as Sample_1-Sample_6. The samples with the slot located at upper layer position were Sample_1-Sample_3,

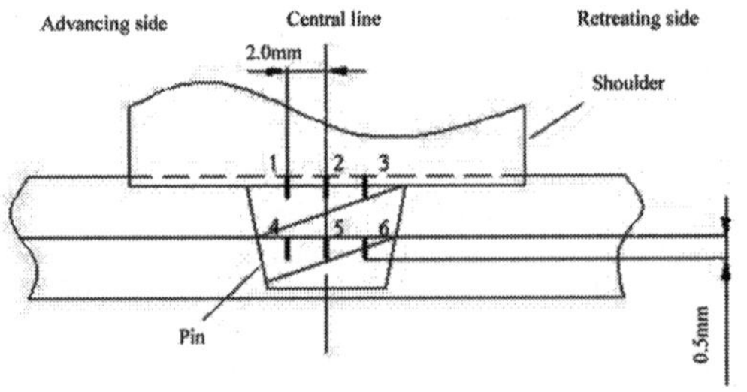

Figure 1. Illustration of the position of the groove.

from advancing side to retreating side. The samples with slot located at middle layer positions were Sample_4-Sample_6. The horizontal distance between the grooves at the upper and middle layers was 2 mm. Both clockwise and counter-clockwise tool rotation directions were applied.

As a result, 12 different samples were achieved with one groove in each sample. To distinguish between the clockwise group and the counter-clockwise group, the samples were named as Sample_1c-Sample_6c and Sample_1cc-Sample_6cc respectively. The groove and the FSP direction were in the same direction, so the final distribution of reinforced particles would be relatively uniform through the length of plate. The small cross-sectional area of the grooves was used to indicate the flow situation of specific position more accurately.

A rotation speed of 1200 rpm and feed speed of 100 mm/min were adopted during FSP. The tool tilt angle was 2.5° . A tool with a screwed pin was designed and produced as shown in Figure 2.

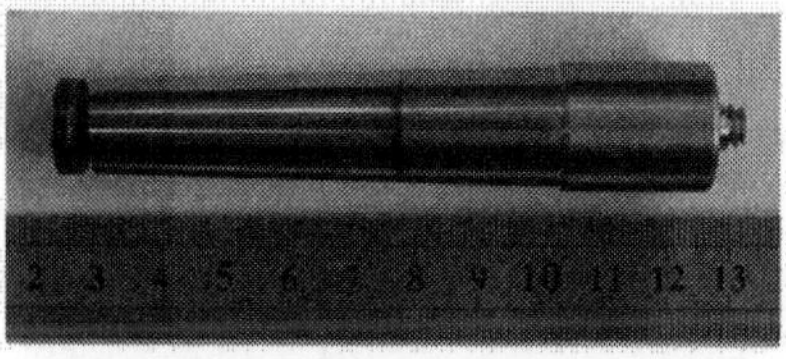

Figure 2. Photo of friction stir tool.

Transverse sections of as-processed plates were mounted and mechanically polished and examined by optical microscopy (OM).

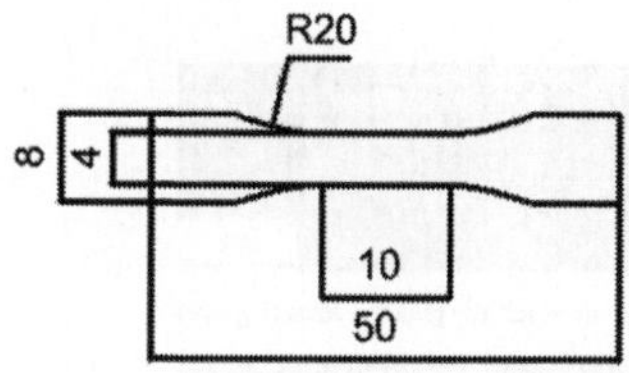

Figure 3. The shape and size of the tensile specimen.

To validate the most suitable location found from the above experiments, SiC-AZ63 MMC was fabricated. Tensile testing was performed on a WDW3020 multifunction testing machine at room temperature. The tensile test specimen was cut by a wire cutting machine along the FSP direction, 2mm below the top surface. The shape of the 2mm thick tensile sample was shown in figure 3.

Results and Discussion

Distribution of SiC with different tool rotation directions and initial positions

An OM photograph is shown in figure 4. The black part is the base metal, and the white part the SiC particles. The white horizontal line is the interface between the two plates. The white bold vertical line is the original position of the SiC, which has been magnified. "A" stands for advancing side and "R" stands for retreating side. Some defects are evident in the sample.

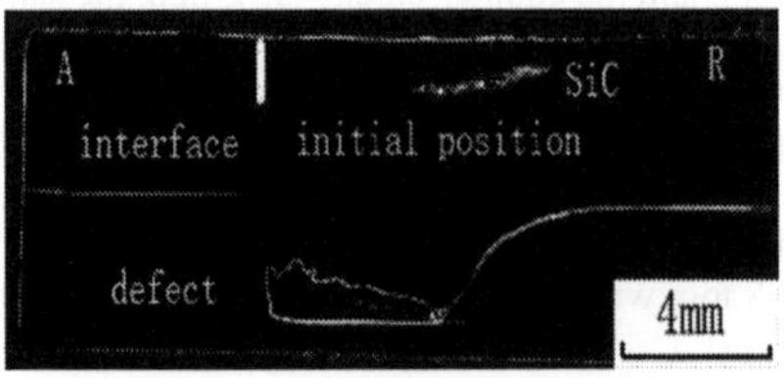

Figure 4. Typical OM photo of a transverse section of an FSP sample.

The SiC particle distributions with the tool rotation in clockwise and counter-clockwise directions were quite different, so the results are shown separately.

<u>Distribution Of SiC With Clockwise Tool Rotation.</u> Samples with Sample_1c-Sample_3c indicated the flow situation when particles were initially placed in the upper layer, as shown in figure 5. No matter where the SiC particles were initially placed, all the SiC particles were finally distributed on the retreating side of the upper layer. It is difficult to observe particles at the middle layer or the bottom layer. The initial SiC area was 0.1 mm^2, but after FSP, the particles dispersed in an area of about 2 mm^2. In the upper layer, the material flow was mainly affected by the shoulder. The pin had almost no effect on the upper layer of the plate.

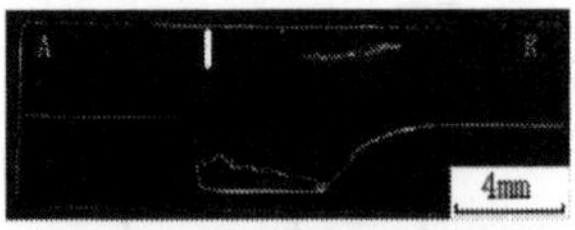

(a) No.1(particles initially placed at advancing side)

(b) No.2(particles initially placed at central line)

(c) No.3(particles initially placed at retreating side)

Figure 5. Distribution of particles initially placed at the upper layer, clockwise tool rotation.

Samples with Sample_4c-Sample_6c indicated the flow situation when particles were initially placed in the middle layer, as shown in figure 6. The SiC particles initially placed on advancing side had a distribution area of about 4 mm^2, and moved 1 mm upward. The SiC particles initially placed on the center line formed a distribution area of about 5 mm^2, most in the margin of the tool on advancing side. The SiC particles initially placed on the retreating side formed a continuous area of about 8 mm^2.The particles dispersed from advancing side to retreating side.

In general, the particles of the middle layer were mainly affected by the screwed pin and the final distribution area was significantly larger than that when particles were initially placed in the upper layer.

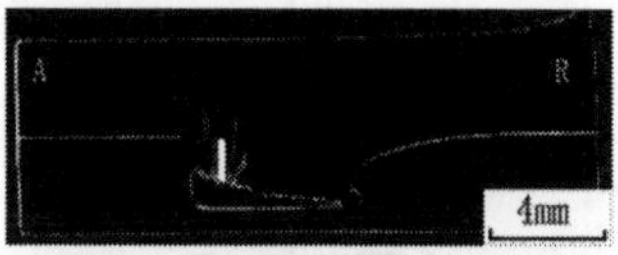

(a)No.4(particles initially placed at advancing side)

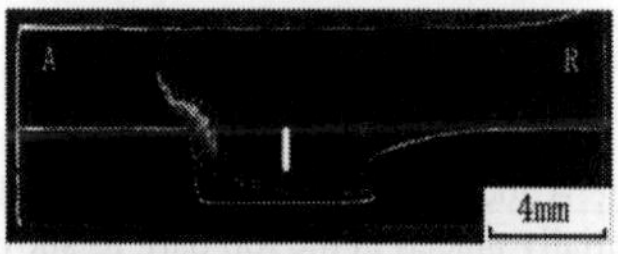

(b)No.5(particles initially placed at central line)

(c)No.6(particles initially placed at retreating side)

Figure 6. Distribution of particles initially placed at middle layer, clockwise tool rotation.

The defect could be observed in all the OM photographs in figures 5 and 6. Figure 6 showed that the material in the middle layer was affected by the pin and the particle distribution area was relatively large. This was especially true for the particles initially placed on the retreating side. The particles distributed from advancing side to retreating side after FSP and formed the largest area. In these clockwise experiments, the screw drove the material flow from bottom to top and formed defects with an average area of 5mm^2. Defects were formed in the bottom layer since there was insufficient material flow from other locations to fill the cavity, as illustrated in figure 7. The defect might not be easy to fix when it is located at the bottom. This would be a significant problem for manufacturing MMC's by FSP. Despite the defect, when reinforced particles were initially placed at the upper layer of the plate, all the reinforced SiC particles flowed to the retreating side no matter the initial particle placement, i.e., on the advancing side, central line, or retreating side. This led to a lack of reinforce particles on the advancing side.

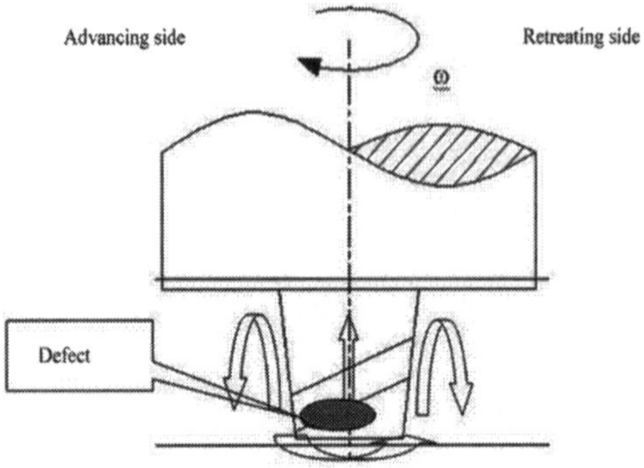

Figure 7. Material flow behavior for clockwise FSP.

Distribution Of SiC With Counter-Clockwise Tool Rotation. Sample_1cc-Sample_3cc indicated the flow when particles were initially placed in the top part, as shown in figure 8. The SiC particles initially placed on the advancing side formed an area about of 1 mm^2 near the central line of the top part. The SiC particles initially placed on the center line formed an area of about 4 mm^2 and moved 3 mm towards advancing side. The SiC particles initially placed on the retreating side formed an area about 3 mm^2 and distributed from advancing side to retreating side of the upper layer.

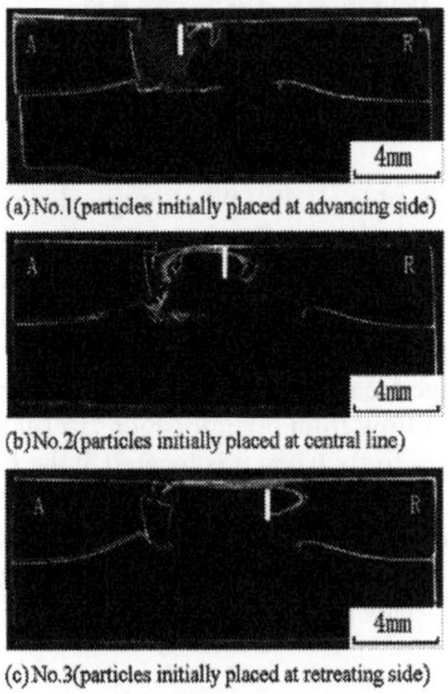

Figure 8. Distribution of particles initially placed at the upper layer, counter-clockwise tool rotation.

Sample_4cc-Sample_6cc indicated the flow when particles were initially placed in the middle layer, as shown in figure 9. The OM photographs revealed that the reinforced particles were very well dispersed. Regardless if the SiC particles were initially placed on advancing side, retreating side, or central line, they all moved 1 mm downward and distributed almost uniformly through the nugget zone. The area is fairly large, nearly $12mm^2$.

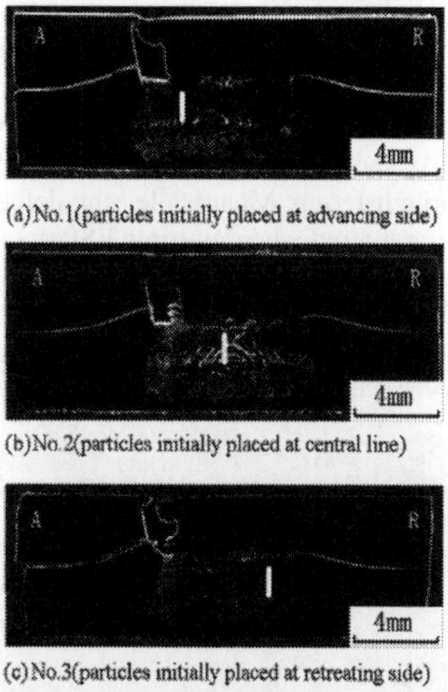

(a) No.1(particles initially placed at advancing side)

(b) No.2(particles initially placed at central line)

(c) No.3(particles initially placed at retreating side)

Figure 9. Distribution of particles initially placed at middle layer, counter-clockwise tool rotation.

Figure 8 indicated the upper layer material was mainly affected by the shoulder. The material was difficult to push down when in the upper layer. Figure 9 showed that the tracer material in the middle layer flowed with a different routine compared with that in the upper layer. The particles were stirred and distributed uniformly in a fairly large area. It could also be found that all the counter-clockwise samples had defects on the advancing side near the top surface. The material had a trend of being pushed downward by the screwed pin during FSP. The existence of the SiC powder reduced the ability of the material to flow. Accordingly, the material from the retreating side and the bottom could not flow to the advancing side to fill the hole. This resulted in a hole in advancing side of the top surface. When the defect is near the top surface, it would be not be as difficult to fix by multi-pass FSP as stated in reference [15]. So the fabrication of MMCs would not be affected.

It can be concluded from the above results that when the particles are put in middle layer of the plate and the rotation direction of the tool is opposite to the screw direction of the pin (counter-clockwise), it is easy to distribute the SiC particles in a fairly large and uniform area. Though defects might be generated during the first pass of FSP, they could be repaired by multiple FSP passes.

Validation of flow mechanism by fabrication of SiC-AZ63 MMC

A SiC reinforced Mg based metal matrix composite (MMC) was fabricated by FSP to validate the flow mechanism. Commercial magnesium alloy AZ63 was selected as the base metal. A 2mm×2mm groove was cut by a milling machine before FSP and subsequently filled with SiC powder. The groove was larger than the tracer experiment with the intention to increase the content of SiC reinforced particles. Multi-pass FSP was applied to make sure there would be no defect in the sample. Plates without a groove were also multi-pass FSPed in the same position as comparison group.

The OM pictures of the base metal, comparison group, and composite are shown in figures 10 (a), (b) and (c) respectively. As shown in figure 10 (c), SiC particles were distributed uniformly in a large area with no clustering. Meanwhile, the defects on the surface in the counter-clockwise situation were repaired when multiple FSP passes were applied.

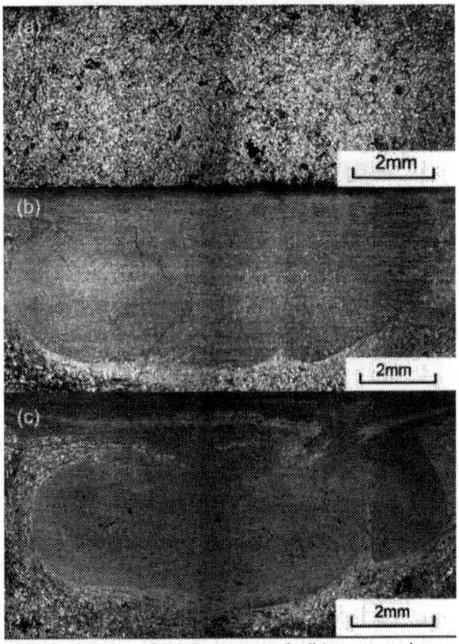

Figure 10. The OM picture of (a) base metal, (b) comparison group, and (c) composite.

Typical tensile curves of the base metal, comparison group, and composites are shown in Figure 11. The ultimate tensile strength of the composite was improved to 312 MPa, compared with 160 MPa of the base metal, and 263MPa of the comparison group.

The validation experiment proved that when SiC particles are filled in the optimized position, the MMC produced by FSP can produce a uniform microstructure and excellent mechanical properties.

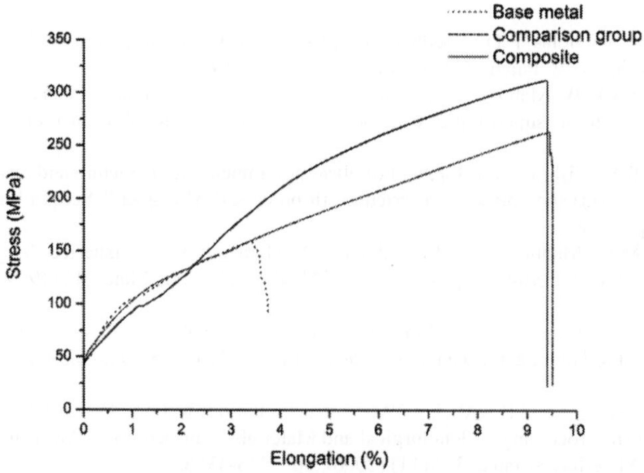

Figure 11 Tensile properties of MMC, comparison group and base metal.

Conclusions

Experiments have been carried out to study the flow behavior with SiC particles as tracer material during FSP. The influence of the FSP tool's rotation direction and initial position of the SiC powder was investigated. The main conclusions are as follows:

1. The FSP tool's rotation direction has a significant influence on the dispersion of SiC particles. The counter-clockwise rotation direction generates a more uniform SiC distribution compared with the clockwise rotation direction.
2. When reinforced particles are placed near the top surface, the final distribution area is relatively small. When reinforced particles are placed in the middle layer, the final distribution area is much larger. When the particles are put in middle layer of the plate and the rotation direction of the tool is counter-clockwise, the final distribution area is fairly large and uniform.
3. Fabrication of a metal matrix composite was achieved by placing the reinforced material in the optimized position. The reinforced particles dispersed uniformly in a relatively large area with no defect observed. MMC's produced by FSP showed excellent tensile properties.

Acknowledgement

This research was supported by the National Natural Science Foundation of PRC under contract number 50875146 and the National Science and Technology Major Project of the Ministry of Science and Technology of China under project No. 2012ZX04012-011.

Reference

1. Thomas, W.M., Nicholas, E.D., Needham, J.C., Church, M.G., Templesmith, P., Dawes, C.J., Friction stir welding. G.B. Patent Application No.9125978.8 1991
2. R.S. Mishra and M.W. Mahoney, "Friction stir processing: A new grain refinement technique to achieve high strain rate superplasticity in commercial alloys", Material Science Forum, 357(3) (2001), pp. 507-512.
3. Z.Y. Ma and R.S. Mishra, "Development of ultrafine-grained microstructure and low temperature (0. 48 Tm) superplasticity in friction stir processed Al-Mg-Zr.", Scripta Materialia, 53(1) (2005), pp. 75-80.
4. R.S. Mishra, M.W. Mahoney, S.X. McFadden, N.A. Mara and A.K. Mukherjee, "High strain rate superplasticity in a friction stir processed 7075Al alloy.", Scripta. Mater, 42 (1999), pp. 163-168.
5. P.B. Berbon, W.H. Bingel, R.S. Mishra, C.C. Bampton and M.W. Mahoney, "Friction stir processing: A tool to homogenize nanocomposite aluminum alloys", Scripta. Mater, 44 (2001), pp. 61-66.
6. Z.Y. Ma, S.R. Sharma and R.S Mishra, "Microstructural modification of as-cast Al-Si-Mg alloy by friction stir processing", Metallurgical and Materials Transactions A-Physical Metallurgy and Materials Science, 37A(11) (2006), pp. 3323-3336.
7. C.J. Hsu, C.Y. Chang and P.W. Kao, "Al-Al3Ti nanocomposites produced in situ by friction stir processing", Acta. Mater, 54 (2006), pp. 5241-5249.
8. Y. Morisada, H. Fuji, T. Nagaoka and M. Fukusumi, "MWCNTs/AZ31 surface composites fabricated by friction stir processing", Materials Science and Engineering A, 419 (2006), pp. 344-48.
9. Y. Morisada, H. Fuji, T. Nagaoka and M. Fukusumi, "Effect of friction stir processing with SiC particles on microstructure and hardness of AZ31", Materials Science and Engineering A, 433 (2006), pp. 50-54.
10. C.J. Lee, J.C. Huang and P.J. Hsieh, "Mg based nano-composites fabricated by friction stir processing", Scripta. Mater, 54 (2006), pp. 1415-1420.
11. W. Wang, Q.Y. Shi, P. Liu, H.K. Li, T. Li, "A novel way to produce bulk SiCp reinforced aluminum metal matrix composites by friction stir processing". J. Mater. Pro. Tech, 209 (2009), pp. 2099-2103.
12. M. Yang, C. Xu, C. Wu, et al, "Fabrication of AA6061/Al_2O_3 nano ceramic particle reinforced composite coating by using friction stir processing", Journal of Materials Science, 45(16) (2010), pp. 4431-4438.
13. B.C. Liechty, B.W. Webb, "Flow field characterization of friction stir processing using a particle-grid method". J. Mater. Pro. Tech, 208 (2008), pp. 431-43.
14. K. Colligan, "Material Flow Behavior during Friction Stir Welding of Aluminum", Weld. J, 78 (1999), pp. 229S-237S.
15. L.M. Ke, L. Xing, F.A. Huang, "Two-dimensional flow of plasticized materials in friction stir welded joints", Trans. China. Weld. Ins, 26 (2005), pp. 1-4.
16. B. London, M. Mahoney, W. Bingel, M. Calabrese, R.H. Bossi, and D. Waldron, "Material Flow in Friction Stir Welding Monitored with AlSiC and Al-W Composite Markers", *Friction Stir Welding and Processing II*, Edited by K.V. Jata, M.W. Mahoney, R.S. Mishra, S.L. Semiatin, and T. Lienert, TMS, March 2-6, 2003, pp. 3-12.

Friction Stir Welding and Processing VII
Edited by: Rajiv Mishra, Murray W. Mahoney, Yutaka Sato, Yuri Hovanski, and Ravi Verma
TMS (The Minerals, Metals & Materials Society), 2013

Processing, Microstructure and Mechanical Property Correlation in Al-B_4C Surface Composite Produced via Friction Stir Processing

Mageshwari Komarasamy[1], Rajiv S. Mishra[1], John A. Baumann[2], Glenn Grant[3], Yuri Hovanski[3]

[1]Center for Friction Stir Processing, Department of Materials Science and Engineering, University of North Texas, Denton, TX 76203, USA

[2]The Boeing Company, St. Louis, MO 63166, USA

[3]Pacific Northwest National Laboratory, Richland, WA 99352, USA

Keywords: Friction stir processing, Surface composite, Modulus enhancement.

Abstract

Friction stir processing (FSP) was employed to prepare surface composite (SC) composed of B_4C particles in 5024 Al matrix. The processing parameters, such as hole pattern and geometry, and the number of FSP passes, were optimized to obtain uniform powder distribution. The micrographs revealed a homogeneous distribution of the particles with good interfacial bonding. The hardness of the composite was uniform across the processed region which again indicates the uniformity of powder distribution. The modulus of the surface composite was measured using strain gage and showed a significant improvement.

Introduction

Friction stir processing (FSP), a variation of friction stir welding (FSW) [1], is an important tool in surface and subsurface modification to improve the components performance. The fabrication of SC using FSP was first introduced by Mishra et al. [2] and they studied the Al-SiC SC with different volume fractions of SiC particles, tool traverse rate and target depth. They have shown the optimum condition of tool traverse rate and target depth at which very good bonding between the SiC particles and the substrate was observed. The surface composites studied so far encompasses different base metals, like Al [2,3,4,5,6,7,8,9,10], Cu [11,12] and Mg [13,14,15], and reinforcement phases, like SiC, B_4C, TiC, Al_2O_3 and multi-wall carbon nanotubes (MWCNTs) for various requirements. Kurt et al. [3] studied the effect of tool rotation rate and traverse speed on 1050Al-SiC surface composite and observed that higher hardness were reached at higher tool rotation rate combined with low traverse speed. Shfiei -Zarghani et al. [4] studied 6083Al-50 nm Al_2O_3 surface nano-composite and observed the decrease in average Al_2O_3 cluster size, Al matrix grain size and increase in hardness with the increase in the number of FSP passes. Asadi et al. [13] produced AZ91-SiC surface composite with varying penetration depth, tool rotation rate and traverse speed, and optimized these parameters to obtain uniform powder distribution and higher hardness in the processed region. Barmouz et al. [9] studied the Cu-SiC with varying size and volume fraction of SiC particles and demonstrated the effect of particles size on the fracture behavior. They reported that the SC with nano sized SiC particles showed better wear resistance than micron sized SiC particles. Mahmoud et al. [16] studied the effect of FSP tool pin shape (circular with and without threads, square and rectangular) and size (pin diameter of 3, 5 and 7 mm) on the homogeneity of the powder distribution. Though the square shaped pin resulted in higher homogeneity and finer grain size, the tool wear rate was high and hence the incorporation of the tool debris into the processing zone, which can be undesirable.

Increasing the number of FSP passes [4,5,9] reduces the particle cluster size and increases the homogeneity of the powder distribution, which thereby enhanced the properties. The reinforcement powders were added into the base material in the form of paste prepared by mixing the powders with evaporating liquid like methanol [1,2,4] and then dried before FSP, and also filled into grooves [3,9,14], dimples [6] and holes [4]. There are different variants in the location of the hole, such as side hole filling [17] and top hole filling [4]. In side filling, the holes are drilled along the length (or width) of the plate and in top filling, the holes are drilled along the thickness of the plate. Though the side filling reduces the powder loss, the drilling of the holes across the length posed the limitation on the length of the hole. So top filling is usually preferred with changing the number of holes and even the pattern of holes. To hold the reinforcing powders inside the base metal and to avoid powder loss, the region is covered either with a cover sheet of the same base metal or closed with a pinless FSP tool.

The properties of the most often used reinforcing particles are given in Table I. Though the presence of ceramic particles inherently increases the strength and wear resistance of the metal matrix or surface composites, the main focus is not to increase the weight of the composite system tremendously, as these components are of major interest to aerospace industries.

Table I. Properties of a few often used reinforcing particles.

Ceramic powder	Young's modulus, GPa	Density, g/cm^3
Al_2O_3	380	3.95
SiC	400	3.1
TiC	440	4.9
B_4C	460	2.5

In this study we have attempted to produce 5024 Al alloys with B_4C as a reinforcement phase to improve the component strength and stiffness, using FSP. Boron carbide was chosen as reinforcing phase because of its higher modulus and also its lower density compared to the other commonly used reinforcement phases like SiC and Al_2O_3, as discussed in Table I. The process of producing SC using FSP is discussed later in this paper. The distribution of the reinforcing particles and both the hardness, and modulus measurements are also presented with further discussion.

Experimental Procedure

Commercially available B_4C powders (6 μm average particle size) and 5024 Al (Al-Mg-Sc) alloy were used in this study. Holes of 1.6 mm depth and 2 mm diameter were first drilled into the 5024 Al alloy using a mini-CNC machine. The hole pattern was designed to obtain the maximum efficiency of particle incorporation in the processed region and the volume fraction of the reinforcing phases can be controlled by changing the number of the rows of holes in the processing zone. The staggered hole pattern which showed improved homogeneity is shown in Figure 1. The 5024 Al alloy plates were thoroughly cleaned before powder filling. The B_4C powder was pressed into the holes to maximize the green density of powders.

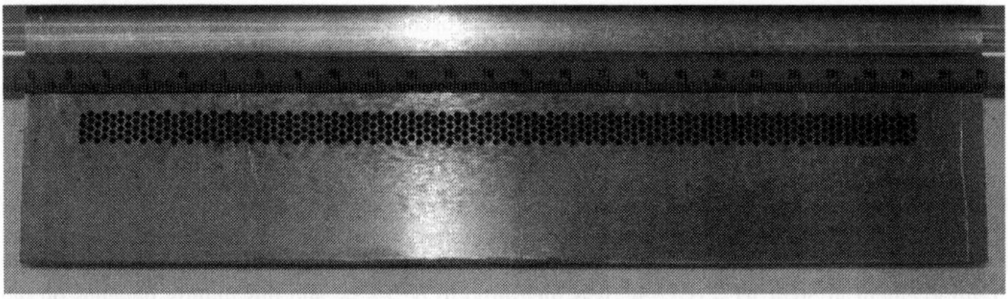

Figure 1. Top view of the staggered holes of eight rows in 5024 Al base metal and filled with B_4C powder.

First, the holes were closed by capping pass wherein a pinless FSP tool was used. The processing parameters used for this capping pass were 600 rpm tool rotation rate, 4 ipm tool traverse speed and zero tool tilt. The number of capping passes can be more than one, depending upon the surface inspection for unclosed holes. Then the plates were processed using FSP tool with pin. The processing parameters for the stirring passes were 1000 rpm tool rotation rate, 2 ipm tool traverse speed and 2.5° tool tilt angle. After each pass, the plates were rotated by 180 degree, so that the advancing side of the previous pass became the retreating side in the next pass and vice-versa. This was done to ensure a better powder distribution in the processed region and to reduce powder agglomeration in the advancing side. The plate was subjected to 4 passes to obtain homogeneous powder distribution. The process parameters were optimized to give uniform powder distribution, better interfacial bonding and mixing of powders. The parameters were held constant for all of the stirring passes. The friction stir processed material was then cold rolled to 2 mm thickness (~ 60% thickness reduction) with 0.5 mm being the composite region.

The transverse cross section was polished down to 1 micron using diamond suspension. Optical micrographs were taken at various magnifications for analyzing the powder distribution at different regions in the nugget. The volume fraction analysis was carried out using ImageJ software. The modulus of the surface composites was measured using strain gage during tensile testing. The test sample was taken along the processed region so that the entire gage region composed of the processed region. The sample was polished to 1 μm surface finish using diamond suspension. The strain gage was installed onto the composite side of the material. The micro strain response of the material in the elastic regime was recorded for the applied load using Vishay P3 strain indicator, which was attached to the mini-tensile testing machine. The modulus of only the composite layer was obtained by removing all of the base metal by polishing. The Vickers microhardness tester was used for hardness measurement in the transverse section across the processed region in both the as-processed and cold rolled conditions. The hardness testing was done at 200 g load for 10 s dwell time.

Results and Discussion

The macrostructures of the vertical cross section in both as- processed and after cold rolling conditions are shown in Figures 2 (a) and (b), respectively. The bright contrast in the processed region shows the presence of B_4C particles and is uniform throughout, which shows the homogeneous nature of powder distribution at macroscopic scale. There are no dark patches in the processed region which clearly indicates the absence of the processing defects. The ratio of the composite layer thickness to the base metal in the SC was maintained at 1:3 after cold rolling to ensure optimum strength and ductility. The optical microstructure of the surface composite is shown in Figure 3. For the above mentioned process parameters, perfect interfacial bonding between the surface composite and 5024 Al base material exists and is shown in Figure 3(a). Figures 3 (b) and (c), show the powder distribution, which is uniform and no clustering of the particles were observed. However, some regions showed different levels of powder incorporation, in terms of volume fraction of B_4C. The overall volume fraction of B_4C reinforcement in 5024 Al – B_4C SC was measured to be 12 %, and this mainly depends on the number of rows of holes. For example, the volume faction can be lowered by having four rows instead of eight, by keeping the tool geometry and other process parameters constant. In all the SC prepared by FSP, perfect interfacial bonding between surface composites and the metal substrate was observed, and the homogeneity of the powder distribution always increased with the number of FSP passes.

(a) (b)

Figure 2. The macrostructure of the surface composite along the vertical cross section, (a) after processing, and (b) after cold rolling.

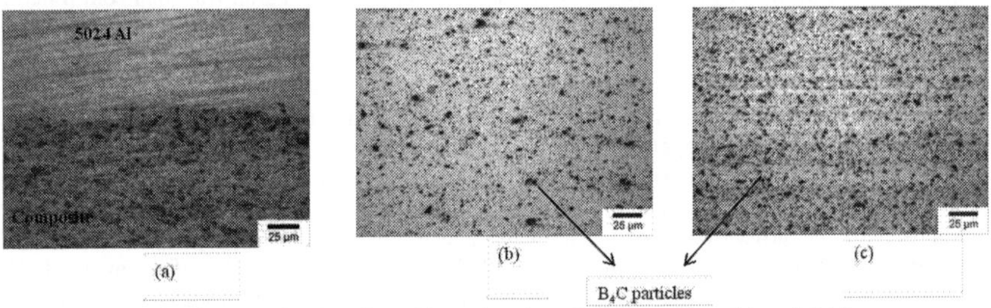

Figure 3. Optical micrographs showing, (a) perfect interfacial bonding, (b) and (c) homogeneous powder distribution with various levels of powder incorporation.

Microhardness Measurements

Vickers microhardness measurements were made in both as FS processed and cold rolled conditions for 5024 Al-B_4C surface composite. The hardness profile of the 5024 Al-B_4C is shown in Figure 4. The average hardness is 96.7±7.2 VHN and 152.7±13.4 VHN in the as-processed and the cold rolled conditions, respectively.

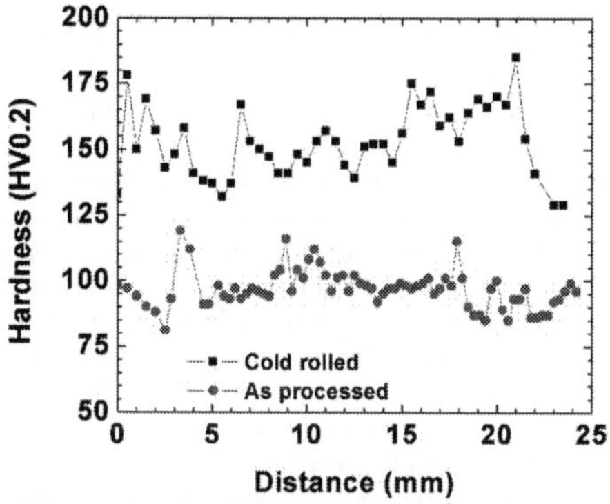

Figure 4. Vickers microhardness profiles of 5024 Al-B_4C surface composite in the as-processed and cold rolled conditions.

As expected, the presence of B_4C particles increased the hardness of the SC. This hardness improvement mainly depends on the type, size, the volume fraction of reinforcing particles incorporated and the FSP parameters. For a given particle type, the hardness of the SC always increases with the number of FSP passes and attains maximum value after three or four passes. The size of the particle governs the type of strengthening mechanisms operative in SC [18], in the case of nano sized particles, it is mainly Orowan strengthening and also the contribution from finer grain size, and in the case of micron sized particles, it will be mainly due to load transfer from the matrix to the particle and the hardening effect from geometrically necessary dislocations (GNDs). The presence of GNDs increase the dislocation density in the material and hence the work hardening capacity of the SC. In addition, 5024 Al alloy is primarily a solid solution strengthened alloy and cold rolling increased the hardness because of work hardening.

Elastic Modulus Measurements

The modulus measurements were done using micro-strain responses as explained before, and approximately five tests were done for better statistical distribution. The modulus of the surface composite is 79±2 GPa and the modulus of the composite alone is 133±2 GPa, as shown in Table II. The modulus of the surface composite is nearly 14% higher and of the composite layer alone is nearly 91% higher than the base metal. In the presence of ceramic particles, the modulus of the SC increased appreciably. This improvement in the modulus is due to the load transfer from the base metal to the reinforcement particles and also from the base metal to the composite region.

Table II. Elastic moduli of the surface composite and the composite layer.

Material	Modulus (GPa)
5024 Al base metal	70
5024 Al-B_4C surface composite	79±2
5024 Al-B_4C composite layer only	133±2

Because of this load transfer, the Al matrix is expected to experience lower stress as compared to the applied stress. As the modulus of the reinforcing phase increases, the load transferred to them increases which ultimately reduces the load on Al matrix. There are only a few studies [19-24] in which the modulus improvement in SC prepared using FSP were measured. In the literature, there are primarily two methods used to measure the modulus of the composites, such as instrumented indentation hardness tester [24] and slope of the normal stress-strain curve [19-23]. So there cannot be one to one comparison in modulus values, because of the variety of reinforcement particles used, volume fraction of the particles, FSP processing parameters and the type of technique employed to measure modulus. The summary of the modulus of composites fabricated using FSP is shown in Figure 5, and in comparison with the current study. In general, increase in the number of FSP passes, and the volume fraction of reinforcement particle enhances the modulus of the composites. The modulus of the 5024-B_4C SC is lower and this can be attributed to the fact that this modulus value is a contribution from both the base metal and the composite layer. The modulus value of the composite layer is nearly double that of the base metal and this is mainly attributed to the presence of higher modulus of the B_4C particles and uniform powder distribution. This also clearly indicates the effectiveness of the B_4C particles on load bearing characteristics. The modulus of the Al-15Ti with almost 50% Al_3Ti phase is somewhat closer to the 5024-B_4C composite modulus values. The size of the reinforcement particle also plays an important role on modulus. Mazaheri et al. [24] concluded that both hardness and modulus values are higher for composites with nano sized particles than for composites with micron sized particles. This is partly because of the shift in mechanism of load transfer at very fine particle sizes.

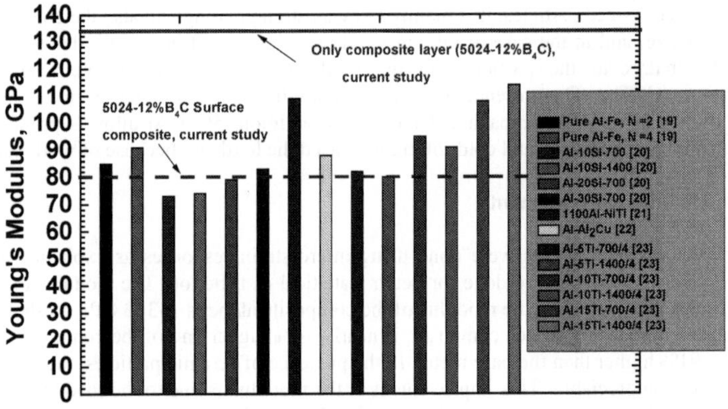

Figure 5. A comparison of modulus values of the SC/composites prepared using FSP.

Conclusion

In the present study, a 5024 – B_4C surface composite was successfully prepared using FSP and the following conclusions were made from the characterization.

1. The uniform distribution of powders in the processed zone and better interfacial bonding between base metal and the composite region were achieved in 5024 Al-B_4C surface composite for the above mentioned processing parameters.
2. The hardness of the surface composites were higher than the base metal hardness and this is mainly attributed to the presence of higher modulus powders, finer grain size, GNDs and increased work hardening.
3. The Young's modulus of the 5024 Al-B_4C surface composite is enhanced for both surface composites and for the composite layer alone. The modulus of the composite layer is almost twice that of the base metal modulus.

References

1. Mishra RS., Ma ZY., Friction stir welding and processing, Materials Science and Engineering R, (2005), 50, pp1-78.
2. Mishra R.S., Ma Z.Y., Charit I., Friction Stir Processing: a novel technique for fabrication of surface composite, Materials Science and Engineering A341, (2003) pp.307-310.
3. Kurt A., Uygur I., Cete E., Surface modification of aluminum by friction stir processing. Journal of Materials Processing Technology, 211 (2011) pp.313-317
4. Shafiei-Zarghani A., Kashami-Bozorg S.F., Zareu-Hanzaki A., Microstructures and mechanical properties of Al/Al_2O_3 surface nano-composite layer produced by friction stir processing, Materials Science and Engineering A500, (2009) pp.84-91.
5. Yang et al., Fabrication of AA6061/Al2O3 nano ceramic particle reinforced composite coating by using Friction stir processing, Journal of Materials Science, 45 (2010), pp 4431-4438.
6. Wang et al., A novel way to produce bulk SiCp reinforced aluminum metal matrix composites by friction stir processing, Journal of Materials Processing and Technology, 209, (2009), pp2009-2103.
7. Qu et al., Improving the tribological characteristics of aluminum 6061 alloy by surface compositing with sub-micro-size ceramic particles via friction stir processing, Wear 271 (2011), pp1940-1945.
8. Alidokht S A et al., Microstructure and tribological performance of an aluminum alloy based hybrid composite produced by friction stir processing, Materials and Design, 32 (2011), pp2727-2733.
9. KashaniBozorg S F., Jazayeri K., Formation of Al/B4C Surface Nanocomposite Layers on 7075 Al Alloy employing Friction Stir Processing, AIP Conf. Proc. 1136,(2009), pp 715-719.
10. Rejil C M et al., Microstructure and sliding wear behavior of AA6360/(TiC + B4C) hybrid surface composite layer synthesized by friction stir processing on aluminum substrate, Materials Science and Engineering A 552 (2012),pp336– 344.
11. Barmouz M., Asadi P., Besharati Givi M.K., Taherishargh M., Investigation of mechanical properties of Cu/SiC composite fabricated by FSP: Effect of SiC particles' size and volume fraction, Materials Science and Engineering A528, (2011) pp.1740-1749.
12. Barmouz M., Besharati Givi M.K., Seyfi J., On the role of processing parameters in producing Cu/SiC metal matrix composites via friction stir processing: Investigating microstructure, microhardness, wear and tensile behavior, Materials Characterization, 62, (2011), pp108-117.

13. Asadi P., Faraji G., Besharati M.K., Producing AZ91/SiC composite by Friction Stir Processing, International Journal of Advanced Manufacturing Technology (2010) pp. 247 – 260
14. Morisada Y et al., MWCNTs/AZ31 surface composites fabricated by friction stir processing, Materials Science and Engineering A, 419, (2006), pp 344-348.
15. Asadi P et al., Experimental Investigation of Magnesium-Base Nanocomposite Produced by Friction Stir Processing: Effects of Particle Types and Number of Friction Stir Processing Passes, Metallurgical and Materials Transactions A, 42A, (2011), pp2830-2832.
16. Mahmoud et al, Effect of friction stir processing tool probe on fabrication of SiC particle reinforced composite on aluminum surface, Science and technology of welding and joining, 14(5), (2009), pp 413-425.
17. M Dixit., Validation of microstructure and property correlation for design of materials, MS thesis, Materials Science and Engineering, Missouri S&T, Rolla, Missouri.
18. Lloyd, D.J., Particle reinforced aluminium and magnesium matrix composites. Int. Mater. Rev. 39 (1994), pp1–24.
19. Lee I S., Kao P W., Ho N J., Microstructure and mechanical properties of Al–Fe in situ nanocomposite produced by friction stir processing, Intermetallics 16 (2008), pp1104–1108.
20. Lee et al., Particle-reinforced aluminum matrix composites produced from powder mixtures via friction stir processing, Composites Science and Technology 71 (2011), pp693–698.
21. Dixit M., Newkirk J W., Mishra R S., Properties of friction stir-processed Al 1100–NiTi composite, Scripta Materialia 56 (2007), pp541–544.
22. Hsu C J., Kao P W., Ho N J., Ultrafine-grained Al–Al2Cu composite produced in situ by friction stir processing, Scripta Materialia 53 (2005), pp341–345.
23. Hsu et al., Al–Al3Ti nanocomposites produced in situ by friction stir processing, Acta Materialia 54 (2006), pp5241–5249.
24. Mazaheri Y., Karimzadeh F., Enayati M H., A novel technique for development of A356/Al2O3 surface nanocomposite by friction stir processing, Journal of Materials Processing Technology 211 (2011), pp1614– 1619.

FRICTION STIR WELDING AND PROCESSING VII

Friction Stir Welding: High Temperature Materials – I

Friction Stir Welding and Proceesing VII
Edited by: Rajiv Mishra, Murray W. Mahoney, Yutaka Sato, Yuri Hovanski, and Ravi Verma
TMS (The Minerals, Metals & Materials Society), 2013

Comparison between Friction Stir and Submerged Arc Welding Applied to Joining DH36 and E36 Shipbuilding Steel

Stephen Cater[1], Jonathan Martin[1], Alexander Galloway[2], Norman McPherson[3]

[1]TWI Technology Centre (Yorkshire) Ltd, Rotherham, S60 5TZ, United Kingdom
[2]University of Strathclyde, Glasgow, G1 1XJ, United Kingdom
[3]BAE Systems Surface Fleet, Glasgow, G51 4XP, United Kingdom

Friction, FSW, submerged arc, steel, DH36, E36, butt, shipbuilding

Abstract

With the impending development of FSW tools for steel with useful lifetimes, attention has turned to the mechanical properties of the welds that can be made in a range of industrially significant steels. This work reports on a comparative study undertaken to examine the use of friction stir and submerged arc welding on DH36 and E36 shipbuilding steels. The study made an assessment of the distortion induced in fabricating plates by the two welding techniques, and provides initial comparative data on weld tensile strength, toughness and fatigue life. In each case, friction stir welding was shown to outperform submerged arc welding.

Introduction

Friction stir welding is an established process widely used for the fabrication of safety critical structures in aluminium, magnesium and copper alloys across many market sectors. It is recognised that in these materials the process brings benefits in terms of weld integrity, durability, fatigue life and reduced cost when compared with conventional fusion welding techniques. Until recently, the transfer of this capability into the steel sector was stalled by the relatively poor performance and high cost of the tools required. This situation is now changing, with composite pcBN/W-Re FSW tools becoming available that are capable of producing industrially useful lengths of welds in steel. Friction stir welding (FSW) is therefore a joining technique of interest to Shipbuilders who are attempting to identify low distortion joining methods that will allow cost-effective fabrication of steel sheets. Using conventional arc welding methods causes significant amount of distortion that relate to increased costs associated with fit-up, fabrication and installation. However any adopted welding technique must also produce joints with at least comparable tensile, fatigue and impact performance.

The work detailed below shows the developmental progression of applying FSW to butt welding of two shipbuilding steels and assessing the joint performance. The two steels selected are DH36 and E36 which are carbon manganese niobium steels with a minimum yield strength of 360N/mm^2 and a minimum impact requirement of 36J at -20°C and -45°C respectively.

There has been previous work in this area (1) which concluded that there was a capability to produce single pass full penetration FSW welds in 6.4mm thick steel plate. The steel used was not C-Mn-Nb but a C-Cr-Mo-V steel. Increasing the travel speed resulted in a progressive increase in weld metal hardness to a maximum level of 350Hv, which is close to the acceptable maximum. The all weld metal mechanical properties appeared to be acceptable with yield and tensile strengths well above the base plate. However, the elongation was below specification and was a function of the welding speed. There was no description of the toughness of the steel. On the basis of hardness and elongation it could be predicted that there was probably an issue in achieving the specification requirements. Another publication (2) alludes to a problem meeting the toughness requirements for DH36. However in that same publication (2), there appeared to be a potential to have better toughness when welding HSLA 65. The work did however show that lower distortion was produced from FSW processed material. An additional publication (3) from the same source highlighted the benefits of FSW as being, less in need for fume extraction and having lower distortion.

Summary of Initial Work Programme

An initial study was carried out in 2012 [4] comparing butt welding of DH36 steel using FSW and submerged arc welding techniques. Distortion was found to be lower in friction stir welded steel plates of 4, 6 and 8 mm thickness than in equivalent submerged arc welded (SAW) plates as shown in Figures 1 to 3 and listed in Table 1. Also no issues were identified with weld metal strength, and Charpy impact toughness at -20°C which was found to be comparable but more uniform across the weld area than with the submerged arc welded material. Microstructural observations were linked to hardness, toughness and fatigue test data.

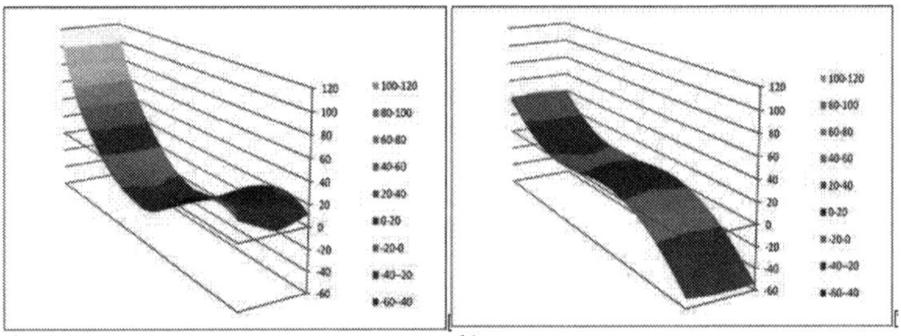

a) b)

Figure 1 Distortion comparison between 4mm thickness DH36 steel plates, 2000 x 400mm, butt welded along the 2000mm edge:
a) SAW;
b) FSW.

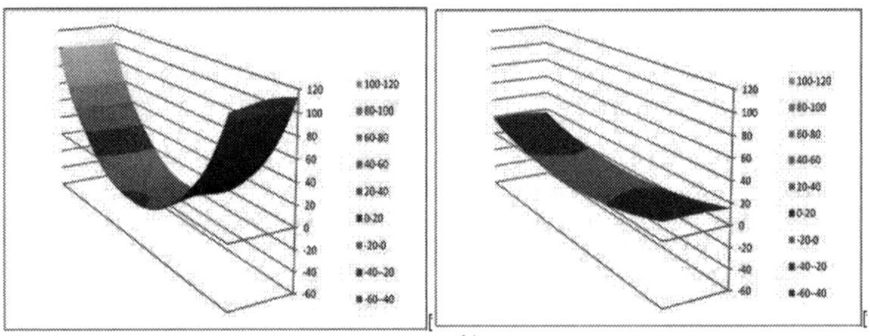

a) b)

Figure 2 Distortion comparison between 6mm thickness DH36 steel plates, 2000 x 400mm, butt welded along the 2000mm edge:
a) SAW;
b) FSW.

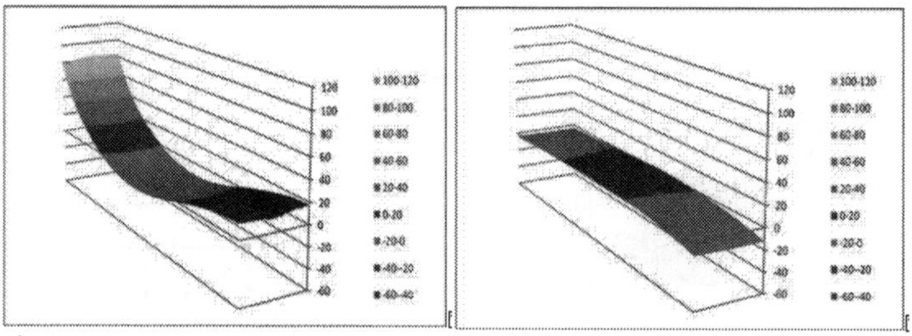

a) b)

Figure 3 Distortion comparison between 8mm thickness DH36 steel plates, 2000 x 400mm, butt welded along the 2000mm edge;

a) SAW;
b) FSW.

Table 1 Summary of distortion in DH36 steel plates butt welded along the 2000 mm edge.

Plate thickness, mm	Distortion over 2000mm submerged arc welded plates	Distortion over 2000mm friction stir welded plates
4	120	60
6	110	20
8	80	15

The fatigue properties reported showed the FSW plates to outperform the SAW plates. A typical result is shown in Figure 4.

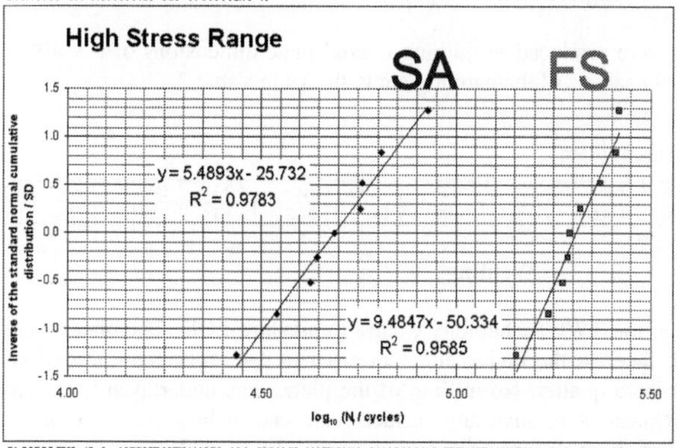

Figure 4 Comparison of high stress fatigue properties of SAW and FSW DH36 steel.

Further work [5] showed that double sided FSW of 8mm DH36 surprisingly gave an increase in the level of distortion from 15 to 25 mm as shown in Figure 5

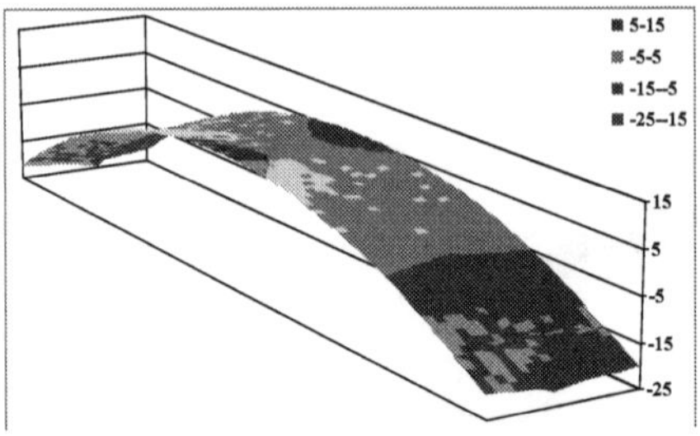

Figure 5 Distortion measurement of 8mm thickness DH36 steel plates, 2000 x 400mm, double sided butt welded along the 2000mm edge;

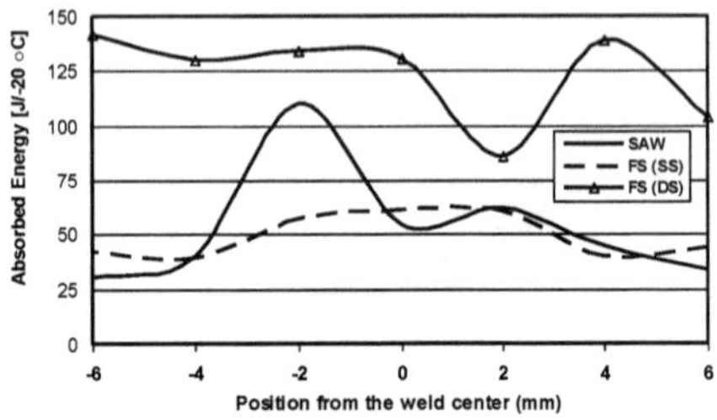

Figure 6 Absorbed energy along the weld joint

This work also showed that a double sided FSW weld has the potential to improve the weld toughness over the single sided FSW and SAW as shown in Figure 6 however further investigation and data was required.

Comparable studies in E36 Steel

To compare these properties and obtain further supporting data a second steel grade is being investigated namely, E36, a tough, moderate strength (355MPa) steel used for marine construction, particularly ship hulls and superstructures, and offshore structures such as oil rigs.

Experimental Approach

Six friction stir welded plates were produced by joining overall plate dimensions of 6 x 400 x 2000mm. The nominal chemical analysis of the parent plate is shown in Table 2.

Table 2 Chemical analysis of parent plate

%C	%Si	%Mn	%P	%S	%N	%Al	%Ti	%Cu	%Cr	%Ni	%Mo	%Nb	%V
0.13	0.31	1.45	0.011	0.001	0.004	0.048	0.02	0.05	0.03	0.04	0.004	0.03	0.004

Welding

The FSW plates were welded at the TWI Technology Centre (Yorkshire), UK. The plates were welded using the 'as received' faces as the faying edges in order to determine the effect of no preparation of the edges upon weld quality. No milling of the plates was undertaken to ensure that they were of uniform thickness or remove any surface scale and inclusions. Some plates were given a light dressing along the weld line with a sanding disc on an angle grinder prior to welding to remove excessive mill scale but no attempt was made to fully remove the scale and take the plate back to bare metal of uniform thickness. This preparation regime was adopted to investigate how robust the FSW process is for industrial applications. A picture of the welding set up is shown in Figure 7

The FSW tools used in this study were manufactured by MegaStir Inc. from a composite ceramic material: refractory metal designated Q70. This material contains 70% by volume of

polycrystalline boron nitride in a tungsten-rhenium binder. The tool design comprises of a stepped spiral probe and scroll shoulder, made from a single piece of pcBN-WRe composite, mechanically held in a metal shank. An argon gas shield was also used, more to protect the tool rather than the surface area of the welded region.
One plate was welded from just one side to give a full penetration through thickness weld whilst other plates were welded from both sides. These double-sided welded plates had different degrees of weld zone overlap as indicated in Table 3. There were two reasons for the manufacture of double-sided plates:

1. FSW is currently limited in the thickness of steel that can be welded by the size of FSW tool that can be manufactured from the pcBN-WRe composite and thus for the foreseeable future welds in steels thicker than 8mm will need to be made as double sided welds
2. There are indications from the initial work reported by McPherson *et al* that double sided welds may have superior properties to single sided welds.

TWI Image SYF 19673-5150

Figure 7 Image showing two plates of E36 steel clamped into the welding jig prior to welding.

As far as possible, the welds used similar welding parameters based upon those reported to give good welds in the initial work programme when welding DH36. For all the double sided welds, the second weld was made on the reverse of the plate in the opposite traverse direction to the first weld in order to even out the asymmetry of the FSW process. (That is, the advancing side of the second weld was over the retreating side of the first weld).

Table 3 Details of weld set up and preparation for each plate

Plate Ref	Type of weld	Overlap zone, mm/%	Plate preparation note
Plate 1	Double sided, 6mm, full penetration from both sides	6/100	Joint line lightly sanded prior to welding
Plate 2	Double sided, 6mm first then 4mm on reverse.	4/66	Joint line lightly sanded prior to welding
Plate 3	Double sided, 4mm then 4mm.	2/33	Minimal surface preparation - wipe with Scotchbrite
Plate 4	Double sided, 4mm first	1/16.6	No surface preparation

	then 3mm on reverse		
Plate 5	Double sided, 3mm first then 4mm on reverse	1/16.6	Minimal surface preparation
Plate 6	Single sided, 6mm, full penetration	0/0	Surface lightly prepared with sanding wheel

Testing

All the plates were measured for distortion. Plates 1, 3 and 6 were subjected to further tests as shown below. These plates were selected to represent the range of tool penetration from single-sided through thickness to double-sided welds having full and partial tool zone overlaps.

- Cross weld tensile strength
- Weld centreline Charpy toughness
- Fatigue testing

Distortion measurement

The plates were assessed for distortion using a laser measurement scanner at the University of Strathclyde. The plates were marked with the position of 4 identical reference points coinciding with the position of the mounting points, to ensure that all the plate measurements were consistent with each other. The plates were then lifted on to the measurement platform and aligned on to the four mounting points.
The overhead travelling laser was then aimed at the bottom left corner of the plate and programmed to measure the position of the plate surface over a series of 6370 longitudinal and 2530 lateral steps. (The welded plates were 2000mm long by 400mm wide.) It then took around 10 minutes for the laser to scan each plate. After the laser had finished scanning each plate, the data were stored in a file compatible with subsequent analysis in Excel. Distortion was defined as the deviation from a plane bounded by the four reference points. Prior to welding, all the plates were flat.
The measured distortions are presented in Table 4 below, and a typical distortion map is presented in Figure 8.

Table 4 Measured plate distortion.

Plate	Overlap zone, %	Type of weld	Maximum distortion, mm
1	100	Double	9.5
2	66	Double	17.5
3	33	Double	10.5
4	16	Double	22
5	16	Double	22
6	0	Single	20.2

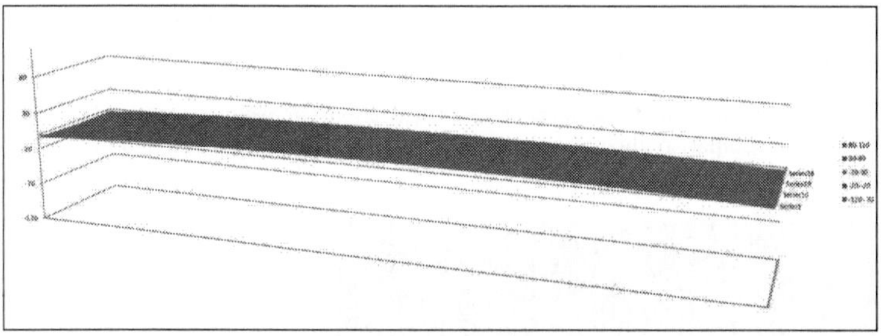

Figure 8 Typical distortion map of E36 plate. Maximum measurement in middle of plate.

All the plates distorted by arching upwards in the centre of the plate with the last weld uppermost.
The results obtained in this study confirmed the earlier work reported by McPherson et al indicating that FSW induces far less distortion in carbon steel weldments than does submerged arc welding. As may be seen from Table 4, all plates exhibited a total distortion of less than 22mm over the 2000m length. By way of comparison, the FSW 6mm thickness DH36 plate had a similar magnitude of distortion (20mm) but the SAW plate was significantly more distorted, almost 110mm. It is considered that the primary reason for this is the lower thermal energy input to a friction stir weld and the lack of volume changes that occur during the melting and solidification cycle of a fusion weld. There may be a further, though lesser, contributory effect from the fact that plates welded by friction stir techniques are clamped very rigidly in place throughout the welding process. The different microstructures that can be generated by FSW may change the volumetric variations and thus residual stresses present in the weld zone, perhaps further reducing distortion. Finally, any residual stresses in the weld will also be influenced by a second pass of the tool in double welded samples though from the limited number of samples in this study it is not possible to identify with certainty the presence and magnitude of this effect.

Tensile testing

To determine the strength of the welds made, cross weld tensile tests were performed on samples from the welded plates in accordance with EN ISO 6892-1:2009. The testing was performed by Exova (UK) Ltd.

Table 5 Tensile strength of friction stir welds

Plate	Degree of overlap	UTS, MPa	Comment
1	100	561	Parent break
3	33	576	Parent break
6	0	574	Parent break

All the tensile test specimens failed in the parent plate. The tested samples had an average strength of 570MPa indicating that the welds were at least as strong as parent material and in all cases exceeded that minimum material strength by a significant margin. The presence of defects in some of the welds tested did not appear to have too great a detrimental effect upon weld

strength. In those cases where surface oxides were not removed from the faying edges before welding, it appears that the tool broke up these layers and distributed them as small particles throughout the weld zone.
The degree of grain refinement that can take place in a friction stir weld is clearly shown by Figure 9 which shows a comparison between the grain sizes in the parent material and the TMAZ of a weld in E36 steel.

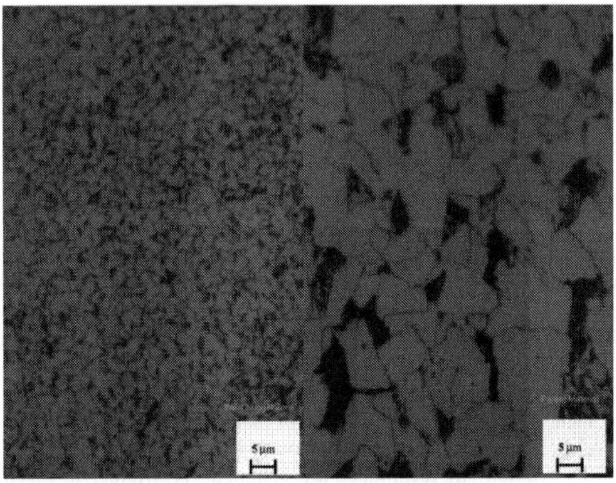

Figure 9 Comparison of grain size in the TMAZ and parent material in E36 steel.

Charpy toughness

The toughness of the selected welds was determined by Charpy impact testing, performed according to ISO 148-1:2009 (KV2) at -45°C. Test samples 10 x 5 x 2mm were taken from the weld centreline, and also from positions offset from either side of the weld centreline by 2, 4 and 6mm. FSW is an asymmetric process and thus it was deemed appropriate to establish the toughness properties of the weld across the full width of the weld zone. The Charpy impact toughness data for three of the E36 steel plates are plotted in the Figure 10. The data are for the plates with a single sided weld (0% overlap), a double sided weld with a 33% and 100% stir zone overlap.
For the three tested, the toughness was greater in the thermomechanically affected zone (TMAZ) than in the heat affected zone (HAZ). The toughness distribution appears to be asymmetric.
The data indicate that the Charpy impact toughness at -45°C in the single side welded plate, ie full penetration with no overlap zone, is similar to that expected from unwelded parent material at around 50J. Both the plates that had been welded from both sides (33% and 100% overlap of the weld zones) exhibited a higher toughness level, close to 100J.

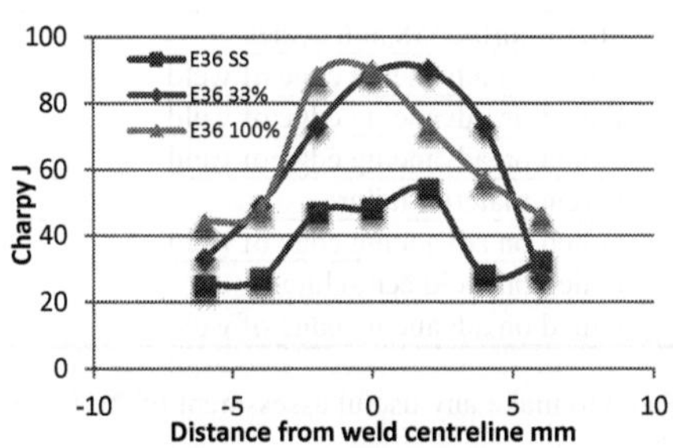

Figure 10 Charpy impact toughness profile at -45°C for FSW E36 steel plates 1, 3 and 6

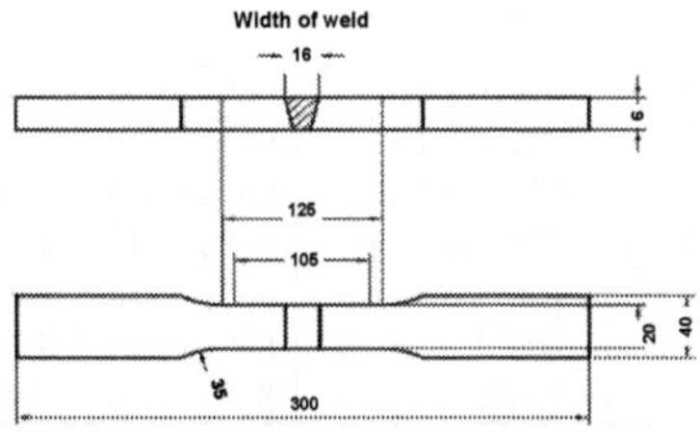

Figure 11 Dimensions of fatigue test piece.

This may be explained in part by the microstructural changes wrought by the FSW process. FSW has produced a microstructure with a much finer grain size in the thermomechanically processed zone and the second pass will have had a tempering effect upon the material in the first weld zone. The initiation and propagation of crack through the complex microstructure present in these E36 friction stir weld zones will therefore require more energy than in coarse grained microstructures typical of a conventional fusion weld.

Fatigue testing

Fatigue testing was performed at the University of Strathclyde. Two fatigue testing machines were used, one an Instron 8801 and the second an Instron 8802. These feature precision-aligned, high-stiffness load frames that encompass a broad range of static and dynamic test applications. The two machines differ in that the 8801 has an axial force capacity of 100kN whilst the 8802 has a capacity of 500kN.

Standard 'dog bone' fatigue test samples were cut from the selected welded plates using an automatic milling machine and their cut edges de-burred. The form and dimensions of the samples are shown in Figure 11. Seven samples were prepared per plate. Three samples were used for both high and low cycle fatigue (HCF and LCF respectively) and one was used to determine the yield stress of the material so that the parameters for the fatigue tests could be determined accurately for each material. It was decided that the LCF test would be run at 95% of the first yield stress and that the HCF test would be run at 60% of the first yield stress. However, after considering the timescale of the project and given that the first batch of HCF tests ran on for over a million cycles, it was decided that a more appropriate stress level would be at 80% of the first yield stress. Fatigue testing of the selected specimens is currently underway. The available results are shown in Table 7 below.

Table 7 Preliminary fatigue test results for E36 steel specimens.

Plate	Stress	Cycles to failure	Comment
E36 - 0% overlap	60% yield	1,800,000	No failure. This sample was then cycled at 80% yield and failed at the advancing edge after a further 17,000 cycles
E36 - 0% overlap	60% yield	1,000,000	No failure. This sample was then cycled at 80% yield and failed at the retreating edge

			after a further 58,566 cycles
E36 - 0% overlap	60% yield	374,797	Failed on advancing edge of weld
E36 - 0% overlap	95% yield	170,000	Failed on advancing edge of weld
E36 - 0% overlap	95% yield	100,000	Failed on advancing edge of weld
E36 - 33% overlap	95% yield	495,815	Parent material failure
E36 - 100% overlap	80% yield	280,000	Failed on advancing edge of weld
E36 - 100% overlap	80% yield	877,108	Failed on weld centreline
E36 - 100% overlap	80% yield	396,849	Failed on advancing edge of weld

Insufficient fatigue data have yet been accumulated to make any useful assessment of the fatigue performance of the welds generated in this work.

Conclusions

1. Single/double sided friction stir welding of both E36 and DH36 shipbuilding steel produces lower angular and longitudinal distortion than the sub arc welding of DH36.
2. Single sided FSW in 6mm thickness E36 produces similar distortion to that witnessed in single sided FSW of DH36 steel.
3. The FSW welds in E36 were at least as strong as the parent material.
4. The presence of oxide inclusions in some of the welds tested did not appear to have a detrimental effect upon weld strength.
5. Performing the fully penetrating second pass in the double side friction stir welding process creates an improvement in toughness which is likely to be related to microstructural differences.
6. From the data generated in this study supports the evidence that FSW it is a superior process to the conventional SAW process.

Acknowledgments

The authors would like to acknowledge FSW group at TWI in performing the FSW, Stuart Campbell (AMRL) in Strathclyde University in helping measuring the plate distortion. BAE Systems-Marine is acknowledged for supplying the material and funding the experimental work and C.McKechnie for producing the SAW plates and the analysis of the fatigue data by Colin Davies of Tata Steel Europe.

References

[1] Reynolds, A.P., Tang, W., Posada, M. and DeLoach, J. "Friction stir welding of DH36 steel," Science and Technology of Welding and Joining, Vol.8, No.6 (2003), pp. 455-460.

[2] Posada, M., Nguyen, J.P., Forrest, D.R., DeLoach, J.J. and DeNale, R. "Friction stir welding advances joining technology," AMPTIAC Quarterly, Vol. 7, No.3 (2003), pp. 13-20.

[3] Posada, M. and Roush, S. "Friction stir welding – a promising new technique for joining metals," Seaframe, Vol. 1, Issue 2 (2005), pp. 17-18.

[4] McPherson, N.A., Galloway A. M., Cater S. R "A comparison between single sided friction stir welded and submerged arc welded DH36 steel thin plate", International Conference on Trends in Welding Research 2012.

[5] McPherson, N.A., Galloway A. M., Cater S. R., Osman M. M. "A comparison between single sided and double sided friction stir welded 8mm thick DH36 steel plate", International Conference on Trends in Welding Research 2012.

Friction Stir Welding and Processing VII
Edited by: Rajiv Mishra, Murray W. Mahoney, Yutaka Sato, Yuri Hovanski, and Ravi Verma
TMS (The Minerals, Metals & Materials Society), 2013

FRICTION STIR WELDING OF PIPELINE STEELS

Murray Mahoney[1], Sam Sanderson[2], Zhili Feng[3], Russell Steel[2], Scott Packer[1] and Dale Fleck[2]
1) Advanced Metal Products, West Bountiful, Utah 2) MegaStir, Orem, Utah, and 3) Oak Ridge National Laboratories, Oak Ridge, Tennessee

Key Words: Friction Stir Welding, Arc Welding, Pipeline Steels, Consistent Full Penetration, NDE, Mechanical Properties

Abstract

In prior FSW studies, *consistent full penetration* in pipeline steel has proved to be a difficult goal when using a *portable* FSW system capable of operation in the field [1]. In a previous study, metallography, mechanical testing (tensile and Charpy impact), and workmanship testing (root bend) demonstrated that full penetration can be achieved through much of the weld length, and when full penetration is achieved, mechanical properties are excellent. However, at times there remained sections of the weld that did not exhibit full penetration. As an example, metallography and root bend tests of the weld root have shown full penetration at many locations around the pipe circumference while from the same weld, locations are identified where remnant faying surfaces remain and full penetration was not achieved. Further, some welds exhibit full penetration accompanied by a continuous oxide path that remains at the weld root. Conversely, if the FSW tool penetrates into the support anvil, anvil material is drawn into the weld nugget. Thus, different approaches need to be developed to assure consistent full penetration without anvil contact.

In the study reported herein, two approaches to achieve consistent full penetration are being evaluated. These methods include 1) root arc welding followed by a partial penetration friction stir weld and 2) use of a sacrificial anvil. Arc welding of the root prior to FSW builds on a technique developed at ExxonMobil whereby an internal root arc weld can be used to provide support for the FSW process. After the internal root weld is made, the butt joint surfaces remain and a partial penetration friction stir weld penetrates into the arc weld root pass. In addition to providing structural support to resist FSW loads, with overlap of the two weld nuggets, there is no possibility of a root lack of penetration defect. This technique takes advantage of the efficiency afforded by internal root welding for onshore pipeline construction. Arc welding of the root is a commercialized widely used practice. For the arc weld / friction stir weld approach, mechanical properties and root bend test results are reported. The sacrificial anvil approach uses a small insert in the structural anvil where metal of the same chemistry as the pipeline material is used as the insert material. In this approach, the FSW tool penetrates into the sacrificial anvil thus achieving consistent full penetration. Removal of the small sacrificial anvil may or may not be required.

Introduction

The pipeline industry has made wide use of fusion welding for pipeline construction. With fusion welding, the larger the pipe diameter or the thicker the pipe wall, welding time increases due to the greater volume of weld metal. For onshore pipelines, particularly in remote locations, it is important the welding be as economical as possible because of the large expense related to deployment of workers and equipment to the pipeline right of way. For offshore pipelines, there are high costs associated with the lay barge and again welding needs to be as economical as possible. Other factors such as availability of qualified welders, pipe thickness, welding productivity, quality of the weld, automation of the welding process, etc., also influence the selection of the welding process.

In a recent study on pipe girth welding, friction stir welding (FSW) was shown to significantly reduce pipeline construction costs compared to gas metal arc welding [2]. Approximately 7% savings in pipeline construction costs were estimated for onshore, large diameter pipelines. However, for onshore construction, use of an internal mandrel does not appear to be an economic construction scenario. Alternatively, an internal GMA root weld can be used to support the normal loads associated with FSW [3]. This root weld approach used in concert with FSW was shown to be economical for onshore pipeline construction. In addition, ~25% savings in pipeline construction costs were estimated for offshore pipeline installation using the J-lay method coupled with FSW [3]. However, to achieve these savings, FSW needs to be capable of thick wall welding in one pass. In smaller diameter pipe, arc welding of the root may not be possible and an internal mandrel may be required.

Thus, potentially, there are significant economic benefits to apply FSW to pipeline construction with two different approaches. That is, to use FSW for onshore pipeline construction, an internal mandrel is not feasible but for offshore construction an internal mandrel likely will be required. Further, the economic benefits for offshore pipeline construction are greater for larger wall thickness pipe. These economic benefits are based on the most recent economic analysis and current pipeline fabrication methods.

Based on the potential savings reported for pipeline fabrication, two FSW approaches were investigated each with the potential to achieve consistent full penetration. For onshore pipeline construction, the ExxonMobil approach using an internal root arc weld combined with FSW was evaluated to determine if defect free, consistent full penetration welds could be made without additional internal support. Most importantly, the ability of the root arc weld to support and prevent collapse of the weld zone during FSW needs to be demonstrated. For offshore pipeline construction, where an internal anvil will be required to support FSW loads, a different approach to achieve consistent full penetration is required. For offshore pipeline fabrication, a sacrificial anvil approach is investigated. Results for both approaches are reported herein.

Experimental Procedures

Pipe Materials

The pipeline steel was X42 carbon steel with dimensions 32 cm (12.75 inch) outer diameter by 0.5 inch (12 mm) thick wall pipe. The nominal composition is 0.28 C (max), 1.3 Mn (max), 0.03 P (max), and S (max). This family of pipeline steels is based on a C-Mn formulation with low overall alloy content. The number 42 indicates the minimum yield strength in ksi.

FSW Tool Design and Tool Material

For all friction stir welds, polycrystalline cubic boron nitride (pcbn) based tools were used. The tool material selected for this study was designated as Q60. Q60 is pcbn ceramic based (60 wt/% cBN) with a metal WRe (40 wt/%) binder [4]. The tool design is the standard convex scroll shoulder step spiral (CS4) shown in figure 1 [5]. The radius on the tool tip increases metal flow at the weld root while the shoulder scroll and step spiral on the pin enhance metal flow at their respective locations.

Figure 1. Photograph of a standard convex scroll shoulder step spiral (CS4) pcbn tool.

Root Arc Weld: The root arc weld was made using an automated gas metal arc weld with ER 70S-6 filler rod. The weld joint design is shown in figure 2. Subsequent metallography showed the arc weld depth to be 5.2 mm (0.205 inches). Thus, to achieve overlap of the two weld nuggets, the FSW depth needed to be >6.7 mm (0.264) inches).

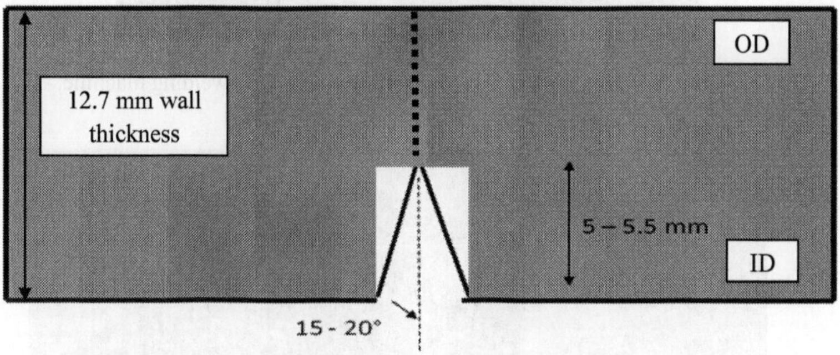

Figure 2. Weld joint design for the internal root arc weld.

FSW Pipe Weld Procedure

Figure 3 is a photograph of the MegaStir portable FSW pipe welding machine used to weld pipe sections for this study. Pipe welds were made with the FSW tool rotating around the stationary pipe (5G welding) versus the case where the pipe is rotated underneath a stationary tool (1G welding). To eliminate the presence of a tool exit hole, a tool run-off ramp was attached during welding. Figure 4 illustrates a cross-section of the run-off ramp. Figure 5 shows the weld sequence. That is, once the weld has started, a) the run-off ramp is attached to the pipe allowing for a short weld overlap distance, b) the tool approaches the run-off ramp, c) the tool traverses onto the run-off ramp with a change in weld parameters to increase the heat input, and d) the run-off ramp is removed by grinding. Weld parameters selected for the pipe welds included:

- Plunge: 500 rpm at a plunge rate of ~0.5 ipm (12.7 mm/min).
- Welding: 275 rpm at 4 ipm (1.7 mm/sec) with variable normal load, i.e., when the weld appearance was not as desired, the normal load was either increased or decreased as necessary to maintain a fixed weld crown width.
- Run-off Ramp: 500 rpm at 1.16 mm/sec (2.75 ipm) with a 2,720 Kg (6,000 lb.) normal load. These ramp parameters result in a considerably hotter weld than for the equilibrium pipe weld parameters. This was necessary because heat transfer ahead of the weld tool is not efficiently transferred to the run-off ramp due to the interface between the pipe and the run-off ramp.

Figure 3. Photograph of the MegaStir portable pipe welding machine.

Figure 4. Cross-section of the run-off ramp.

Figure 5. a) The run-off ramp is attached to the pipe allowing for a small amount of weld overlap, b) the tool approaches the run-off ramp, c) the tool traverses onto the run-off ramp with a change in weld parameters to increase the heat input, and d) the run-off ramp is removed by grinding.

Sacrificial Anvil FSW Procedure

A sacrificial anvil concept for flat plate is schematically illustrated in figure 6. As shown, figure 6a illustrates the workpiece (green) positioned on top of a structural anvil (red) with a thin sacrificial anvil (yellow) inserted into the structural anvil beneath the weld joint. Figure 6b illustrates the FSW tool penetrating through the thickness of the workpiece and into the sacrificial anvil. Following weld completion, figure 6c shows the workpiece welded to the sacrificial anvil and the structural anvil removed. This illustration is for flat plate. Preparations are in-progress to perform the same weld procedure on pipe.

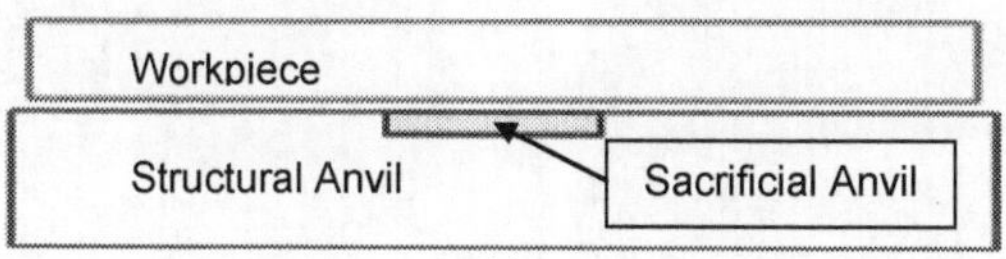

(a)

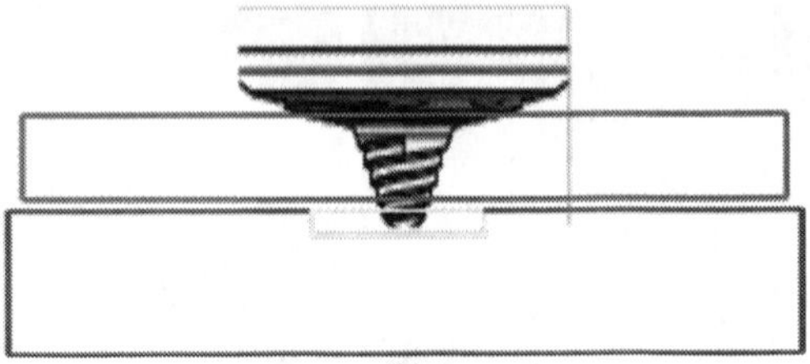

(b)

(c)

Figure 6. Schematic illustration of FSW with a sacrificial anvil.

Results and Discussion

Combined Root Arc Weld and Friction Stir Weld

Friction stir welding was completed on four pipe weld sections all containing a 5.2 mm (0.204 inch) deep root arc weld. During FSW, the weld interface in front of the FSW tool can open and if this separation becomes too great can result in volumetric defects in the weld nugget. However, the root arc weld held the weld interfaces flush throughout FSW. This is an additional benefit associated with arc welding the weld joint root. Figure 7 is a photograph illustrating a completed weld trial showing the good surface finish on a friction stir welded pipe, i.e., very little flash, no undercut, and little oxidation. Figure 8 shows output parameters of normal load and temperature for a typical weld run. As shown, the peak temperature was ~930°C (1706°F) at normal loads of 55 MPa (8,000 lbs.).

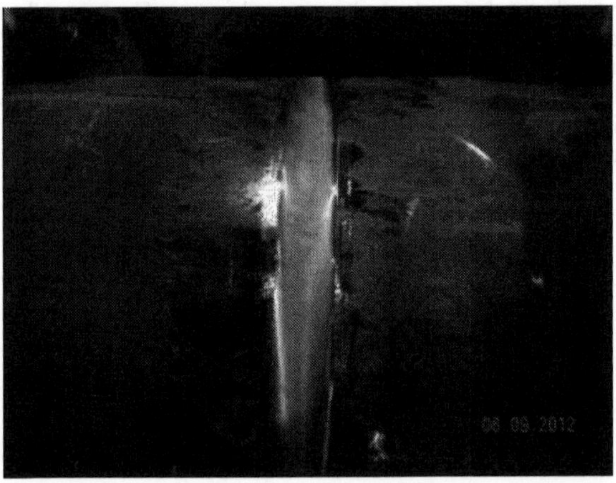

Figure 7. Photograph of a completed weld trail using the ExxonMobil root arc weld approach.

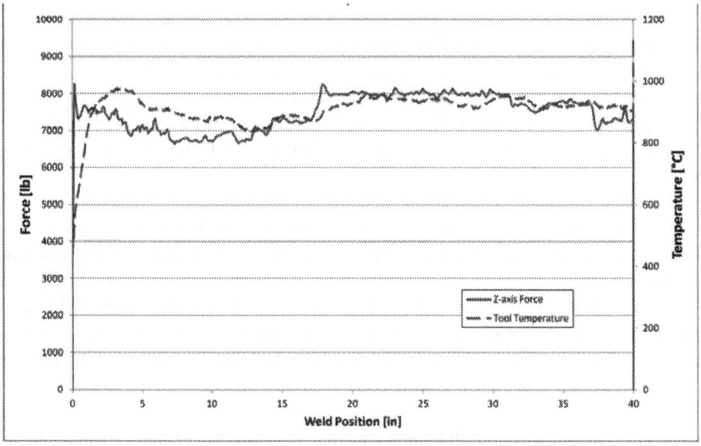

Figure 8. Output parameters of normal load and temperature to complete the pipe weld shown in figure 7.

Figures 9a and 9b are macrographs illustrating typical weld cross-sections using the root arc weld as support for the subsequent friction stir weld. As shown, the friction stir weld nuggets overlap with the root arc welds and all weld nuggets are defect free. However, most importantly, for the boundary conditions evaluated, i.e., 1) weld parameters of 275 rpm at 1.7 mm/sec (4 ipm), 2) a root arc weld depth of 5.2 mm (0.205 inch), 3) X42 steel pipe, 4) a CS4 tool design with a penetration depth of ~7.6 mm (0.3 inch), and 5) the normal loads and temperatures shown in figure 8, there was either no distortion of the pipe wall (figure 9a) or very little (arrow in figure 9b) on the inner diameter surface. Clearly, the volume of arc weld metal was sufficient to support the FSW loads. This result validates the ExxonMobil method of using an internal arc weld as the support anvil.

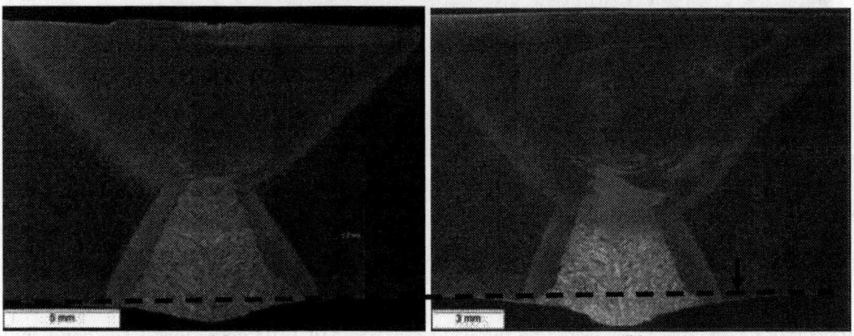

(a) (b)

Figure 9. Cross-sections of the weld zone in a pipe weld illustrating 1) overlap of the arc and friction stir weld nuggets, 2) defect free welds in all weld nuggets, 3) penetration depth (5.2 mm) of the arc weld, and 4) no distortion on the weld inner diameter surface in (a) and very little in (b).

The limits of the boundary conditions for this weld approach need to be explored. For example, it would be beneficial to know how small the arc weld volume could be to support FSW loads and how thick a pipe wall section could be welded with an arc weld depth of 5 to 6 mm (currently the approximate limit of automated internal pipe weld systems) without collapsing the weld zone. These boundary conditions should be explored as a function of weld parameters and weld material.

Further, the ExxonMobil arc weld approach eliminates both the potential for a lack of penetration weld defect at the weld root or contact of the FSW tool with a support anvil. That is, unless the arc weld misses the weld joint, the weld joint at the weld root is consumed by the arc weld prior to FSW and there can be no root lack of penetration defect. In addition, the FSW tool can be designed, both pin width and depth, and using conventional process control, such that overlap of the two weld nuggets should be achieved easily and no internal defect created.

Hardness results for the weld nugget shown in figure 9b are illustrated in figure 10. These are Vickers hardness values obtained at a load of 500 g. As shown, the cast microstructure in the root arc weld is the hardest location in the weld zone with hardness values ranging from ~150 to 180 HV. The friction stir welded nugget is softer with hardness ranging from ~125 to ~140 HV. Adjacent to the friction stir weld nugget, the TMAZ is slightly harder, i.e., ~142 to ~148 HV. Beyond the TMAZ is the HAZ with hardness values >148 HV. Unfortunately, the metallographic sample was not sufficiently large to include the parent metal and additional hardness tests need to be performed.

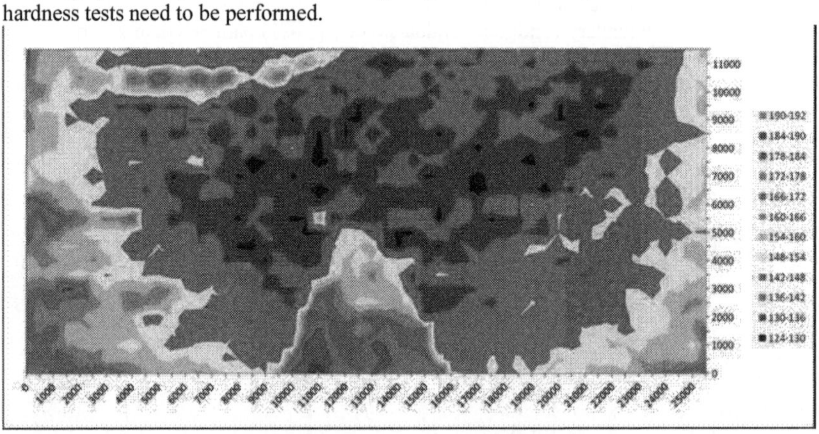

Figure 10. Hardness map for the weld nugget shown in figure 9b (HV at a load of 500 g).

Root Bend Tests were performed on the weld root per API 1104 guidelines. The bend angle for all tests was 180° with a one inch bend radius. Samples were selected from different locations around the pipe circumference. All root bends passed.

Mechanical Properties were established using transverse tensile tests per API 1104 guidelines. In all tests, the tensile strength was >427 MPa (>62 ksi), the failure was ductile, and the failure location was in the parent metal. This failure location implies the parent metal hardness was either less than that of the friction stir weld nugget or the support provided by the arc weld was sufficient to strengthen the weld nugget beyond that of the parent metal. The arc weld support can be either dimensional, i.e., a greater wall thickness, or higher strength.

Sacrificial Anvil

Initially, flat plate weld studies were used to evaluate the potential for using a sacrificial anvil. Metallography in figure 11 shows a weld cross-section with a lap interface. As shown, following FSW, the weld interface is straight at the intersection of the lap joint and weld nugget, i.e., there is no uplift typical of friction stir lap joints in aluminum alloys when using a butt joint tool design. However, the weld penetration is relatively shallow and likely the step spiral on the tool pin did not sufficiently engage the lap interface to create uplift.

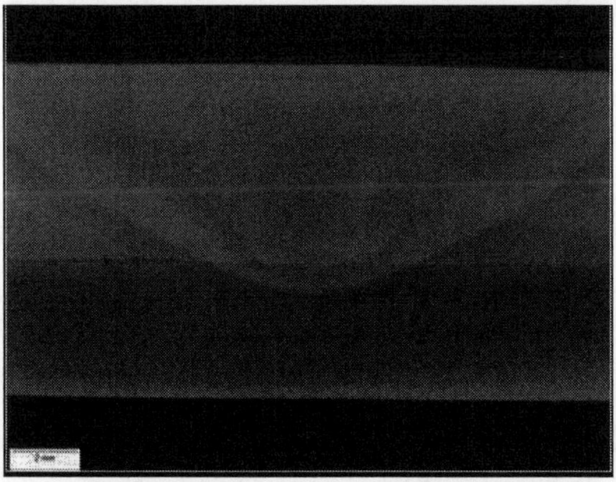

Figure 11. Metallography illustrating a friction stir welded lap interface in steel.

To achieve greater weld penetration without uplift, it may be necessary to use a specially designed lap joint tool. Figure 12a illustrates a tool specially designed to prevent uplift during FSW of lap joints [6]. Rather than features that create vertical metal flow, the feature at the pin tip creates horizontal metal flow. With this lap joint tool design, the pin tip feature must penetrate into and be positioned at the lap interface. Figure 12b illustrates a metallographic cross-section using the lap joint tool design in figure 12a showing no uplift and deeper weld penetration. These weld results on flat plates have been reported previously but are included herein for completeness and to demonstrate the approach[1]. The eventual goal is to demonstrate the capability to perform similar welds on pipe.

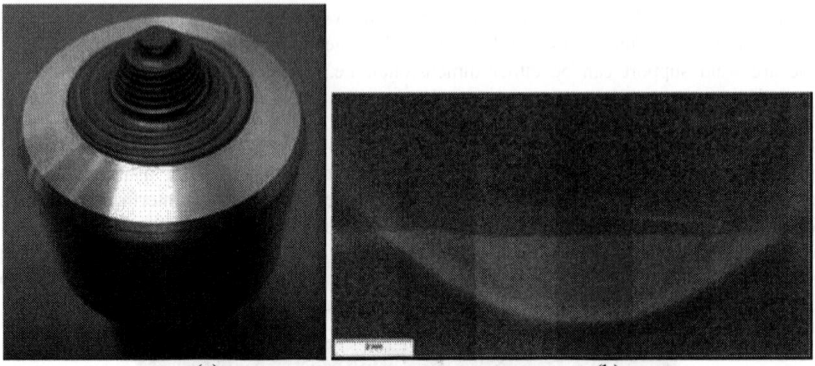

(a) (b)

Figure 12. a) A friction stir tool designed to create horizontal metal flow at the tool tip as opposed to vertical flow at the lap interface , and b) Metallography illustrating a friction stir welded lap interface using the tool shown in figure 12a.

With the sacrificial anvil approach, the potential for a lack of penetration defect will be significantly reduced. That is, with the conventional anvil approach, tool positioning must be relatively precise to achieve consistent full penetration without contacting the structural anvil. Conversely, if the tool is allowed to penetrate beyond the pipe wall thickness into a sacrificial anvil, figures 6 and 12b, the dimensional tolerance for consistent full penetration can be large.

Conclusions

Studies by others have illustrated the significant potential economic advantage to construct pipelines using friction stir welding [2] The economy of savings can be realized for onshore pipeline construction if no removable internal mandrel is used and for offshore pipeline construction of thick wall pipe using a removable internal support. The ability to friction stir weld both onshore and offshore while eliminating or minimizing the potential for lack of penetration root defects has been demonstrated. For onshore pipeline construction, an internal arc weld serves as the support anvil to resist FSW loads. For offshore FSW, a sacrificial anvil approach is currently being evaluated.

Based on the results presented herein, for X42 pipeline steel, and using an orbital FSW system, the following may be concluded:

- For the FSW boundary conditions investigated, an internal arc weld of depth 5.2 mm (0.20 inch) can support the normal loads associated with FSW.
- Transverse mechanical properties in the overlapping friction stir and arc welds are good with ductile failures occurring in the parent metal.
- Root bend tests and metallography illustrate full and consistent weld penetration.

Acknowledgements

The authors would like to acknowledge the support of the US Department of Energy, Contract No. 6400009575 and technical guidance by ExxonMobil personnel A. Wasson and D. Fairchild.

References

1. M. Mahoney, S. Sanderson, Z. Feng, R. Steel, S. Packer, and P. Higgins, "Approaches to Consistently Achieve Full Penetration in Pipeline Steel Welds Using Friction Stir Welding", Proceedings of the 9th International Friction Stir Welding Symposium, Huntsville, Alabama, May 2012.
2. A. Kumar, D.P. Fairchild, M.L. Macia, T.D. Anderson, H.W. Jin, R. Ayer, N. Ma, A. Ozekcin, and R.R. Mueller, "Evaluation of Economic Incentives and Weld Properties for Welding Steel Pipelines Using Friction Stir Welding", Proceedings of the Twenty-first (2011) International Offshore and Polar Engineering Conference, Maui, Hawaii, USA, June 19-24, 2011,vol.4, pp. 460-467.
3. D. P. Fairchild, S. J. Ford, A. Kumar, N.E. Nissley, N. E. Biery, and M.I. Macia, "Butt Weld and Method of Making Using Fusion and Friction Stir Welding", US Patent No. 7,874,471 B2, Jan. 25, 2011.
4. Q. Liu, R. Steel, J. Peterson, S. Horman, M. Collier, D. Marshall, J. Davis, and M. Mahoney, "Advances in Friction Stir Welding Tooling Materials Development," Proceedings of the Twentieth (2010) International Offshore and Polar Engineering Conference, Beijing, China, ISOPE, 2010, vol. 4., pp. 298-302.
5. C. Sorensen and B. Nielsen, "Exploring Geometry Effects for Convex Scrolled Shoulder, Step Spiral Probe FSW Tools," *Friction Stir Welding and Processing V*, Edited by R. Mishra, M. Mahoney, and T. Lienert, TMS 2009, San Francisco, California, February 2009, pp. 85-92.
6. Christian B. Fuller, Murray W. Mahoney, and William H. Bingel, "Friction Stir Weld Tool and Method", Patent No. US 6,994,242, June 9, 2005.

Friction Stir Welding and Proceesing VII
Edited by: Rajiv Mishra, Murray W. Mahoney, Yutaka Sato, Yuri Hovanski, and Ravi Verma
TMS (The Minerals, Metals & Materials Society), 2013

MICROSTRUCTURE AND PROPERTIES OF FRICTION STIR PROCESSED HY80 STEEL

Garth W Young[1], William C Stewart[2], Murray W Mahoney[3], Russell Steel[4], Jon Babb[4], Sarath K Menon[5] and Terry R. McNelley[5]

[1] NAVFAC ESC, 1100, 23rd Road, Port Hueneme, CA 93043, USA
[2] US Navy, NROTC Placement Officer, 250 Dallas St, Pensacola, FL 32508, USA
[3]Consultant, [4]Megastir Technologies, Provo, Utah, USA
[5] Naval Postgraduate School, Center for Materials Science and Engineering, Department of Mechanical and Aerospace Engineering, Monterey, CA 93943, USA

Keywords: HY80 steel, friction stir processing, underwater welding, inclusions

Abstract

Microstructures and mechanical property changes associated with Friction Stir Processing (FSP) of HY-80 steel both dry and under seawater were examined. FSP on HY-80 plates employed a PCBN / tungsten rhenium tool operating at 400 rpm and 2 ipm. Microstructural characterization of the as-received HY-80 plate emphasized the differences in the distribution of constituent phases and inclusions in the rolling, transverse and normal planes using optical and scanning electron microscopy. Microstructural details of transverse sections as well as the plan section at the tool extraction sites of processed plates were also evaluated. Mechanical properties of these samples were evaluated by tensile tests, microhardness tests and Charpy V-notch impact tests. Residual hydrogen content was also evaluated. Stir zones exhibited untempered bainitic / martensitic microstructures with minimal hydrogen pick up but distinct property gradients from stir zone to base metal. The influence of a post-FSP tempering treatment also is summarized.

Introduction

HY-80 steel is a high strength steel used in the quenched and tempered condition to achieve mechanical properties desired in many U.S. Navy applications, such as ship and submarine hulls, as well as some pressure vessels. The resulting tempered martensitic microstructure though strong, is quite susceptible to Hydrogen Assisted Cracking (HAC) [1]. When HY80 steel structures are fabricated by conventional fusion welding techniques, there is a potential for hydrogen to be liberated from the water molecule and to diffuse/dissolve into the fusion zone. Hydrogen diffusion in weld metal cannot be prevented. However, extensive site preparation and strict environmental controls can be used to restrict the amount of diffusible and residual hydrogen in the resulting weldment. The material preparation requires pre- and post-heating of the materials to be welded, filler electrode controls such as baking and storing and control of moisture and hydrocarbons. Due to these strict environmental controls, performing conventional fusion welding underwater or in otherwise 'wet' environments would require expensive site preparation and, likely, prolonged time [2]. An alternative to conventional fusion welding is friction stir welding (FSW) and friction stir processing (FSP) which are solid state processes for the joining and processing of a wide variety of metallic materials. FSW/P has been implemented in the U.S. Navy in such applications as joining aluminum material on Littoral Combat Ships along with the surface treatment of nickel aluminum bronze propellers.

In this work, a series of studies on dry as well as underwater FSP of steels are being carried out [3-6]. Results of experiments on FSP of HY-80 steel plates both under dry as well as seawater conditions are discussed. In particular, the influence of FSP on microstructural modifications, microhardness, and Charpy V-notch impact resistance are discussed.

Experimental

A steel plate was obtained from the Naval Surface Warfare Center – Carderock Division with dimensions 0.25 inches (6.4mm) thick and 26.5 inches (673 mm) wide by 43.5 inches (1105 mm) long. Pre-welding analysis was conducted by Anamet, Inc. and conformed to HY-80 steel specification. The plate was cut into three sections for base metal analysis and dry and wet (underwater) friction stir processing. Each plate measured approximately 26.5 inches (673 mm) long by 14.5 inches (368 mm) wide. One plate was sectioned to produce Charpy V-notch samples and tensile test samples in both the rolled and transverse directions. The second plate was cut into two sections each measuring approximately 26.5 inches (673 mm) by 7.25 inches (184 mm). One section each was used for dry and wet (underwater) FSP. Initially, the plate was sand blasted to remove mill-scale. FSP was conducted by Advanced Metal Products and MegaStir Technologies in Provo, Utah. A dry FSP run of approximately 25 inches (635mm) was completed at 400 RPM and 2 IPM. An initial plunge load of greater than 15,000 pounds (68,000 N) was applied and decreased to 10,000 pounds (45,500 N). The MS80 tool material was polycrystalline cubic boron nitride (pcbn) consolidated with a metallic binder. All FSW was performed parallel to the long dimension of the plate. Due to the size limitations of the underwater chamber, two approximately 10 inch 9254 mm) underwater weld runs were completed for a total length of approximately 20 inches (508 mm). Figure 1(a) shows the underwater FSP setup employed and a typical result of an FSP run is shown in Figure 1(b).

Figure 1: (a) Photograph showing the progress of underwater FSP of HY80 steel plate (b) typical appearance of an underwater FSP carried out at 400rpm/2ipm

Results

Residual hydrogen concentration was determined (Luvak Inc.) by vacuum hot extraction method (ASTM E 146-83) to be 1.01 mL/100gm and 2.58 mL/100gm in the stir zone regions of the dry and wet runs respectively while that of the base material was determined to be 2.13 mL/100gm. There is virtually little or no hydrogen pickup in both stir zones when comparison is made to the base metal. Furthermore, the hydrogen pickup is greater in the 'wet' sample than in the dry sample although all of these hydrogen levels are relatively low in comparison to the specifications of MIL-E-23765 that restricts the maximum hydrogen content in weldments to be 2-5 mL/100gm [7]. The hydrogen content probably decreases in the dry FSW/P due to the material being heated and the escape of hydrogen into the atmosphere. It appears that the hydrogen content of the underwater FSW/P is approximately 20% higher than the base material. However the error in the vacuum hot extraction process is 5 to 10%. Thus it is difficult to conclude that the slightly increased hydrogen content in the underwater FSW/P is due to the production of atomic hydrogen by dissociation of water and its diffusion in the steel.

Microstructures of the as-received HY-80 plate in three orthogonal surfaces showed distinct banding in the ND/TD as well as the ND/RD planes but not in the TD/RD surfaces. Figure 2 shows the optical micrographs illustrating this feature. Scanning electron microscopy and energy dispersive x-ray spectroscopy analysis (EDS) clearly showed that the banding was associated with the inclusions in the material and EDS confirmed that both oxide/silicate inclusions (spherical morphology) as well as MnS (elongated stringers) were present in the steel. During rolling, the inclusions were preferentially elongated in the rolling direction with a rod or pancake like cross section. Often, these inclusion stringers were hundreds of microns long and thus expected to be extremely deleterious to mechanical properties and were confirmed by the Charpy V-notch test results [5].

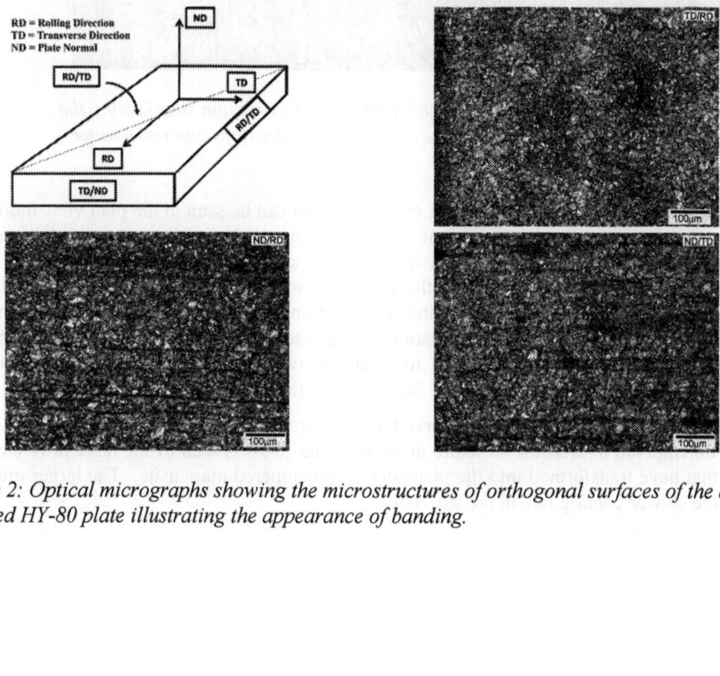

Figure 2: Optical micrographs showing the microstructures of orthogonal surfaces of the as-received HY-80 plate illustrating the appearance of banding.

A summary of the microstructural characterization of the HY80 plates subjected to both dry and underwater FSP are presented here. 'Plan view' montages of the pin extraction site for the dry and underwater FSW/P are shown in Figures 3(a) and (b), respectively.

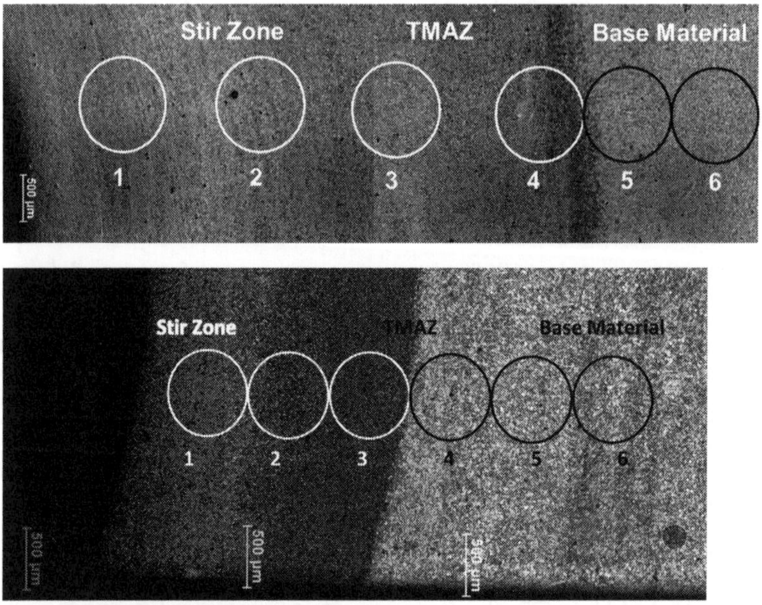

Figure 3: Montage of optical micrographs from the tool extraction site showing the microstructure ahead of the tool traverse direction from dry (top) and underwater (bottom) FSW/P sample.

The flow pattern of the material moving around the tool can be seen in the plan view montage of material processed under dry conditions. A distinct change in etching response delineates the location ahead of the tool where transformation to austenite takes place on heating due to approach of the tool. Comparison of the montages in Figures 3(a) and (b) reveals that this transformation occurs a distance of approximately 5mm ahead of the tool under dry conditions and 2.2 mm under wet conditions, indicating a steeper temperature gradient ahead of the tool for FS under water. Higher magnification micrographs were taken from the stir zone (zone 1), through the TMAZ (zone 4) and into the base material (zone 5 & 6).

The microstructures from the zones marked 1 in Figure 3 are presented in Figure 4 and show a coarse untempered martensitic structure in the stir zone. This is seen by the large prior austenitic grains that have transformed into the plate-shaped untempered martensite. The larger grains are due to the slower cooling rate in the stir zone.

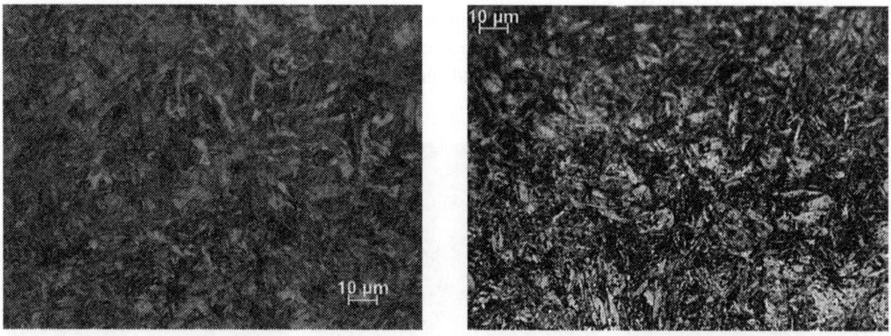

Figure 4: Optical micrographs from dry (left) and underwater FSP (right) obtained from zone 1 from the dry and underwater FSP samples shown in Figure 3 showing untempered martensitic structure.

The microstructure undergoes a change when approaching the TMAZ. The microstructure changes from a tempered martensitic structure to a refined, smaller grained untempered martensitic structure in the zones marked 4. As zones 5 and 6 are approached, the microstructure of the base material, composed of tempered martensite, is observed. This transformation occurs due to the change in temperature produced by FSW/P and cooling rates. A key observation made in this work was the absence of change in the shape of the grains near the TMAZ/SZ boundary unlike that reported in aluminum or bronze alloys [8, 9]. The grains remain equi-axed and not elongated due to deformation. Clearly, deformation of the steel occurs primarily in the fully austenitic condition indicating that the gradient in the thermal field is less sharp than the deformation field, similar to those in aluminum alloys [8] and nickel aluminum bronze [9]. Since the austenite undergoes martensitic transformation during the fast cooling subsequent to FSP, the effects of deformation of the austenite is completely masked in the microstructures observed.

Figure 5: Optical micrographs illustrating the morphology of the inclusions in the base material and the stir zone of dry processed material

Apart from these microstructural developments, FSP has a strong influence on the morphology of the inclusions. For example, Figure 5 illustrates the morphology of the inclusions in the base plate and the stir zone (SZ). The long inclusions were completely broken up into fine spherical shaped inclusions within the SZ. Details of the morphology of the inclusions, especially the fine spheroidal ones in the stir zone are better seen in the secondary electron images shown in Figure 6. It appeared that the inclusions were well-dispersed in the SZ as well. The results from both the dry FSP as well as the underwater FSP did not show any noticeable differences in the morphology and redistribution of the inclusions.

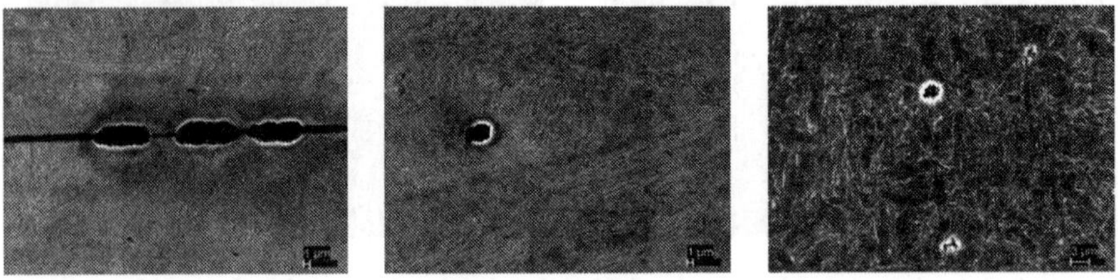

Figure 6; Secondary electron images showing the morphology of inclusions in the base material (left), SZ of dry FSP (center), and SZ of underwater FSP (right)

The morphology of the inclusions has a strong influence on the impact resistance as a function of plate orientation and thus highly anisotropic impact resistance behavior is exhibited by the hot-rolled HY-80 plates. Charpy V-Notch Impact testing was carried out to determine the Ductile-to-Brittle Transition Temperature (DBTT) of the base material to establish a baseline to compare the processed material. Charpy V-Notch tests revealed the DBTT to be between -192°C and -54.5°C. The results of these tests showed a substantial difference in the behavior of the base plate in the transverse and longitudinal directions. This is as anticipated from the elongated inclusions present in the hot-rolling direction of these HY80 steel plates and as expected with the substantially reduced impact resistance in the transverse samples [5].

Microstructural examination of the FSP samples showed that during FSP (both dry and underwater) temperatures in the austenitic phase regions were approached and martensitic transformation occurred during subsequent cooling. This is illustrated in the micrographs shown in Figure 7 comparing the microstructure of the base plate with those obtained during FSP both under dry and underwater conditions. It appeared that the martensitic structure was retained in the FSP samples.

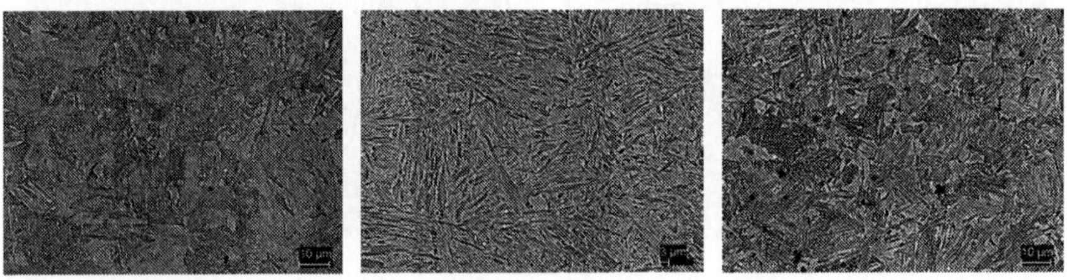

Figure 7: back-scattered electron images from (left) base plate and within the stir zones of dry (middle micrograph) FSP and underwater FSP (right micrograph) showing the typical appearance of the microconstituents.

Figure 8 compares the microhardness data from the dry and underwater FSP samples along with the corresponding low magnification optical micrograph montages. One noteworthy feature is the uniformity of the microhardness data in the underwater FSP sample suggesting microstructural homogeneity throughout the stir zone. In addition, it is also clear that the TMAZ in the dry FSP sample is substantially wider while the TMAZ in the underwater FSP sample is very narrow.

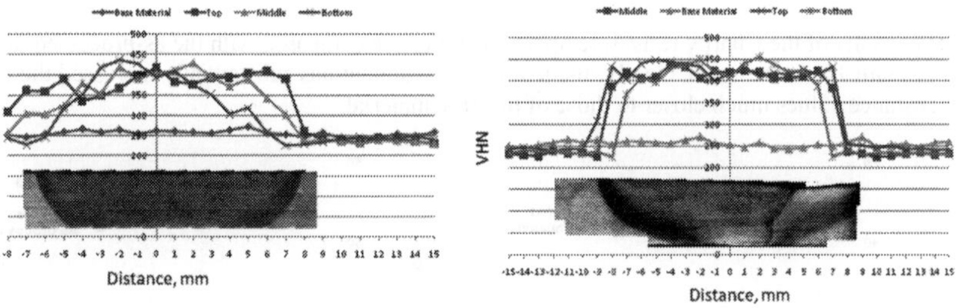

Figure 8: Microhardness data and the corresponding optical montages from dry FSP (left) and underwater FSP (right) samples.

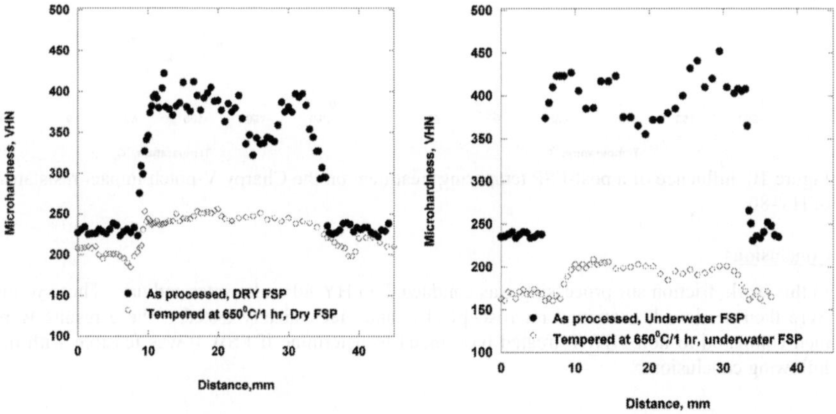

Figure 9: Variation in microhardness across the FSP regions in the dry (left) and underwater (right) samples before and after tempering.

FSP leads to austenitization of the SZ material and subsequent formation of martensite within these regions, and as a result, the microhardness within the SZ is substantially higher. Indeed, as expected, the Charpy impact resistance of the SZ material is significantly lower than that of the base material as well though the DBTT is not significantly altered. Thus, tempering of the SZ material is required to minimize any variation in mechanical properties between the base material and the SZ. The FSP'ed samples were tempered at 650°C for 1 hour as part of our investigation. Figure 9 shows the effect of tempering on the microhardness variation of both the dry and underwater FSP samples. Clearly, for both these processing conditions, the substantial increase in the hardness within the SZ has been practically eliminated and only a minimal difference between the base material and the SZ is observed. Importantly, the entire sample including the base material (which was already provided in the heat-treated condition and hence tempered once) has now been tempered. Thus, the base material hardness values reflected in Figure 9 corresponds to a doubly-tempered HY-80. Presumably, the carbides in the base material may have coarsened substantially due to over-tempering leading to the behavior seen here.

The results of the Charpy tests of tempered samples are compared with the as-processed material in Figure 10. Here again, clearly, the tempering treatment has substantially increased the impact resistance values much closer to those of the base material.

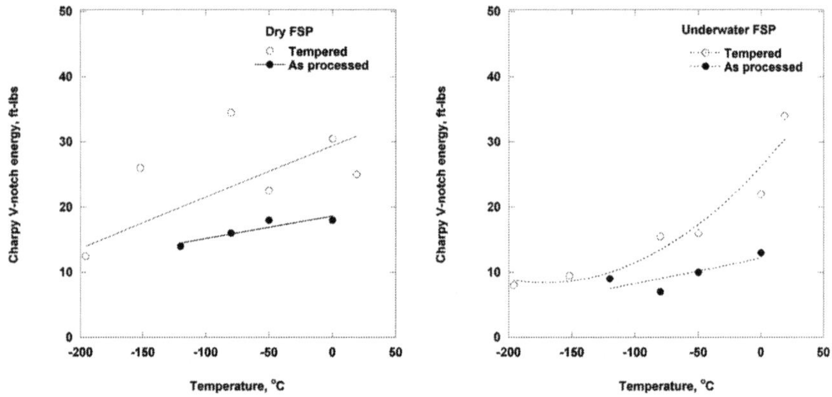

Figure 10: influence of a post-FSP tempering treatment on the Charpy V-notch impact resistance of HY-80

Conclusions

In this work, friction stir processing was conducted on HY-80 underwater and dry. These welds were then sectioned, examined microscopically, and mechanically tested. These results were then compared to non FSW/P treated base metal to determine if FSW/P was feasible with the following conclusions:

1. Friction Stir Processing/welding of HY-80 grade steels in both dry and underwater conditions is feasible while maintaining the hydrogen content within allowable levels.
2. The MnS inclusions appearing as elongated stringers in the hot rolled HY-80 plate are broken up and distributed due to the thermomechanical processing of FSW/P.
3. Material is fully austenitized during FSW/P and transforms to martensite with the cooling rates established during both dry and underwater FSW/P.
4. Deformation during FSW/P of HY-80 did not result in the elongation of the grains in the TMAZ region unlike that reported in aluminum alloys and Nickel Aluminum Bronze.
5. The martensite undergoes partial tempering in the dry FSW/P resulting in a lower hardness than that of underwater FSW/P in which the microstructure is untempered martensite and hence harder.

6. These microstructural features and the observation of some cleavage steps in FSW/P samples indicate that post-FSW/P cooling rates must be controlled or subsequent tempering employed to control properties of the resulting welds.
7. Charpy impact resistance, tensile testing, and microhardness data is in conformity with the microstructural observations mentioned above.

Acknowledgement

Financial support for this work was provided by the Office of Naval Research under contract number N0001410WX21362 with Dr William Mullins as Program Officer.

References

[1]. K. Masubuchi and D. C. Martin, "Mechanisms of Cracking in HY-80 Steel Weldments," *Welding Journal,* vol. 41, pp. 375S-384S, 1962.
[2]. J. H. Nixon, "Underwater repair technology," pp. 108, 2000.
[3]. LT N Overfield, "Feasibility of Underwater Friction Stir Welding of Hardenable Alloy Steel," MS thesis, Naval Postgraduate School, December 2010.
[4]. LT C Stewart, "Feasibility of Underwater Friction Stir Welding of HY-80 Steel," MS thesis, Naval Postgraduate School, March 2011.
[5]. G. W. Young II, "evaluation of friction stir processing of HY-80 steel under wet and dry conditions", MS thesis, Naval Postgraduate School, March 2012.
[6]. Norman Overfield, Murray Mahoney, Russell Steel, Jon Babb, Sarath Menon and Terry McNelley, "Friction stir welding (FSW of a hardenable alloy steel in dry and wet environments"; in Friction Stir Welding and Processing VI, Eds., R. Mishra, M. W. Mahoney, Y. Sato, Y. Hovanski and R. Verma, John Wiley & Sons, , Hoboken, New Jersey, pp. 59-64, 2011.
[7]. Military specification, MIL-E-23765/2E (SH), 22 April 1994.
[8]. Garcia-Infanta, J. M., Zhilyaev, A. P., Carreno, F., Ruano, O. A., Su, J. Q., Menon, S. K., & McNelley, T. R. , "Strain Path and Microstructure Evolution during Severe Deformation Processing of an As-Cast Hypoeutectic Al-Si Alloy", *Journal of Materials Science, 45*(17), 2010, 4613-4620.
[9]. S. Swaminathan, J.Q. Su, S.K. Menon and T.R. McNelley, "The Effect of Concurrent Straining on Phase Transformations in NiAl Bronze during the Friction Stir Processing Thermomechanical Cycle", *Metallurgical and Materials Transactions A*, Vol. 42A, 2011, pp. 2420-2430

Friction Stir Welding and Processing VII
Edited by: Rajiv Mishra, Murray W. Mahoney, Yutaka Sato, Yuri Hovanski, and Ravi Verma
TMS (The Minerals, Metals & Materials Society), 2013

MECHANICAL PROPERTIES AND MICROSTRUCTURAL CHARACTERIZATION OF A MULTILAYERED MULTIPASS FRICTION STIR WELD IN STEEL

Yong Chae Lim[1], Samuel Sanderson[2], Murray Mahoney[3], Dongxiao Qiao[1], Yanli Wang[1], Wei Zhang[1], and Zhili Feng[1]

[1] Materials Science & Technology Division, Oak Ridge National Laboratory
One Bethel Valley Road; Oak Ridge, TN 37831, USA

[2] MegaStir Technologies LLC,
333 West 2230 North; Provo, UT 84604, USA

[3] Consultant
1161 W. Sunburst Lane; Midway, UT 84049, USA

Keywords: friction stir weld, Multilayer, High strength low alloy steel, Mechanical properties

Abstract

Multilayered multipass friction stir welding (MM-FSW) makes it possible to use FSW to fabricate thick-section structures. In this work, MM-FSW was demonstrated on a high strength low alloy steel; ASTM A572 Grade 50. Three steel plates with thicknesses of 0.18", 0.18", 0.24" respectively were stacked and friction stir welded together to form a 0.6" thick welded structure. The welded plate was sectioned into rectangular bars transverse to the weld direction for tensile testing to evaluate mechanical properties. Digital image correlation (DIC) was employed to map the local strain fields during tensile testing. The initial failure was found to occur simultaneously at the bottom and middle layers away from the weld zone. The top layer failed last in the base metal. The failure locations were consistent among different samples tested. Also, Charpy V-notch impact tests were conducted for weld metal, heat affected zone, and the base metal at each layer as a function of temperature. The weld microstructures were characterized using optical and electron microscopy and micro-hardness mapping.

Introduction

Friction stir welding (FSW) is a unique solid-state joining process that uses the advantages of solid-state joining to fabricate continuous linear welds. A specially designed tool rotates and traverses along the joint line, creating frictional heating that softens a column of material underneath the tool. The softened material flows around the tool through extensive plastic deformation and is consolidated behind the tool to form a solid-state continuous joint. In the current form of FSW, the length of the pin limits the thickness of workpiece that can be friction stir welded. For example, in single sided welding, the depth of weld is about the same as the pin length. For double-sided welding, it is about twice of the pin length.

The power required to spin a friction stir tool increases significantly when the process is applied to thicker and/or harder materials. Consequently, current friction stir welding practices are generally best suited for welding aluminum or other comparatively soft metals with thicknesses of less than approximately one inch (2.5 cm). What is needed therefore are improved friction stir welding processes that may be applied to thicker and/or harder materials.

In this work, an initial attempt to friction stir weld thick-sectioned steel by means of multilayer and multipass friction stir welding (MM-FSW) is demonstrated [1].MM-FSW makes it possible to eliminate the limitations on thickness of structures, allows for the addition of different materials as filler to tailor the composition and properties of the weld region, and creates new alloys by solid-state thermomechanical processing. The study is inspired by the superior mechanical properties obtained in friction stir welding of high-strength line-pipe steels [2], as well as potential applications for high-pressure storage vessels [3] and ballistic resistance structures. Since the overall strength of any weld depends on the distribution of properties across the zones affected by the welding operation, the present work was designed to develop and characterize the mechanical properties of MM-FSW of a high strength low alloy (HSLA) steel. Transverse tensile tests were performed on the welded sample for mechanical properties. Local strain fields were also recorded by a digital image correlation (DIC) technique during tensile testing. Also, Charpy V-notch impact tests were performed for stir zone (SZ), heat affected zone (HAZ), and base metal (BM) zone at each layer, including top, middle, and bottom layer, as a function of temperature. Microhardness at SZ, HAZ, and BM at each layer was also measured. Microstructures were evaluated using optical microscopy (OM) and scanning electron microscopy (SEM).

Materials and Experiment

In the present work, a three layered structure with five multipass friction stir welds was made using flat HSLA steel plates (ASTM A572 Grade 50) with the total thickness of 0.6 inch (15 mm). The nominal chemical composition of the A572 steel is given in Table I.

Table I. Typical chemical composition of ASTM A572 Grade50

Element	C	Mn	P	S	Si	Cu
Weight %	0.23 (max)	1.35 (max)	0.04	0.05	0.3 (max)	0.2 (min)

Figure 1 shows a schematic of each step for MM-FSW. For demonstration purposes, a stepped assembly was machined to accommodate MM-FSW. Subsequently, plates are inserted into the assembly and friction stir welded to achieve fabrication of a multi-layered structure. Initially, a bottom or root weld was made to join the two stepped machined plates with a standard butt joint geometry, as shown Figure 1(a). Next, the first inserted weld plate was placed above the root weld and it was welded using a butt/lap or corner weld joint. Finally, the second weld plate was inserted above the first inserted plate and joined in the same manner. The reason why it is called 'butt/lap' is that the additional metals were inserted into the premachined workpiece for the middle and top layers. Therefore, the weld geometry was a mixed butt and lap joint. The addition of a lap feature to the weld joint is important during FSW. The samples were welded using the MegaStir stationary FSW system.

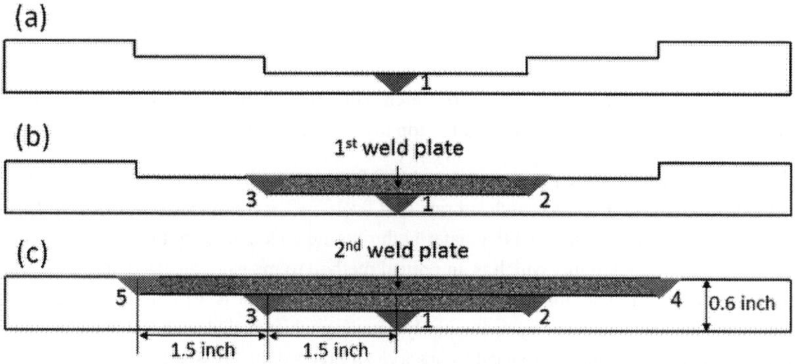

Figure 1. Cross-sections showing (a) Initial butt weld joint design. (b) Insertion of the first weld plate above the root weld. (c) Insertion of the second weld plate above the first weld plate and the final welded structure. Each thickness of top and middle layer was 0.18 inch (4.6 mm) and the bottom layer was 0.24 inch (6.1 mm). The total thickness of MM-FSW is 0.6 inch (15.2 mm). The tool rotation was counter clockwise and into the page.

After initial weld trials to establish weld parameters, defect free weld assemblies were made on flat workpieces. For FSW, the plunge was performed at 1100 rpm followed by a 6 second dwell at 250 rpm. Following the dwell, the tool rotation rate was maintained at 250 rpm and the tool travel speed ramped up to 3 to 4 ipm (1.29 to 1.69 mm/s) within one inch of travel. FSW was performed using load control with normal loads ranging from 7,200 to 8,000 lbs. Also, in order to minimize weld defects at the lap feature of the weld joint, such as hooking, friction stir welds were made with the tool direction such that the tool advancing side was on the lap side of the weld joint. For FSW, a tool with a concave scroll shoulder step spiral pin (CS4 tool design) was used and was fabricated from an MS80 grade of PCBN.

The MM-FSW steel plate was sectioned into rectangular bars transverse to the weld direction for tensile testing to evaluate overall mechanical properties. Specimens for the transverse tensile tests were prepared by electrical discharge machining (EDM). As shown in Figure 2, dimensions of the tensile test samples were 0.5 inch (12.5 mm) in width with a uniform gauge length of 8 inch (203.2 mm) and thickness of 0.6 inch (15.24 mm). Similarly, tensile tests were completed using base metal to compare the mechanical properties with the MM-FSW steel samples.

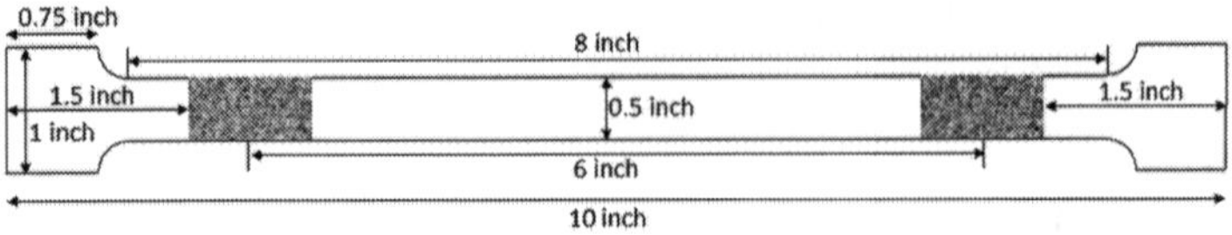

Figure 2. Schematic of the tensile test specimen with dimensions (top view). Two weld zones (textured) at the top layer are illustrated.

Mechanical tensile tests were performed on a MTS 22kip servo-hydraulic machine driven by an MTS model 407 controller at a crosshead rate of 4.0 $x10^{-3}$ s^{-1} at room temperature. In order to study the deformation and strain of the material during the tensile testing, DIC (Correlated Solutions, VIC-3D) technique, which is an optical method to measure displacement and strain, was employed. Prior to DIC measurements, random speckle patterns were made on the cross section of the tensile specimens by sequentially spraying white and black paint. Two digital cameras of the stereo system were used to observe and record the cross section view of the tensile sample during the tensile test. Samples were illuminated by a fiber light and the recorded data analyzed by the software (VIC-3D 2010).

Based on ASTM E23 specification, Charpy V-notch impact tests were conducted to study the fracture toughness of the SZ, HAZ, and BM at each layer. Due to the different thickness at each layer, sub-sized samples were prepared for impact testing with the following dimensions (10 x 4 x 55 mm). Charpy tests were performed at six different temperatures for the SZ, i.e., -50, -30, -20,-10, 0, and +20 °C. Similarly, samples located in the HAZ and BM were tested at -50, 0, 20 °C. Triplicate samples were tested at each temperature at Laboratory Testing Industry (LTI) facility.

Vickers hardness of each weld sample was measured by Leco microhardness tester (LM 100AT) with 200µm spacing, 200g of load and 13 seconds of dwell time. The measured hardness data was transferred to Origin software (Origin Pro 8.1) for mapping.

To reveal the microstructure, the samples were cut, mounted, and polished using 6, 3, and 1µm diamond paste solutions and etched using a 5% Nital solution. An optical microscope (Nikon EPIPHOT) and field emission scanning electron microscope (FE-SEM) (Hitachi S4800) were used to characterize microstructures of the weld samples.

Results and Discussion

A low magnification optical image of the final weld assembly is presented in Figures 3(a) and (b). Because the assembly is small and the welds are partial penetration, there is considerable bowing or distortion. The overall weld length was about 33 inch (838.2 mm). Five multipass welds were used to make the three layer total thickness of 0.6 inch (15.24 mm). Figure 3(b) shows a cross sectional view of the MM-FSW steel sample.

Figure 3 (a) Top view of finished weld (b) cross sectional image of MM-FSW steel sample.

Figures 4(a) and 4(b) show optical macrographs of the friction stir welds at the bottom and top layers respectively. As shown in Figure 4(a), the root weld is a defect free full penetration weld. Figure 4(b) presents the top layer weld, showing a minimum of interfacing hooking. Hooking is minimized by positioning the tool advancing side on the lap interface side of the corner joint. Hooking is typically caused by the upward flow of metal at the sheet interface when the tool is penetrated into the bottom workpiece [4]. This interface defect has been found with lap joints coupled with FSW. The authors also tried a corner weld with the FSW tool retreating side positioned on the lap interface. Even though the weld nugget was free from volumetric defects, it showed uplift of the interface (not presented here). These are just a few metallographic examples but they are representative of all weld locations, i.e., hooking was consistently prevented/minimized when the tool advancing side was positioned on the lap interface. Thus, to minimize the possibility of this type flaw, the advancing side of the tool should be positioned on the lap interface.

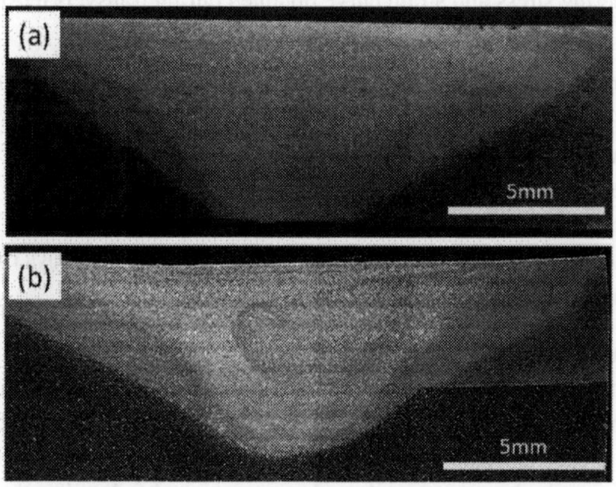

Figure 4. (a) Macrograph illustrating full penetration and a defect free weld nugget for the bottom butt weld. (b) Macrograph showing a defect free weld nugget for a top layer corner joint. The metal insert is on the right side, resulting in butt/lap joint geometry. The corner joint was welded with the FSW tool advancing side positioned on the lap interface.

Figure 5 plots the engineering stress and strain graph obtained during tensile testing of MM-FSW steel samples. Each sample showed relatively consistent results. Averaged yield stress (σ_{yp}= 60 ksi [419 MPa]) and ultimate tensile strength (σ_{ult}= 76.3 ksi [533 MPa]) were also obtained from Figure 5. These strengths were approximately the same as the base workpiece (σ_{yp}= 62 ksi and σ_{ult}= 76.6 ksi). This result is consistent with results of previous researcher's [5]. A fracture elongation of at least 14% was reached at the initial failure for MM-FSW samples. A fracture elongation of ~17.5% was obtained for the final failure for both samples. The average elongation at fracture for the base metal was found to be 43.6%.

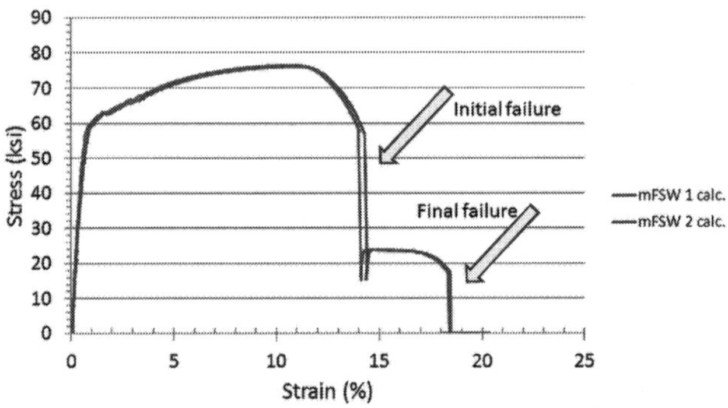

Figure 5. Engineering stress and strain curve obtained during transvers tensile testing of MM-FSW steel samples.

Two failures occurred during the tensile testing from the recorded video images and the obtained engineering stress-strain curves. Considerable necking occurred distant from the weld nugget with eventual failure in the parent metal. This initial failure simultaneously occurred at the bottom and middle layer. Subsequently, the top layer was failed last. Both the initial and final failure locations for each sample were consistently found in the parent metal. This test result, i.e., failure in the base material, is consistent with other FSW joined steel alloys reported by previous researchers, i.e., the weld joint is stronger than the base metal [6,7].

DIC images were used to observe local strain fields in transverse weld tensile samples. Figure 6 shows a fractured tensile specimen (a) and corresponding DIC strain maps near initial (b) and final fracture (c) loads. Results show the base metal had higher strains (around 0.4) than the weld zone (ranged from 0.025 to 0.05) right before fracture, approximately 8 to 16 times higher than the weld zone. This result also corresponds to the experimental failure locations both for the initial and final stages of fracture. These results indicate that the stir zone (SZ) and HAZ have higher yield strength than the parent metal.

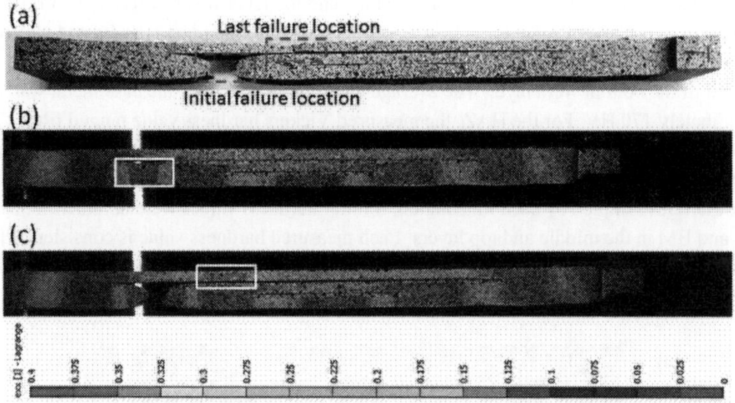

Figure 6. (a) Cross section view of tensile specimen showing initial and final failure locations. (b) DIC image presenting strain level right before the initial failure at middle and bottom layers. (c) DIC image displaying the strain map right before the final failure at the top layer. Failure locations had high strain levels and both failure locations were found in the base metal.

Charpy V-notch impact tests were conducted to study to fracture toughness of the SZ, HAZ, and BM of each layer. Figure 7 summarizes the Charpy impact test results for each zone with different layers at different test temperatures. Generally, toughness of the SZ was much higher than the base metal at all temperatures. Similarly, it is observed that the toughness of HAZ was also higher than the parent material. This substantial increase in the absorbed impact energy can be related to the grain size of the SZ and HAZ. Similar observations were obtained for aluminum alloys in previously reported work [8]. The grain size of each zone will be discussed below with the microstructure images.

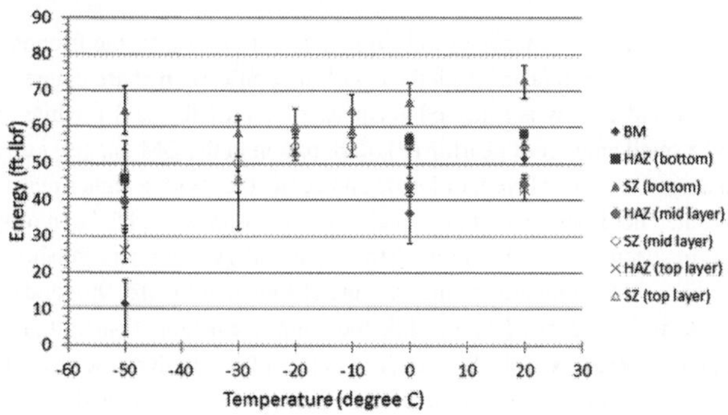

Figure 7. Charpy V-notch impact test results for each layer at different temperature ranges. Toughness of WM and HAZ was typically higher than the parent material.

Microhardness of each weld at the bottom, middle, and top layer was measured to correlate with the strength of each weld zone. Figures 8(a) and (b) show microhardness results and plotted data as a function of the x position (along the weld width) at three different through thickness positions of the bottom weld layer. The average measured hardness of the base metal was approximately 170 HV. For the HAZ, the measured Vickers hardness value ranged from 175 to 185 HV. In the SZ, the highest hardness value is shown, ranging from 195 to 220 HV. Variations in hardness results can be associated with differences in the grain size and different microstructures sampled by each indentation. Similar hardness values were obtained for each SZ, HAZ, and BM in the middle and top layers. Each measured hardness value is consistent with the tensile test results. That is, as expected, hardness of the weld is much higher than that of the parent metal.

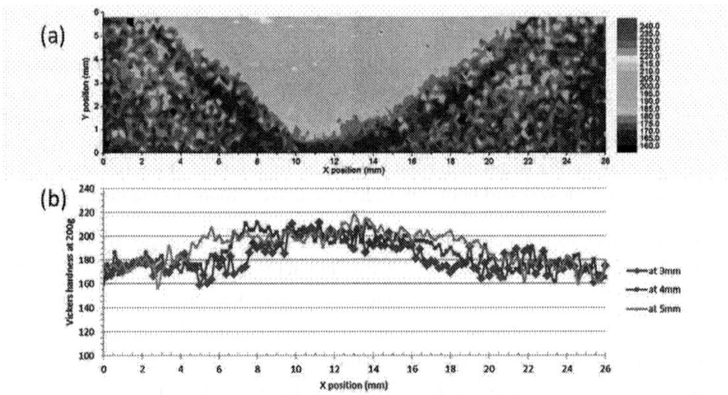

Figure 8. (a) Mapped Vickers microhardness value for the bottom layer (butt joint) weld area, and (b) cross sectional hardness plot at three different locations in through the plate thickness.

Figure 9 shows OM and SEM images of weld zone microstructures for the friction stir welded butt joint area, i.e., the bottom layer. With the aid of the iron-carbon phase diagram, the base metal with 0.23 wt% of C alloy is expected to consist of mainly ferrite (F) (white region in the OM image) and a small amount of pearlite (P) (dark region in the OM image), i.e., a lamellar structure consisting of alternate layers of ferrite and cementite. With a higher magnification SEM image (not presented here), the size of the lamellar layer was measured to be approximately 100 nm with a uniform spacing of slightly more than 100 nm. Figure 9(a) and (b) show the refined grain structures in the SZ. Compared to the base metal shown in Figures 9(e) and (f), the size of refined grain in SZ is the order of 10 μm, while the grain size is more than 100 μm for the parent metal. This grain refinement is typically observed in the SZ due to dynamic recrystallization during FSW [5]. In addition, mixtures of a refined ferrite grain structure and fine/randomly distributed pearlite were apparent in the SZ. In the HAZ, the grain size was slightly larger than the grain size in the SZ due to grain growth, as shown in Figures 9(c) and (d). Grain growth in the HAZ can be affected by exposure to a lower peak temperature during FSW. That is, the fine microstructure in the HAZ can be due to less time for the ferrite to transform from austenite because of low heat input from FSW, but sufficiently high for grain growth. Similarly, there is evidence for ferrite and a fine pearlite mixture in the HAZ in both OM and SEM images (white dashed box). This observation can be related to the measured hardness of the SZ, HAZ, and BM. Variations in hardness with grain size follow a Hall-Petch relationship, i.e., the hardness value decreases with increasing grain size [9].

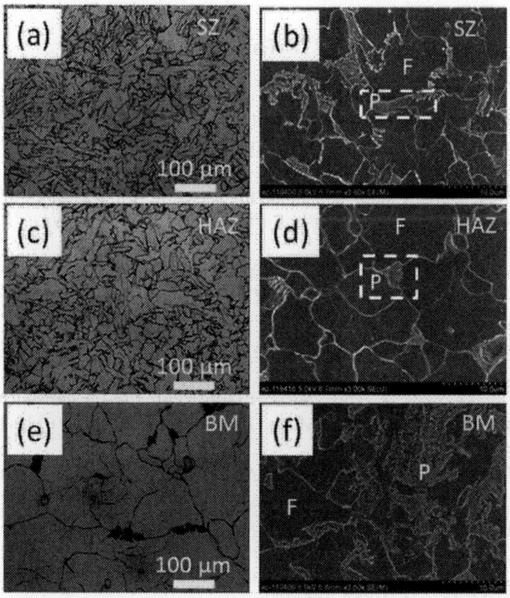

Figure 9. Microstructures of SZ, HAZ and BM at the bottom layer. Optical microscope (OM) images (a), (c), (e); and scanning electron microscope (SEM) images (b), (d), (f). Note the white region is ferrite and dark region is pearlite in OM images.

Conclusions

In conclusion, the MM-FSW technique was developed and demonstrated for thick layer joining of a high strength low alloy steel. Defect free welds were produced using optimized FSW parameters. Mechanical properties of transverse tensile tests of MM-FSW samples were close to those of the base metal and the fracture location was found in the base material. Strengths in the SZ and HAZ were typically higher than the base metal based on the microhardness and fracture toughness test results. Taken together with all the mechanical tests and hardness measurements, the present work indicates that the overall mechanical strength of MM-FSW steel samples were comparable to the parent metal.

Acknowledgements

The authors would like to acknowledge the financial support of the US Department of Energy, Energy Efficiency and Renewable Energy, Advanced Manufacturing Office. ORNL is managed by UT-Battelle, LLC for the U.S. Department of Energy under Contract DE-AC05-00OR22725.

References

1 Z. Feng, S. A. David, D. A. Frederick, "Multiple pass and multiple layer friction stir welding and material enhancement processes." USA patent 7762447 (2010)

2 Z. Feng, R. J. Steel, S. M. Packer, S. A. David, "Friction Stir Welding of API Grade 65 Steel Pipes," in *2009 ASME Pressure Vessels & Piping Division Conference.* PVP2009-77248.

3 Z. Feng, J. Wang, W. Zhang, "Vessel Design and Fabrication Technology for Stationary High-Pressure Hydrogen Storage," in *2012 DOE Hydrogen and Fuel Cells Annual Merit Review Proceedings.* http://www.hydrogen.energy.gov/pdfs/review12/pd088_zhang_2012_o.pdf

4. H. Badarinarayan, Q. Yang, Z. Zhu, "Effect of tool geometry on static strength of friction stir spot-welded aluminum alloy," Int. J. Mach. Tools Manuf. 49 (2009) 142-148.

5 . T. J. Lienert, W. L. Stellwag, JR., B.B. Grimmett, R.W. Warke, "Friction stir welding studies on mild steel - Process results, microstructures, and mechanical properties are reported," Welding J. 82 (2003) 1s-9s.

6. W. M. Thomas, P. L. Threadgill, E. D. Nicholas, "Feasibility of friction stir welding steel," Sci. Technol. Weld. Join. 4 (1999) 365-372.

7. J. Fujii, L. Cui, N. Tsuji, M. Maeda, K. Nakata, K. Nogi, "Friction stir welding of carbon steels," Mater. Sci. Eng. A 429 (2006) 50-57.

8. K.S. Arora, S. Pandey, M. Schaper, R. Kumar, "Microstructure Evolution during Friction Stir Welding of Aluminum Alloy AA2219," J. Mater. Sci. Tech. 26 (2010) 747-753.

9. W.B. Lee, J. W. Kim, Y. M. Yeon, S.B. Jung, "Evaluation of the microstructure and mechanical properties of friction stir welded 6005 aluminum alloy," Mater. Sci. Technol. 19 (2003) 1513-1518.

Friction Stir Welding and Processing VII
Edited by: Rajiv Mishra, Murray W. Mahoney, Yutaka Sato, Yuri Hovanski, and Ravi Verma
TMS (The Minerals, Metals & Materials Society), 2013

EFFECT OF WELDING PARAMETERS ON THE MICROSTRUCTURE AND MECHANICAL PROPERTIES OF A FRICTION STIR WELDED 11CR-FERRITIC/MARTENSITIC STEEL

Yutaka S. Sato[1], Hiroyuki Kokawa[1], Yasuhide Yano[2], Yoshihiro Sekio[2]

[1]Department of Materials Processing, Graduate School of Engineering, Tohoku University; 6-6-02 Aramaki-aza-Aoba, Aoba-ku; Sendai 980-8579, Japan
[2]Japan Atomic Energy Agency; 4002 Narita-cho; Oarai-machi, Ibaraki 311-1393, Japan

Keywords: Friction stir welding, Ferritic/martensitic steel, Mechanical properties, Microstructure

Abstract

PNC-FMS is a newly developed 11Cr-ferritic/martensitic steel with good swelling resistance designed for the wrapper tubes of fast reactors. Because fusion welding of PNC-FMS significantly reduces mechanical properties through the formation of a brittle microstructure, friction stir welding (FSW) was attempted as a solid state welding process. FSW was applied to PNC-FMS at 100 to 300 rpm using a Q60 tool, and defect-free welds were obtained. The stir zones had a fine microstructure with ferrite and martensite with the grain size and fraction of martensite increasing with tool rotational speed. Since all welds were overmatched, all transverse tensile specimens failed in the base material. Based on mechanical properties in the stir zone, welds produced at 100 rpm exhibited both higher strengths and elongation than the base material. This study showed that FSW at lower rotational speed produced stir zone microstructures having better mechanical properties in PNC-FMS.

Introduction

Ferritic/martensitic (F/M) steels are considered as a future material for the long-life core of fast reactors (FRs) and also for the blanket of fusion reactors due to their good swelling resistance. An 11Cr F/M steel, PNC-FMS, has been newly developed for wrapper tubes for the Japan Sodium-cooled FR [1,2]. When PNC-FMS is used in the FR application, metallurgical joining is required. In preliminary studies, conventional fusion welding processes, such as gas tungsten arc welding and electron beam welding, have been applied to this steel [3]. These fusion welding methods resulted in a significant reduction of mechanical properties through the formation of a brittle microstructure in the weld, i.e., coarse δ-ferrite and hard martensite. Since these undesirable phases are formed by holding at high-temperatures followed by rapid cooling during welding, solid-state welding processes with low heat-input should be favorable for joining PNC-PMS. In this study, friction stir welding (FSW) [4] has been attempted as a solid-state welding process for this steel.

Studies using FSW to join 11Cr ferritic steels for nuclear applications, such as PNC-FMS, have been limited, although some papers have shown preliminary results on FSW of commercial 9-12Cr ferritic steels. Ahn et al. [5] examined the microstructure and mechanical properties of a friction stir welded 409L ferritic stainless steel, and reported that mechanical properties of the weld were similar to those of the base material (BM). Lakshimanarayan et al. [6] showed that friction stir welded 409M stainless steel exhibited superior fatigue properties to the BM. Cho et al. [7] reported formation of a fine grain structure with a high fraction of low-angle boundaries in the stir zone (SZ) of friction stir welded 409 stainless steel. These previous studies [3-5] have provided fundamental knowledge on the mechanical properties and microstructure in friction stir welded 9-12Cr ferritic steels, but the effect of welding parameters on mechanical properties, associated with the microstructure in friction stir welds, is not fully understood.

The objective of this study is to systematically examine the relationship between welding parameters, mechanical properties, and the resulting microstructure in fiction stir welded PNC-FMS. Since the rotational speed of the welding tool significantly affected the weld microstructure, more so than other welding parameters, the effect of rotational speed on the mechanical properties and microstructure in the weld was evaluated in this study.

Experimental procedures

The material used in this study was 6 mm thick PNC-FMS. Chemical composition and heat treatment conditions of PNC-FMS are shown in Table 1. Examination of the phase transformations revealed that the A_1 and A_3 temperatures of this steel are 1083 K (810°C) and 1153 K (880°C), respectively. Bead-on-plate FSW was applied to PNC-FMS plate at tool rotational speeds between 100 rpm and 300 rpm using a Q60 tool consisting of 60% cBN (cubic boron nitride) and 40% W-Re alloy. Appearance and dimensions of the Q60 tool are presented in Figure 1. The tool travel speed and plunge depth were constant at 1.0 mm/s and 4.2 mm, respectively. Ar shielding was employed to prevent oxidation of the plate surface during FSW.

Table 1. Chemical composition and heat-treatment condition of PNC-FMS used in this study.

C	Si	Mn	P	S	Ni	Cr	Mo	W	V	Nb	N	Fe
0.14	0.07	0.67	<0.005	0.0034	0.54	10.38	0.40	1.79	0.19	0.060	0.061	Bal.
Normalized at 1323 K for 2.4 ks (40 min) and then tempered at 983 K for 2.4 ks (40 min).												

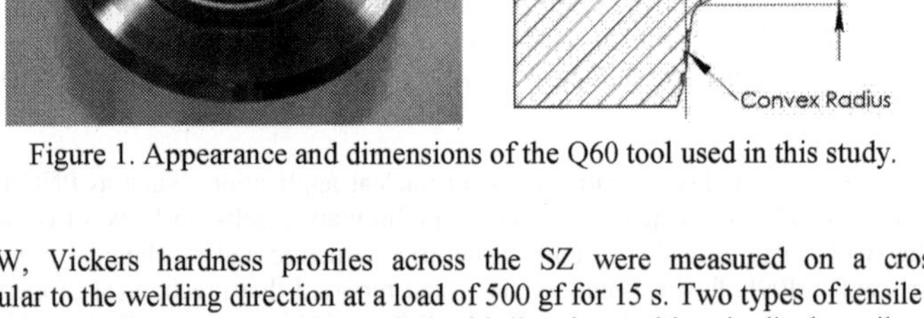

Figure 1. Appearance and dimensions of the Q60 tool used in this study.

After FSW, Vickers hardness profiles across the SZ were measured on a cross section perpendicular to the welding direction at a load of 500 gf for 15 s. Two types of tensile tests were conducted, i.e., tensile tests transverse to the weld direction and longitudinal tensile tests of the SZ. Configurations of these tensile specimens are shown in Figure 2. The transverse tensile tests were carried out in air using a screw-driven tensile testing machine at an initial strain rate of 5.0×10^{-5} s^{-1}, which was changed to 1.3×10^{-3} s^{-1} after yielding. The test temperatures were 298 K (25°C), 823 K (550°C), 873 K (600°C) and 923 K (650°C). The longitudinal tensile tests of the SZ were also conducted in air using the same machine at a constant crosshead speed of 6.35×10^{-3} mm/s (an initial strain rate of 8.3×10^{-3} s^{-1}) at room temperature.

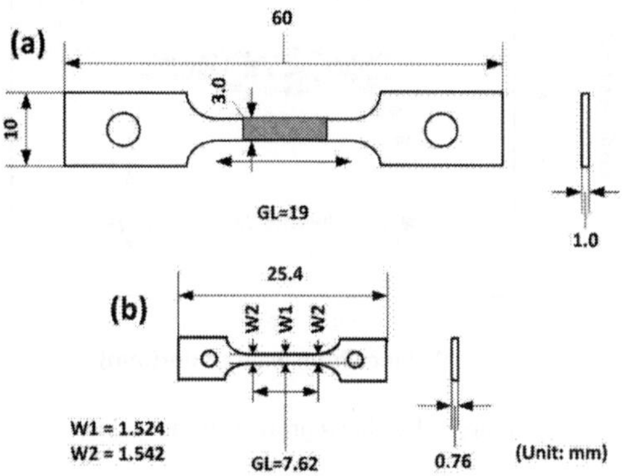

Figure 2. Configurations of (a) transverse tensile test specimen and (b) longitudinal tensile test specimen.

Microstructures in the SZ were examined by optical microscopy, scanning electron microscopy (SEM) and electron backscatter diffraction (EBSD) methods. Samples for optical microscopy and SEM were mechanically polished with Emery papers and 1 μm alumina paste, and then etched in aqua regia. Martensite and ferrite were easily identified in the SEM images because etching created distinctly different contrasts. The fraction of martensite was quantified by point counting on the SEM image. Samples for EBSD analysis were electrolytically polished in a 10% perchloric acid + 90% acetic acid solution. Grain boundary maps were drawn using crystallographic data obtained by the EBSD method. In the maps, high-angle boundaries (HABs) with misorientation over 15° and low-angle boundaries (LABs) with misorientation between 2° and 15° were expressed by thick black lines and thin gray lines, respectively. Grain size was also quantified by a mean linear intercept method for HABs in the grain boundary map.

Results and discussion

Mechanical properties

Vickers hardness profiles of the welds are shown in Figure 3. Hardness of the BM is about 300 Hv. Hardness increases sharply in the heat affected zone (HAZ) moving toward the SZ, and the SZ exhibits a much higher hardness than the BM, reaching about 530 Hv. Hardness of the SZ is hardly affected by the rotational speed.

Tensile strength and elongation of the weld as a function of temperature are shown in Figure 4. All welds exhibited roughly the same strengths as the BM at all temperatures and mostly failed in the BM because the welds are overmatched. Elongation of the weld was smaller than that of the BM due to localized deformation of the BM in the specimen gauge section [8].

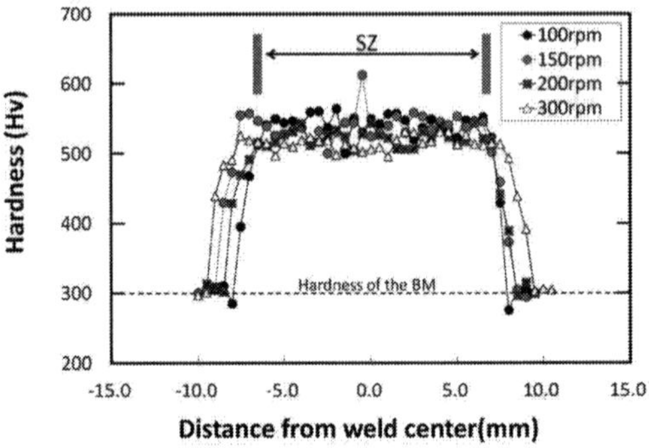

Figure 3. Hardness profiles of the welds.

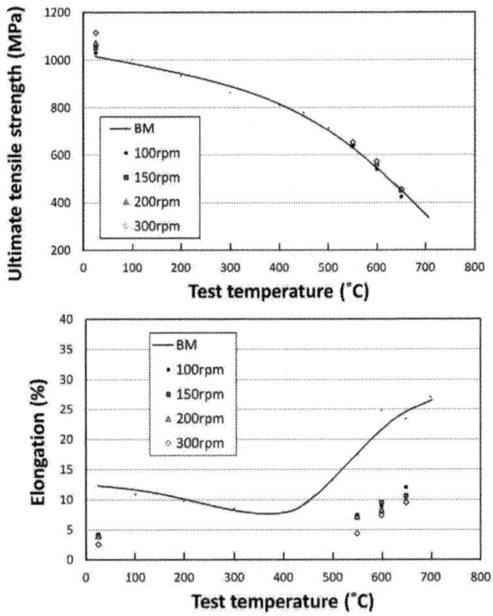

Figure 4. Temperature dependence on ultimate tensile strength and elongation of the weld obtained by transverse tensile tests.

The effect of rotational speed on tensile properties of the SZ is presented in Figure 5. All SZs show an ~1.8 times higher ultimate tensile strength (UTS) than the BM, and the elongations were roughly the same as that of the BM. The SZ produced at 100 rpm exhibits higher strengths and elongation than the BM.

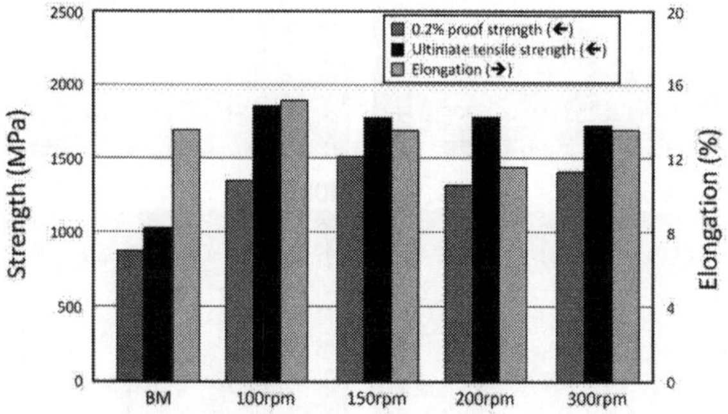

Figure 5. Tensile properties of the SZs.

Microstructure

Optical-microscopy and SEM images of the BM are shown in Figure 6. The BM has a lath-shaped microstructure with carbide precipitation along the lath boundaries, i.e., the BM consists of tempered martensite. Optical micrographs of the SZs produced at the different rotational speeds are shown in Figure 7. Although the microstructure observed in the SZ produced at 100 rpm is unclear at this magnification, the SZs produced at 200 rpm and 300 rpm are composed of a lath-shaped microstructure. SEM images of the SZs are shown in Figure 8. All SZs consist primarily of a martensitic structure without carbides, i.e., the quenched martensite. In addition, ferrite phases are also observed in all SZs. The ferrite content decreased with increasing rotational speed.

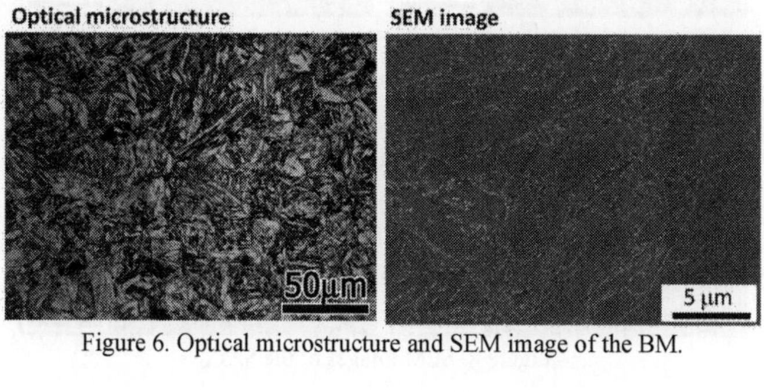

Figure 6. Optical microstructure and SEM image of the BM.

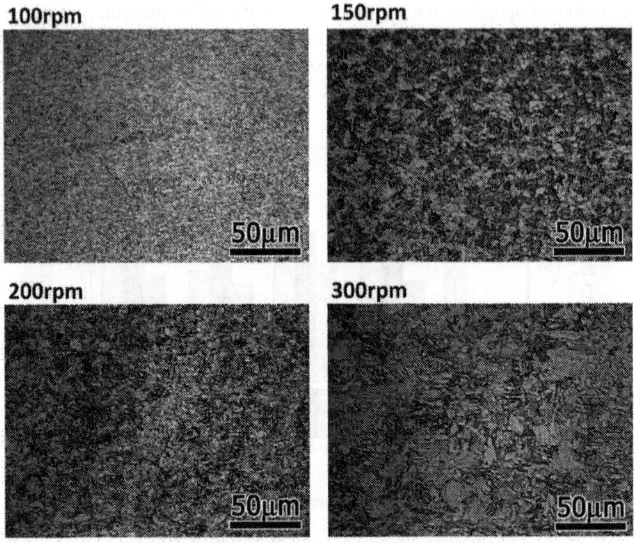

Figure 7. Optical microstructures of the SZs.

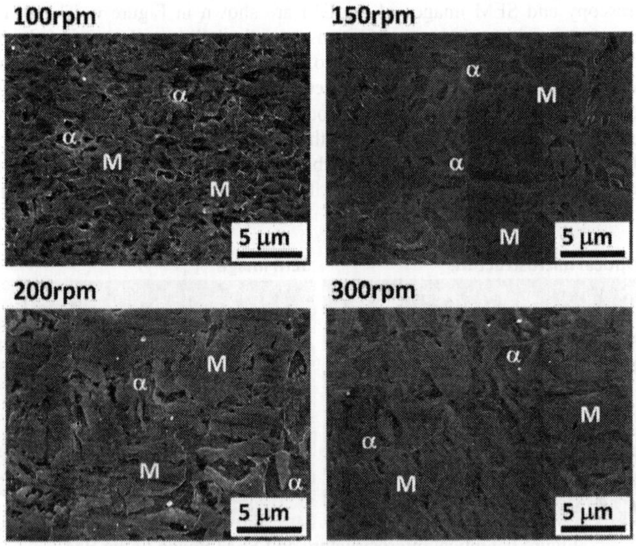

Figure 8. SEM images of the SZs.

Coexistence of the quenched martensite and ferrite in the SZ suggests that FSW was carried out within the temperature range between the A_1 and A_3, because formation of ferrite from austenite during the cooling cycle of FSW would be unexpected due to the high hardenability of high Cr steels [9]. Since the higher rotational speed creates a higher maximum temperature during FSW [10], the ferrite fraction in the (ferrite + austenite) microstructure should be lower at the maximum FSW temperature. This could lead to the decrease in ferrite content with increasing rotational speed. The austenite transforms into quenched martensite during the cooling cycle resulting in a SZ composed of the quenched martensite with ferrite.

Grain boundary maps obtained from the SZs are presented in Figure 9. The lath martensitic structure is observed in the SZs produced at the higher rotational speeds, while the grain structure of the SZ produced at 100 rpm appears equiaxed. The grain size increased with increasing rotational speed. Reconstruction of the prior austenite grain structure, based on crystallographic orientation relationships between martensite and austenite, revealed that the prior austenite grain size in the SZ also increased with increasing rotational speed. Quantitative data of microstructural features in the SZ, i.e., martensite content, grain size and prior austenite grain size, are summarized in Table 2.

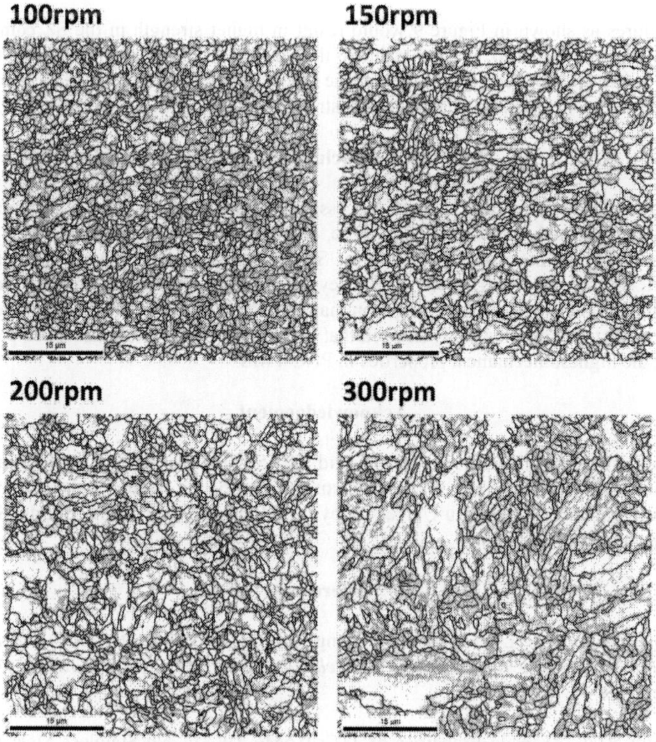

Figure 9. Grain boundary maps obtained from the SZs.

Table 2. Quantitative data of microstructual features in the SZs.

Rotational speed (rpm)	Martensite content (%)	Grain size (μm)	Prior austenite grain size (μm)
100	66	0.6	1.8
150	68	0.7	4.6
200	78	0.8	6.9
300	82	0.9	11.1

Relationship between mechanical properties and microstructure

Microstructures of the SZs consisted of quenched martensite with ferrite. As shown in Table 2, the higher rotational speed resulted in a higher fraction of martensite and larger grain size in the SZ. Although those microstructural factors change with rotational speed in the SZ, hardness values of the SZs were hardly affected by the rotational speed. This result might be due to a compensating effect of martensite fraction and grain size on hardness, because the higher hardndess results from a higher fraction of martensite [12,13] and smaller grain size [9] in steels.

Microstructures as shown in Figure 9 would result in higher strength in the SZ compared with the BM. Elongation in the SZ is comparable to that of the BM and is attributable to the presence of ferrite.. In the SZ produced at 100 rpm, a fine equiaxed microstructure consisting of quenched martensite with ferrite might yield both higher strength and elongation than in the BM.

Conclusions

Friction-stir welds, without defects, were successfully obtained in PNC-FMS. The microstructure consisting of quenched martensite with ferrite, was formed at all rotational speeds evaluated, resulting in higher hardness and strength of the SZ. Elongation in the SZ was comparable to that of the BM, and roughly 100% joint efficiency was achieved. The SZ produced at 100 rpm exhibited both higher strength and elongation than the BM possibly due to the formation of a fine equiaxed microstructure. This study showed that FSW at the lower rotational speed produced the SZ having the highest mechanical properties in PNC-FMS.

Acknowledgement

The authors are grateful to Mr. A. Honda and Mr. S. Fukushi for technical assistance. The present study includes the result of "Friction stir welding of the wrapper tube materials for Na fast reactors" entrusted to Tohoku University by the Ministry of Education, Sports, Science and Technology of Japan (MEXT).

References

1. S. Nomura et al., "Development of long life FBR core materials," *Proceedings of International Conference on Fast Reactor Related Fuel Cycles* (Kyoto, Japan, October 28-November 1, 1991), 7.4-1-7.4-10.

2. A. Uehira et al., "Tensile Properties of 11Cr-0.5Mo-2W, V, Nb Stainless Steel in LMFBR Environment," *Journal of Nuclear Science and Technology*, 37 (2000) 780-786

3. T. Kaito, S. Ohtsuka, and M. Inoue, "Progress in the R&D Project on Oxide Dispeersion Strengthened and Precipitation Hardened Ferritic Steels for Sodium Cooled Fast Breeder Reactor Fuels," *Proceedings on International Conference GLOBAL 2007* (Boise, Idaho, September 9-13, 2007), 37-42.

4. W.M. Thomas et al., "Friction Stir Butt Welding," *International Patent Application No. PCT/GB92/02203.*

5. B.W. Ahn et al., "Microstructure and properties of friction stir welded 409L stainless steel using a Si_3N_4 tool," *Materials Science and Engineering A*, 532 (2012), 476-479.

6. A.K. Lakshminarayanan and V. Balasubramanian, "Assessment of fatigue life and crack growth resistance of friction stir welded AISI 409M ferritic stainless steel," *Materials Science and Engineering A*, 539 (2012), 143-153.

7. H.H. Cho et al., "Microstructural analysis of friction stir welded ferritic stainless steel," *Materials Science and Engineering A*, 528 (2011), 2889-2894.

8. Y.S. Sato and H. Kokawa, "Distribution of Tensile Property and Microstructure in Frction Stir Weld of 6063 Aluminum," *Metallurgical and Materials Transactions A*, 32A (2001), 3023-3031.

9. R.W.K. Honeycombe and H.K.D.H. Bhadeshia, *Steels Microstructure and Properties, second ed.* (Edward Arnold, London, 1995).

10. Y.S. Sato, M. Urata, and H. Kokawa, "Parameters Controlling Microstructure and Hardness during Friction-Stir Welding of Precipitation-Hardenable Aluminum Alloy 6063," *Metallurgical and Materials Transactions A*, 33A (2002), 625-635.

11. Y.S. Sato et al., "Microstructural evolution of ultrahigh carbon steel during friction stir welding," *Scripta Materialia*, 57 (2007), 557-560.

12. Y.S. Sato et al., "Characteristics of Microstructure in Ultrahigh Carbon Steel Produced during Friction Stir Welding," *ISIJ International*, 48 (2008), 71-76.

Friction Stir Welding and Processing VII
Edited by: Rajiv Mishra, Murray W. Mahoney, Yutaka Sato, Yuri Hovanski, and Ravi Verma
TMS (The Minerals, Metals & Materials Society), 2013

FRICTION-STIR-WELDING OF THICK CARBON STEELS USING CO-BASED ALLOY TOOL

Itto Sugimoto[1], Akihiro Sato[1], Seung Hwan C. Park[1], Satoshi Hirano[1], Shinya Imano[1], Yutaka S. Sato[2], Hiroyuki Kokawa[2], Toshihiro Omori[2], and Kiyohito Ishida[2]

[1] Hitachi Research Laboratory, Hitachi, Ltd;
7-1-1 Omika-cho, Hitachi, 319-1292, Japan
[2] Graduate School of Engineering, Tohoku University;
6-6-02 Aramaki-aza-Aoba, Aoba-ku, Sendai, 980-8579, Japan

Keywords: Friction-stir-welding, Tool, Co-based alloy, Carbon steels

Abstract

Friction-stir-welding (FSW) is a solid-phase joining process, and many studies on the application to structural steels have been published. In this study, FSW of thick carbon steel was conducted using heat-resistant moldable Co-based alloy tools. In the weld trials, acceptable appearances were created using a tool with a probe shorter than 9 mm, while a tool with a 12-mm long probe could not successfully weld the materials. The stir zone (SZ) had slightly higher hardness than the base material and fine grains of less than 10 μm in diameter. Comparing the different tool geometries, the tool with 3-flats and a step spiral embedded in the probe effectively reduced the output power of the motor for tool rotation.

Introduction

The application of friction-stir-welding (FSW) has been currently limited to non-ferrous metals with relatively low melting points, including Al and Cu alloys. FSW of various steels has recently been performed to extend the application of FSW to conventional structural parts to avoid large distortions. FSW of high-softening-temperature materials (HSTMs), including steels, must use a tool made of a hard material, such as ceramic or superalloy, to withstand the high temperatures. A Co-based superalloy, strengthened by Co_3 (Al,W) precipitates with a $L1_2$ structure designated as γ', was identified in 2006 as a potential new tool material [1]. It has a coherent γ/γ' microstructure at high temperatures, and exhibits greater high-temperature strength than Ni-based superalloys. Further, tools can be made by casting and the tools are inexpensive compared to sintered polycrystalline cubic boron nitride (PCBN) or W-Re based tools. Sato et al., reported that FSW of various steels, cp-Ti and Ti-6Al-4V was achieved using a Co-based alloy tool with the probe shorter than 2 mm [2]. This study focused on FSW of thick section carbon steels and the effects of the tool geometry on the FSW machine loads.

Experimental details

An FSW machine with a rated motor power of 45 kW and a critical axial force of 10,000 kgf was used to carry out butt welding of the carbon steel. Tools were machined from the Co-based alloy cast by the lost wax process. The different tool geometries and tilt angles of the Co-based alloy tools used in this study are summarized in Table 1. Tools with different probe lengths, i.e., 6 mm, 9 mm, and 12 mm, were prepared. Tilt angles of 1° to 3° were applied to the tools during FSW.

As shown in Figure 1, 6-mm length probes were prepared with (a) normal, (b) 3-flats, and (c) step spiral 3-flats geometries. Carbon steel workpieces, whose chemical composition was Fe-0.17wt%C-0.16wt%Si-0.74wt%Mn-0.012wt%P-0.006wt%S, with a tensile strength of 433 MPa were used.

Table 1 Probe geometries and tilt angles of Co-based alloy tools

No.	Length of probe / mm	Probe geometry	Tilt / °
1	6.0	Normal	3
2	9.0	Normal	2
3	12.0	Normal	1
4	6.0	3-flats	3
5	6.0	Spiral 3-flats	3

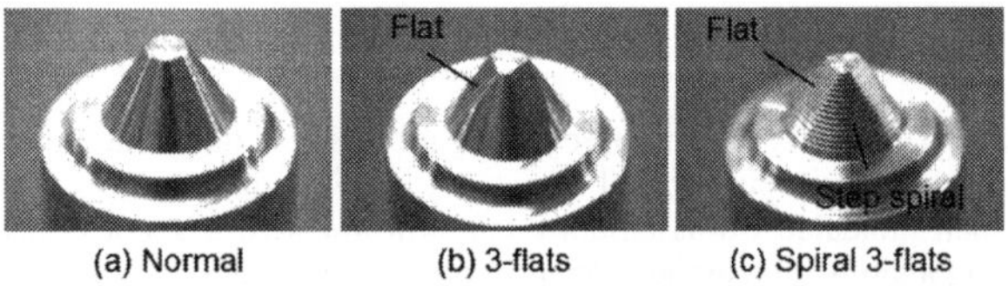

Figure 1. Geometries of 6-mm long probes: (a) normal, (b) 3-flats and (c) spiral 3-flats.

FSW of thick carbon steels

When a normal cone tool with a 6-mm probe was used, the carbon steel workpieces could be butt-welded at a tool rotation lower than 250 rev/min. The appearance and cross-section of the welds produced at the different travel speeds are shown in Figure 2. An acceptable surface weld appearance was obtained at 150 mm/min, as shown in Figure 2-(a). However, a cavity can be seen in the SZ around the advancing side (AS) in the weld cross-section. In contrast, as shown in Figure 2-(b), a high travel speed of 500 mm/min results in incomplete consolidation of the joint. Variation of the output power of the motor with travel speed is shown in Figure 3. Acceptable surface weld appearances were confirmed less than 250 mm/min of travel speed. The output power increased with the travel speed, except for 300 mm/min.

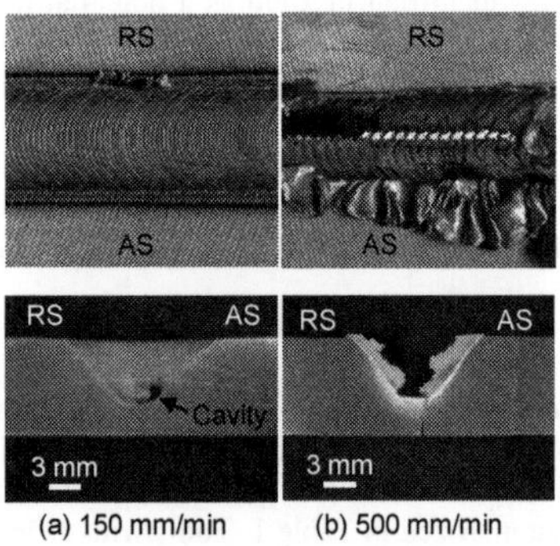

Figure 2. Appearances and cross-sections of the welds produced at two different travel speeds: (a) 150 mm/min, and (b) 500 mm/min.

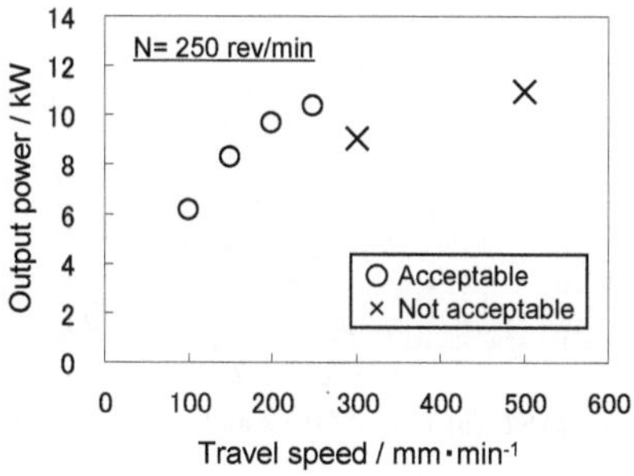

Figure 3. Variation of output power with travel speed.

In this study, the materials were successfully welded using a tool with a probe shorter than 9 mm. The appearance, cross-sections and hardness variations of the weld produced using at 9-mm probe are presented in Figure 4. Although a cavity was still formed, an acceptable surface appearance was obtained. The SZ was hardened up to 185 Hv. Microstructures of the weld and the base material (BM) are shown in Figure 5. The SZ had fine grains less than 10 μm in diameter while the BM contains several regions consisting of the hypoeutectoid microstructure. The effect of probe length on output power of the motor is shown in Figure 6. FSW trials with the 6-mm and 9-mm probes resulted in acceptable weld appearances on the crown surface, while the 12-mm probe did not lead to any successful welds. With the 12-mm probe tool, breakage of the Co-based tool occurred at a rotation speed of 150 rev/min during the plunge. When the rotation speed was 67 rev/min or 100 rev/min, the FSW machine stalled due to overload, i.e., the output power reached 66 kW, significantly higher than the rated power of 45 kW.

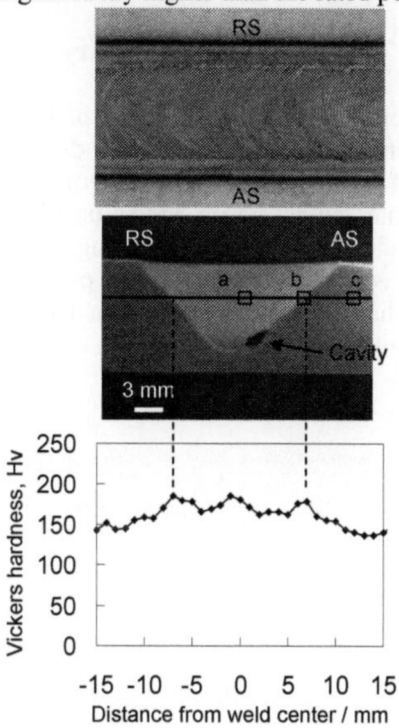

Figure 4. Appearance, cross-section and hardness variation of the weld produced using a 9-mm probe.

Figure 5. Microstructures of (a) SZ, (b) TMAZ / HAZ and (c) BM produced using a 9-mm probe.

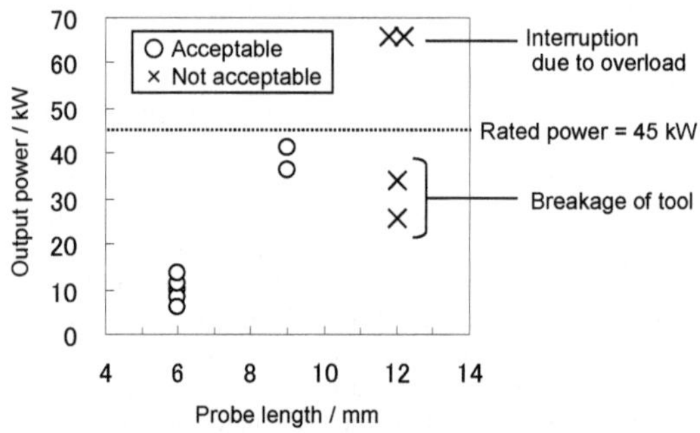

Figure 6. Effect of probe length on output power.

Effects of tool geometry

To enhance stirring of the material and to reduce the load on the FSW machine, the effect of the tool geometry on machine outputs was examined using the 6-mm probe tools with different geometries at a travel speed of 150 mm/min. The effects of rotation speed and probe geometry on the FSW machine outputs, i.e., output power and vertical load, are shown in Figure 7. Both the output power and axial load decreased with increasing rotation speed. The probe geometry strongly affected the output power, and the spiral 3-flats probe resulted in the lowest power. In contrast, the probe geometry had a negligible effect on the axial load, because that load is primarily dependent on the shoulder diameter.

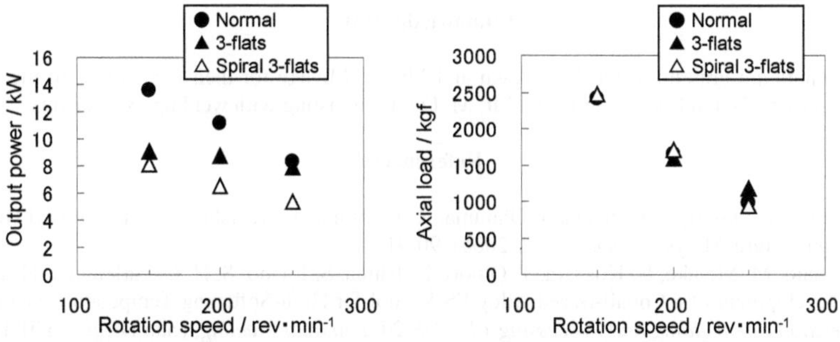

Figure 7. Effects of rotation speed and tool geometry on FSW machine outputs: output power and axial load.

The surface appearance and cross-sections of the welds produced using the tools with a 6-mm probe and with the different probe geometries at a tool rotational speed of 250 rev/min are shown in Figure 8. The weld surface appearance was acceptable for all welds. A probe with only the 3 flats, as shown in Figure 8-(b), produced a bigger cavity around the AS in the SZ than a normal cone probe. In contrast, a probe with both 3 flats and a step spiral produced the smallest cavity in the SZ, as shown in Figure 8-(c), but showed a tendency to form surface undercut.

Acknowledgments

The authors are grateful to Dr. Y. Takaku and Mr. M. Miyake for their evaluations of the Co-based alloy tools and Mr. Y. Fujita and Mr. M. Doi for assisting with welding experiments.

References

[1] J. Sato, T. Omori, K. Okikawa, I. Ohnuma, R. Kainuma and K. Ishida, "Cobalt-Based High-Temperature Alloys", *Science*, 312 (2006), 90-91.
[2] Y. Sato, M. Miyake, H. Kokawa, T. Omori, K. Ishida, S. Imano, S. H. C. Park and S. Hirano, "Development of a Cobalt-Based Alloy FSW Tool for High-Softening Temperature Metals", *Friction Stir Welding and Processing VI. TMS 2011 Annual Meeting*, (San Diego, California, USA), February 27- March 3, 2011.

Friction Stir Welding and Processing VII
Edited by: Rajiv Mishra, Murray W. Mahoney, Yutaka Sato, Yuri Hovanski, and Ravi Verma
TMS (The Minerals, Metals & Materials Society), 2013

ESTABLISHING W-BASED FRICTION STIR WELDING TOOL LIFE FOR THICK SECTION STEEL APPLICATIONS

Michael Eff[1], Brian Thompson[1], Sudarsanam Suresh Babu[2], Todd Leonhardt[3]

[1]EWI; 1250 Arthur E. Adams Dr., Columbus OH 43221, USA
[2]The Ohio State University; 1248 Arthur E Adams Dr., Columbus OH 43221, USA
[3]Rhenium Alloys Inc.; 38683 Taylor Parkway, North Ridgeville OH 44035-6200, USA

Keywords: Tool Wear, Friction Stir Welding, Tungsten, Rhenium, Degradation, Steel

Abstract

Recently, Friction Stir Welding (FSW) has been demonstrated as a solid-state welding technology capable of joining thick section steel in a single pass. This process's ability to reduce cycle time and manpower as opposed to traditional joining methods creates an opportunity to significantly reduce the cost of welding thick section steels. Open literature has recently indicated a 25% cost reduction if FSW can be applied to offshore pipe line construction. This study was initiated to establish a robust tool design capable of joining 19-mm (0.75-in) X-70 steel, maximize tool life, and identify cross weld mechanical properties throughout the life of the tool. This presentation will summarize trends identified during tool life testing and the mechanical properties observed as the tool wore. Over 59-m (193-ft) of weld length was realized with a single W-based tool and the ultimate tensile strength throughout welding was near matching to the base metal.

Introduction

Friction Stir Welding was invented in 1991 to join aluminum alloys [1]. Advancements in the tool material and tool designs have enabled FSW to join high melting temperature materials such as steel and titanium [2.3]. The main limitation in adapting FSW for use in high melting temperature materials has been identifying a tool material capable of withstanding welding temperatures above 1200°C (2192°F), high material flow stresses, and large force loads implicit to the process [4]. Such conditions can cause excessive wear, deformation, or fracture of the tool itself [5, 6]. Despite these challenges, there are considerable benefits associated with the FSW of high melting temperature metals.

Over many years of research published in the open literature, two material types have emerged for the FSW of steels. These include ceramic and refractory metal materials [7]. Ceramic materials include Polycrystalline Cubic Boron Nitride (PcBN) which is second only to diamond in hardness [8]. Due to this high hardness, PcBN as a FSW tool wears minimally during welding but can suffer cracking and premature pin failure due to its low ductility [4]. New ceramic composite tools incorporating tungsten-rhenium (W-Re) as a second phase increase the overall ductility of the material [9]. These new composite metallic/ceramic materials known as grades Q-60, Q-70, and Q-80 have shown the ability to weld a wide range of industrial steels over long weld lengths (40-60-m) in thickness of less than 6.4-mm with improved toughness and durability [9]. Ongoing research to increase the PcBN tool size has allowed the tool material to join thicknesses up to 20-mm (0.78-in) [10]. Refractory materials used as friction stir welding tools include alloys such as tungsten carbide, tungsten lanthanum, and in particular tungsten rhenium. W-Re exhibits high strength at elevated temperatures and remains ductile at room temperatures making it a very durable tool material. Due to its toughness, it has demonstrated an ability to weld steel alloys up to 19-mm (0.75-in) thick in a single pass [11]. However, this toughness at room and elevated temperatures comes with the price of reduced hardness. As a result, W-Re tools experience a higher rate of tool wear than their PcBN counterparts. Through redressing operations, the extent of this tool wear can be managed effectively enabling W-Re to weld a variety of industrial steels, including line pipe grade steels [12].

Currently, there are no published industrial applications for the FSW of steel. However, ongoing research continues to advance the economic and technical case for industrial acceptance. Based upon studies conducted by Kumar et al, FSW offers approximately a 7% savings for onshore pipeline construction and a 25% savings for J-lay offshore applications [13]. The majority of the savings estimated for onshore applications are a reduction in equipment needs and manpower. The drive for cost effective tool material is key to making FSW economically feasible in onshore pipeline manufacturing. Offshore applications see a majority of the savings from a large reduction in cycle time stemming from FSW's ability to complete thick section welds in a single pass. The typical wall thickness used in offshore applications is around 20-mm (0.78-in). The study showed the savings of using the process greatly increased with thicker welding sections, faster travel speeds, and longer tool life [13]. Conversely, high FSW tool cost and short tool life would result in an economic loss for the use of FSW over GMAW.

With the overall goal of making FSW economically feasible in pipeline construction, this study was initiated to establish a robust tool design capable of joining 19-mm (0.75-in) thick X-70 steel. Additionally, this study will quantify the maximum tool life for the particular design selected and will evaluate cross-weld mechanical properties throughout the life of the tool in order to determine long term weld quality. W-Re was chosen as the tool material for this study due its blend of strength and toughness both at room and welding temperatures.

Approach

Materials

Two compositionally identical materials, each with a unique thermomechanical history were examined in this study. An original bar of W-25%Re was manufactured using powder metallurgy. The process steps for this original bar included a cold isostatic press, followed by sintering at 2475 °C (4487°F) and lastly, a hot extrusion. This original bar was then cut in half. One half remained in the as extruded (XT) condition while the other half of bar underwent a heat treatment at 2000°C (3632°F) for a time of one hour, which resulted in a recrystallized microstructure (RX). The microstructures of each bar are displayed in Figure 1.

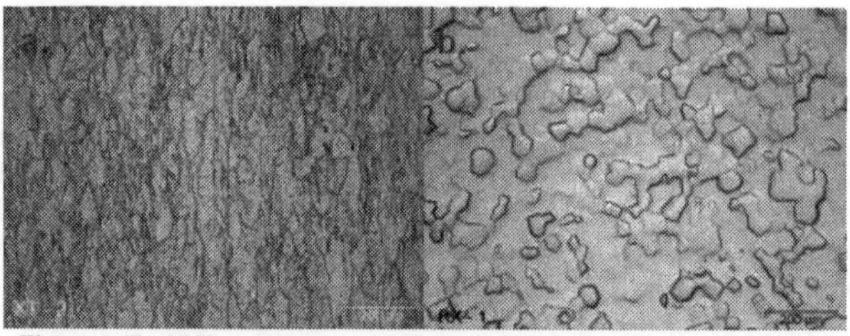

Figure 1. Tool Microstructures (a) Extruded (XT) (b) Recrystallized (RX)

Each half of the bar was approximately 89-mm (3.5-in) in length, with a diameter of 43.2-mm (1.7-in). Both bars were machined into two identical FSW tools based upon the Variable Penetration Tool (VPT) design. One tool was designated as XT for extruded and the other tool was designated as RX for recrystallized. A schematic of the tool design referencing important tool features is displayed in Figure 2. The tool design also incorporated an internal cooling hole. This hole was approximately 12.7-mm (0.5-in) in diameter, 40.6-mm (1.6-in) deep and located on the backside of the tool.

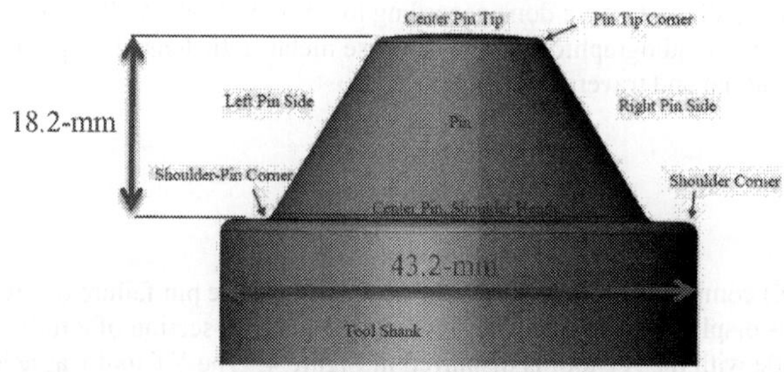

Figure 2. Tool Schematic

X-70 steel was chosen as the substrate for this study. It is characterized as a high strength low alloy (HSLA) steel with a yield strength exceeding 483 MPa (70 ksi). The plates were nominally 19-mm (0.75-in) thick. The material was cut into two separate plate sizes. 18.5-cm (7.25-in) wide by 33-cm (13-in long) plates was used for butt welded mechanical test plates. Larger plates, nominally 0.46-m (18-in) wide by 1.5-m (60-in) long, were used for the tool life trials.

Tool Life Trials

After a brief parameter development study, final parameters of 90-RPM and 76.2-mm/min (3-IPM) were down-selected for the tool life trials. All welds were completed in position control and the heat input of the tool was managed through internal cooling using 50°F chilled water with a flow rate of 6.8-L/min (1.5-GPM).

Each tool life trial began with the joining of a mechanical test plate at 19-mm (0.75-in) thick full penetration. The mechanical test plates were done in a butt joint configuration and were approximately 30-cm (12-in) in length. Following the mechanical test plate, bead on plate welds nominally 1.5-m (60-in) long were made sequentially on the larger plate, 25.4-mm (1.0-in) thick, until a defect was observed in the weld. Between each weld a laser profilometry scan was conducted to record any changes to the tool geometry due to wear. The bead on plate welds were made on a single plate and spaced approximately 5-mm (0.2-in) apart on the plate. Defects were identified, through visual inspection and radiographic inspection. Each weld underwent RT inspection after welding. Once a defect was observed during a 1.5-m (5-ft.) weld, the trial was ended and the tool was redressed to the original pin profile. The tool life trials were then continued until another defect was created or the redressing operation caused the shoulder of the tool to be within 3-mm (0.11-in) of the bottom of the cooling hole. At this point, the tool life trial was determined to be complete. Both the RX tool and the XT tool were tested for extended tool life following this procedure.

Post-Weld Analysis

Radiographs were taken of each weld in accordance with ASME Section V Article 2 and the defect quantity and length was characterized for each weld. Following the radiography, mechanical tests were taken out of the initial test plates from the beginning of each trial. The following mechanical tests were taken out of each plate, two bend tests, one traverse tensile test, one macro cross section, and five CVN tests Each CVN notch was through the thickness and located in the TL orientation (weld centerline) Each CVN test was tested at -20°C (-4°F). All tensile tests were conducted according to ASTM E8. Bend tests were conducted according to AWS B4.0:2007 and all CVN tests were done according to CSA W48-06. CVN test were not taken from welds that failed radiographic inspection. Base metal static tensile properties were evaluated in the longitudinal and traverse directions.

Results and Discussion

Tool Life and Wear

Each tool (XT and RX) completed the tool life trials without complete pin failure or severe tool wear. A typical weld is displayed in Figure 3. A representative cross-section of a fully consolidated weld made with the XT tool is displayed in Figure 4. The XT tool was able to travel an average weld length of 5.9m (19.4 ft.) and was able to be redressed a total of 10 times for a total weld length of 59–m (194-ft) on a single 89-mm (3.5-in) long piece of XT tool stock. The RX tool traveled an average weld length of 4.4m (14.6 ft.) per redress and was redressed seven times for a total 31.2-m (102.3 ft.) of weld length without a defect for a given piece of RX tool stock (Table I).

Figure 3. FSW 0.75-in Thick Steel

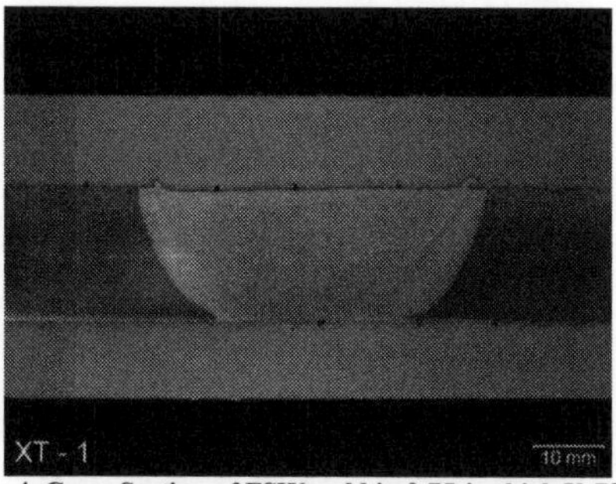

Figure 4. Cross Section of FSW weld in 0.75-in thick X-70 Steel

Table I. Tool Life Results

Material	Average Defect Free Weld Length per Redress	Number of Redresses	Life of Tool Stock	Average Normalized Wear
Extruded	5.9m (19.4 ft.)	10	59m (193ft.)	0.69%/m (0.21%/ft.)
Recrystallized	4.4m (14.6 ft.)	7	31.2m (102.3 ft.)	0.88%/m (0.27%/ft.)

In between each redress, the tool primarily wore along the left and right pin edges (Figure 2). A tool representing the different stages of the tool life trials is displayed in Figure 5. In this study, the majority of weld defects occurred in areas of the weld which corresponded to the areas of the most wear on the FSW tool (Figure 6). This result suggests that the wear of the tools causes the formation of weld defects.

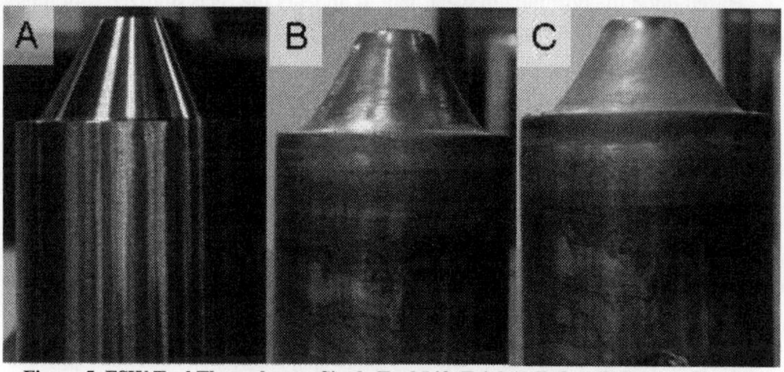

Figure 5. FSW Tool Throughout a Single Tool Life Trial A) Before B) During C) After

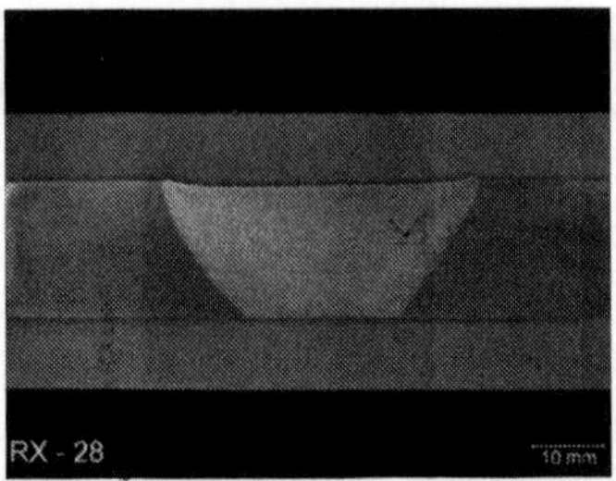

Figure 6. Macro of Defect Weld RX-28

This tool wear pattern is best represented by the laser profilometry line scans displayed in Figure 7, which combines several line scans taken throughout a tool life trial. This type of wear pattern is indicative of the wear that occurred throughout the tool life trials for both tools (RX and RT). The majority of the wear occurred along the right and left sides of the pin and a small amount of wear occurred at the shoulder. To quantitatively capture the wear, the cross-sectional area of the tool was calculated after each weld pass and redress.

The chart displayed in Figure 8 summarizes the normalized reduction in cross-sectional area with each linear meter of welding. The increase in cross-sectional area present at 0.3-m (1-ft.) in some trials is due to a limited amount of material sticking to the tip of the pin after welding of the mechanical test plate. However, the sticking did not occur during the 1.5-m (5 ft.) passes. The cross-sectional areas of the starting tool geometries varied from 629.88-mm^2 to 657.4-mm^2. This variation was found to be normally distributed and likely due to variances in the machining operations; therefore the data was then normalized to help minimize the effect of this variation on analysis. Both tool materials exhibit linear normalized wear. The extruded material typically lost 0.69% of its original cross-sectional area per linear meter of welding (.21%/ft.), while the recrystallized material lost .88% of its original cross-sectional area per linear meter of welding (.27%/ft.). Based upon the normalized wear data, it appears the recrystallized tool material degraded 27.5% more for a given weld length.

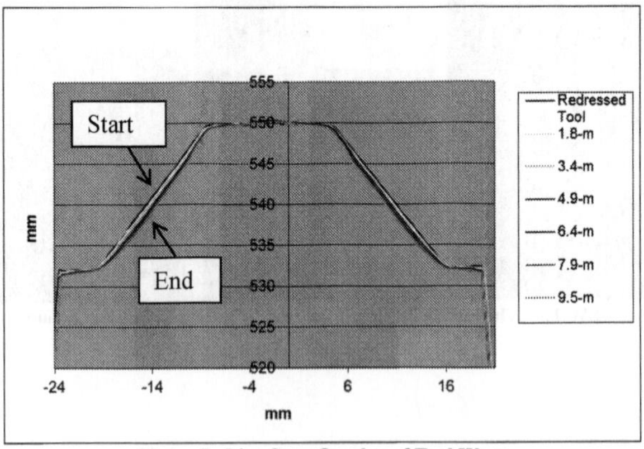

Figure 7. Line Scan Overlay of Tool Wear

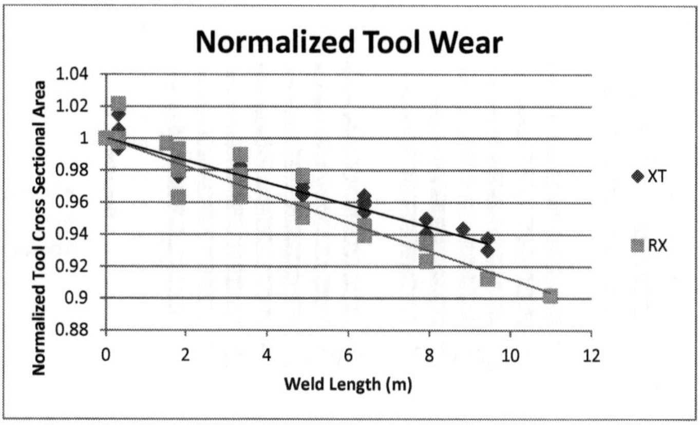

Figure 8. Normalized Tool Wear

Mechanical Test Results

The results of the cross-weld static mechanical tests for each FSW tool are summarized and compared with the base metal values in Figure 9. The base metal exhibited a yield strength of 497-MPa (72.2-ksi) in the transverse direction and 510-MPa (74-ksi) in the longitudinal direction. The ultimate tensile strength exhibited by the base metal was 613-MPa (89-ksi) in the longitudinal direction and 596-MPa (86.5-ksi) in the transverse direction with elongations of 41.79% and 42.09% respectively.

All the cross-weld static tensile tests were transverse to the welding direction and therefore best compared to the transverse base metal properties. On average, the friction stir welds exhibited slightly overmatching ultimate tensile strengths of 610-MPa (89-ksi), and about a 9% loss in yield strength. The elongation is also around 35% in the majority of the tensile tests, representing a 7% loss in ductility. Each of the tensile tests failed within the stir zone. The reason for this failure location is demonstrated in the stress strain curve displayed in Figure 10. The base metal exhibits much more ductility and is stronger than the weld metal at strains above 20%. At 32% strain, where the welds failed, the base metal is over 100 MPa (14.5-ksi) stronger under the given strain and results in a failure in the stir zone.

In general, the failure occurred on the retreating side of the weld. The smallest recorded defect size which noticeably affected properties was 3.6-mm (0.14-in) in the through thickness direction and is displayed in Figure 6. The strength of the welds was not greatly affected but the defects did greatly reduce the elongation of the material. Other welds failed visual inspection but showed no loss in mechanical properties and still failed on the retreating side of the weld. This demonstrates that an acceptable flaw size does exist within a friction stir weld without loss in material performance. In this case, the flaw size was than 2% of the material thickness as it was not detected by RT inspection. Further work is needed to determine the size of an acceptable flaw. All the welds passed the bend test with the exception of RX-10, which exhibited one crack greater than 3.175-mm (0.125-in) in the stir zone, which met failure criteria.

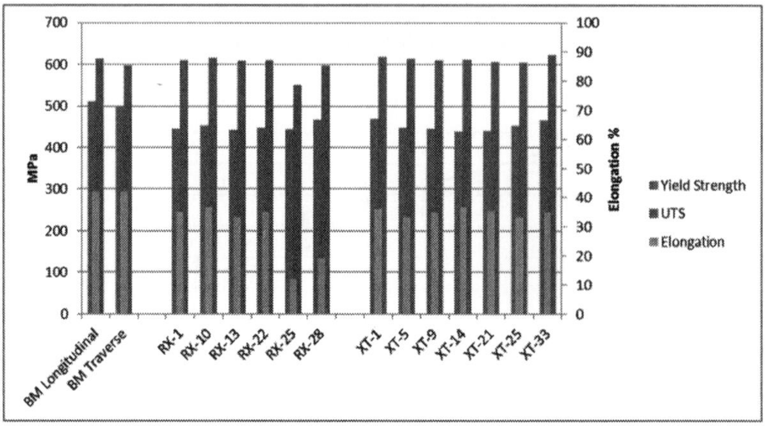

Figure 9. Tensile Test Results

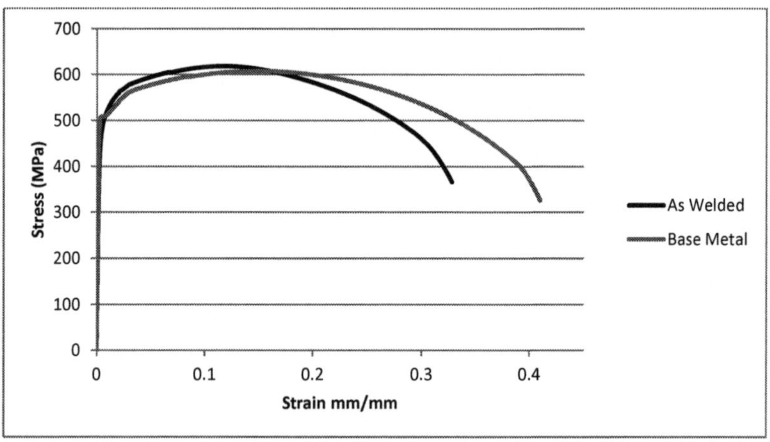

Figure 10. Typical Stress vs. Strain Curves

The weld metal exhibited low toughness values in the range of 19.50- 36.06 J, which is only about 6.5-11.5% of the base metal toughness at -20°C (-4°F). The ductile to brittle temperature was raised in the stir zone, as the toughness values are still very high in the base metal at this temperature. The weld metal is likely exhibiting lower shelf behavior whereas the base metal is likely exhibiting upper shelf behavior. This increase in ductile to brittle transition temperature is consistent with previous work [4, 14]; however other work has demonstrated a reduction in the ductile to brittle transition temperature [15]. Due to this conflicting research and the fact that all of the CVN tests for this work were conducted at -20°C (-4°F), more work is needed to fully characterize the toughness of FSW welds in steel. Overall, the weld mechanical properties remained constant for both tool materials throughout the entire life of the tool.

Conclusions

1. **The XT tool achieved a total tool life of 59-m in 19-mm thick steel:** Further development of the parameters, tool material, and design could extend this tool life beyond 59-m
2. **The RX tool achieved a total tool life of 31.2-m in 19-mm thick steel**:
3. **The extruded microstructure is more resistant to tool degradation:** The extruded material experienced 27.5% less material loss than the recrystallized material. The extruded material was also able to produce 31.5% more defect free weld length than the recrystallized material.
4. **Redressing is an effective method to increase the tool life of a given piece of material:** The tools were able to create repeatable welds throughout the tool life trials, without a loss of mechanical properties. In turn, redressing greatly increased the usable life of a tool and the weld length which a single piece of tool stock may produce.
5. **Mechanical properties of the welds remained consistent throughout the tool life:** The Ultimate Tensile Strength, Yield Strength, Ductility, and Toughness values remained relatively unchanged throughout the tool life.
6. **Defects up to 2% of material thickness were acceptable in this study:** Welds which failed visual inspection did not experience a loss in ductility, yield strength or Ultimate Tensile Strength

References

1. W. M. Thomas., et. al.., “Friction Stir Butt Welding”; International Patent No. PtCt/GB92702203, (1993).

2. A. Kumar, J. Rodelas, and R.S. Mishra, “The Effects of Friction Stir Processing on The Microstructural Evolution and Mechanical Properties of Ti-6Al-4V Alloy,” *TMS Friction Stir Welding and Processing V*, (2009), 45-53.

3. A. Kumar et al., “Research Progress on Friction Stir Welding of Pipeline Steels” Proceedings of the 8th International Pipeline Conference (2010).

4. P.J. Konkol, and M.F. Mruckzek, “Comparison of Friction Stir Weldments and Submerged Arc Weldments in HSLA-65 Steel” *Welding Journal*, 86 (2007), 187-195.

5. B. Thompson and S.S. Babu, “Tool Degradation Characterization in the Friction Stir Welding of Hard Metals” (Master Thesis, The Ohio State University, 2010).

6. Eff, Michael and Babu, S.S., “The Effects of Tool Texture on Tool Wear in Friction Stir Welding of X-70 Steel” (Master Thesis, The Ohio State University, 2012).

7. J. Deflaco and R. Steel, “Friction Stir Process Now Welds Steel Pipe” *Welding Journal*, 5 (2009), 44-48.

8. J. Peterson, “Evaluation of Advanced Tool Material Technology for Friction Stir Welding of 6mm AISI 304L Stainless Steel and 6mm A36 Cold Rolled Mild Steel,” (Paper presented at 8th International Friction Stir Welding Symposium, Timmendorfer Strand, Germany, Session 12 2010).

9. J. Perret et al., “Friction Stir Welding of Industrial Steels” *Friction Stir Welding and Processing VI* (2011), 65-72.

10. Russell J. Steel et al., “Thick Section Friction Stir Welding of Steel” (Presented at 9th International Symposium on Friction Stir Welding Huntsville Alabama, USA 15 May 2012).

11. J. Bernath, , B. Thompson, and N. Ames, “Friction Stir Welding of Steel and Hard Metals-Advancements in Tool Technology and Thick Section Welding” (Presented at 7th International Friction Stir Welding Symposium, Section 03B 2008).

12. J. Seaman, and B. Thompson, “Challenges of Friction Stir Welding of Thick-Section Steel” (Presented at ISOPE 2011), 486-493.

13. A. Kumar et al., “Evaluation of Economic Incentives and Weld Properties for Welding Steel Pipelines Using Friction Stir Welding” *Proceedings of the Twenty-first International Offshore and Polar Engineering Conference* (2011), 460-467.

14. Kumar, A. et. al., “Research Progress on Friction Stir Welding of Pipeline Steels” Proceedings of the 8th International Pipeline Conference (2010).

15. Sam Sanderson and Tracy Nelson, “Effect of Processing Parameters on Friction Stir Welded HSLA-65 Charpy V-Notch Impact Toughness,” *Friction Stir Welding and Processing VI* (2011), 19-24.

ASSISTED FRICTION STIR WELDING OF CARBON STEEL: USE OF INDUCTION AND LASER AS PREHEATING TECHNIQUES

Ana I. Álvarez[1], Víctor Cid[1], Gloria Pena[1], Jose Sotelo[2], David Verdera[2]

[1] ENCOMAT Group, School of Industrial Engineering - University of Vigo, 36310-Vigo, Spain
[2] AIMEN Technology Centre, Joining Technology Plant, Relva 27A, 36410-O Porriño, Spain

Keywords: FSW of steels, induction preheating, laser preheating, assisted FSW

Abstract

The use of FSW in high melting point materials is still restricted due to limited tool life. Due to the high wear involved in the process, tool life is too short and the cost of replacing it is still too high. One of the ways that was explored to increase tool life is the use of preheating techniques to soften the material to be welded. Softening the material would cause a decrease in tool wear and therefore an increase in tool life. In this study, two preheating techniques were tested on the FSW of carbon steel: induction heating and laser heating. The data arising from the study show that the forces taking place during both hybrid processes were importantly reduced compared to those obtained in conventional FSW. Microstructure and mechanical testing of welds made in the three conditions were done showing the influence of each process on weld quality

Introduction

One of the main problems that limit the extension of FSW to high melting point materials is the limited tool life[1]. The tools have to withstand temperatures, loads and stresses much higher than FSW of aluminum, without losing strength or suffering excessive wear. Tool materials must retain sufficient strength at temperatures in excess of 1100°C when welding steels.[2] A big effort has been done in the last decade to develop an appropriate tool material for welding steels. The three materials that best perform with steels are WRe, PCBN and the composite between those two[3],[4]. However, tool wear or unpredictable tool breakage is not completely solved and tool cost is still too high. For this reason, other strategies have been approached to address this problem[5]. One of the ways that was explored to increase tool life is the use of preheating techniques to soften the material to be welded. Softening the material would cause a decrease in tool wear and therefore an increase in tool life[6]. In this study, two preheating techniques were tested on the FSW of carbon steel: induction heating and laser heating. In both cases, the heat source (induction coil or laser spot) is placed in front of the rotating FSW tool, and used to preheat the workpiece at a localized area ahead of the rotating probe, thus plasticizing a volume of the material ahead of the probe[7]. The work piece is then joined in the same way as in conventional FSW process. The high temperature ahead of the rotating tool softens the work piece and enables joining without strong clamping fixtures. Less downward force is required to carry out the process, thus reducing tool wear[8].

Induction heating occurs when a electrically conductive material is placed in a varying electromagnetic field. The magnetic field induces an electric current inside the work piece causing resistive heating of the material. Induction heating is a non-contact method of heat

treatment[9]. On the other hand, laser heating is another non-contact heating method. Placing the laser spot just in front of FSW tool will heat the work piece, thus reducing process forces[6].
The material used to do this research is a carbon steel marine grade. FSW of steels has interest in shipbuilding industry, due not only to the good mechanical properties achieved with this technology but also because of the weld distortion reduction.[10]

Experimental procedure

Assisted-friction stir welds were produced on marine steel plates grade A (similar to ASTM A131) of dimensions 5mm thick, with the welding direction transverse to the rolling direction of the plates. The nominal composition of the steel is (wt%) C max. 0.21, Mn min. 2.5xC, Si max. 0.50, P max. 0.035, S max. 0.035. The ultimate tensile strength of this carbon steel grade is 400-520 N/mm^2, its yield strength 235 N/mm^2 and elongation of 22%. Welds were made in square groove butt-joint configuration of samples typically 350mm in length and 125mm width.

The FSW tool employed in the study was a PCBN tool (MS80), with shoulder diameter of 23.7mm., pin diameter of 8.9mm and pin length of 5mm. Its geometry consisted on a convex scrolled shoulder and conical stepped spiral pin.

Two processes were used to preheat the work piece at a localized area ahead of the rotating tool: induction preheating and laser preheating (induction and laser-assisted friction stir welding).

The induction equipment consisted on a EFD mobile machine, MINAC 25/40, capable of operating in the medium frequency range (10-40 kHz), with a maximum output power of 40KW. The laser preheating was carried out with a 3.3kW direct diode laser (Laserline, model LDL 160-3300), and the FSW equipment was a MTS ISTIR PDS-4 machine.

In order to carry out the assisted process, a conventional friction stir welding machine was modified to incorporate the auxiliary preheating systems. The custom designed heads are shown in figure 1

a) Induction assisted FSW system	b) Laser assisted FSW system

Figure 1: modifications in the conventional FSW head to incorporate the preheating systems

After some previous FSW tests on the same ASTM 131 steel performed by our group, a process parameter window that allowed producing defects-free welds was established. The parameters range is presented in table 1

Rotation speed	400-500 rpm
Travel speed	35-75 mm/min
Angle	0°
Work mode	Force control
Downward force	15-20 kN

Table 1 Parameters window for FSW of marine grade A steel with MS80 tool

In order to limit the test matrix, a fix rotation speed (500 rpm) was selected for all tests and three travel speeds (50mm/min, 150 mm/min and 225 mm/min) were used to compare the welds obtained under conventional friction stir welding (FSW), induction assisted FSW (IA-FSW) and laser assisted FSW (LA-FSW). In all conditions, tool was tilted 0°, under argon gas shielding, and force control was chosen as the operating mode. A backing plate of Inconel 625 was used in order to prevent any adhesion of the root to the backing strip. The paint primer covering the samples was not removed before the welding procedures.

After welding, a qualitative macroscopic inspection of the weld surfaces was performed for surface roughness, presence of surface breaking defects, and side flash visual inspection. Metallographic samples of the cross sections of the welds were prepared according to standard metallographic practice and etched with a 4% nital solution. Etched samples were examined using optical microscopy (OM), in an Olympus GX51 optical microscope, and scanning electron microscopy (SEM) with X-ray energy-dispersive spectroscopy, in a JEOL 5410.

The Vickers microhardness traverse profile of each joint was obtained using a 100-g load and a 15-s dwell time, using a EMCOTEST Durascan.

Face and root bend tests were also performed on machined specimens (250mm x 20mm x 5mm) of the welded samples, using a die block of 20mm diameter (4 x thickness).

Sample Identification	**Welding Conditions**		
	Welding System	**Common parameters**	**Travel speed**
FSW-50	Conventional FSW	Rotation speed: 500 rpm Tilt angle: 0° Argon gas shield Backing plate: Inconel 625	50 mm/min
IA-FSW-50	Induction assisted FSW		
LA-FSW-50	Laser assisted FSW		
IA-FSW-150	Induction assisted FSW		150 mm/min
LA-FSW-150	Laser assisted FSW		
IA-FSW-225	Induction assisted FSW		225 mm/min
LA-FSW-225	Laser assisted FSW		

Table 2: Processing conditions of samples at higher transverse rate

Results and discussion

Process characterization

Table 2 summarizes the welding conditions of the processed samples. Only IA-FSW-225 conditions produced unacceptable weld runs, showing a surface-open tunnel at the advancing side and excessive lateral flash in most of the welds, resulting from the outflow of the plasticized material.

Load data were recorded as function of time for several defect-free welds obtained at a travel speed of 50mm/min in a first set of samples. Representative load curves are presented in Fig. 2.

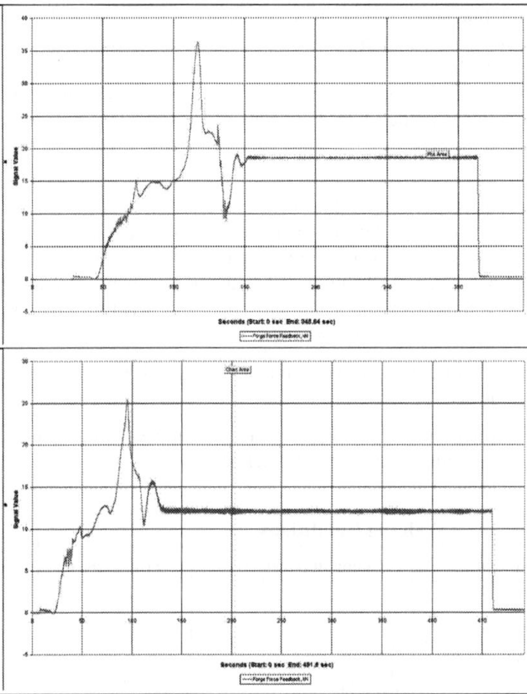

a) Conventional FSW

b) Induction assisted FSW

Figure 2: Load curves measured during plunge and weld stages in samples processed at 50mm/min

During plunging of the tool, the load rises initially when the pin is in contact with the workpiece. As the material is softened a fall in the force is produced, the softened material is displaced until the pin encounters new colder material, and the process is then repeated. Thus the initial load curves exhibit several peaks, with an intense maximum corresponding to the time when the shoulder contacts the top surface of the metal. After some seconds the steady state is reached IA-FSW-150 sample[11].

As can be seen from the figure, the preheating treatments do not clearly modify the shape of the curves, but the load values are substantially reduced.

The maximum forces measured during the plunge and the welding stages, presented in table 3, reveal a remarkable softening of the material in both stages as a consequence of preheating the plates before plunging the FSW tool and ahead the weld run.

Decreases in downward force during the plunge stage close to a 30% and 35% during welding stage are measured for Induction assisted FSW, compared to conventional FSW. In the case of laser preheating, the metal softening causes a reduction in forces of 33% and 40% in both stages.

Sample	Maximum force during plunging (kN)	Max. Force through weld run (kN)
FSW-50	36	18,5
IA-FSW-50	25	12
LA-FSW-50	24	11

Table 3: Forge force measured in samples processed at 50 mm/min

The diminution in the force necessary to reach the same penetration depth and to run the weld when previous preheating has been done implies a reduction in the stress on the tool, decreasing not only its strength requirements but also the whole machine requirements, in terms of strength and rigidity. Furthermore, the tool wear would probably be decreased, having an impact on durability of these tools, one of the main drawbacks when trying to apply FSW to steels.

Microstructural characterization

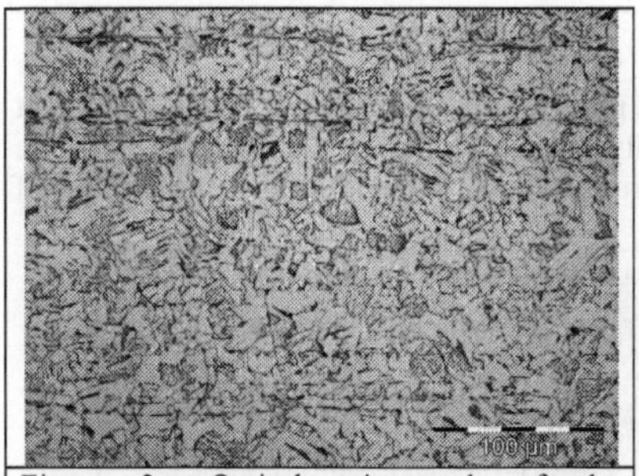

Figure 3.- Optical micrograph of the transverse section of the as received steel (500x)

The as received material is characterized by a mixed polygonal ferrite, Widmanstätten ferrite and pearlite microstructures. At the transverse sections elongated manganese sulfide stringers are found, especially at the central region of the plates. Fig. 3 shows a representative microstructure of the steels. Low –magnification optical micrographs of the cross sections of weldments FSW-50, LA-FSW 50 and IA-FSW 150 are presented in Fig.4. Advancing side (AS) and retreating side (RS) are indicated on the images.

The central band where sulfide inclusions are concentrated is clearly seen, passing through the welding zone suffering relatively small change.

All welded areas display several microstructurally distinct regions including the stir zone (SZ), the heat-affected zone (HAZ) surrounding it and the base material. Only small differences in grain size are detected between the samples processed in the studied conditions. The thermomecanical affected zone (TMAZ) is no clearly found in mild steels, apparently due to a heating over A1 temperature and the allotropic transformations experienced during cooling[10,11].

A more detailed study on the microstructure was performed. Some of the studied zones are indicated on the fig. 4c, that may be used as representative for the purpose of showing the microstructural evolution across the weld run.

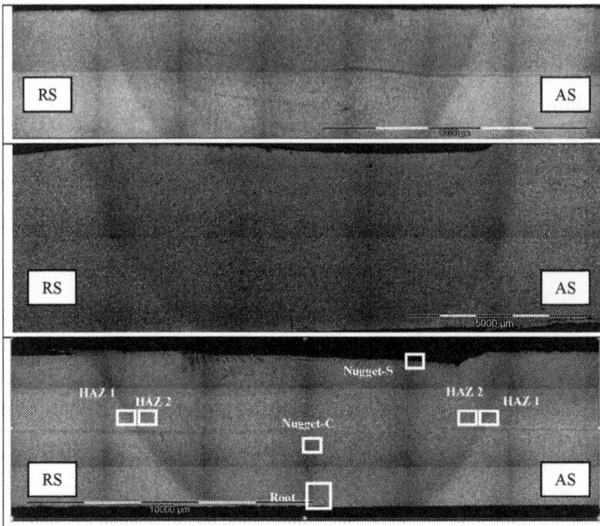

a) FSW-50

b) LA-FSW-50

c) IA-FSW-150

Figure 4.- Cross section of three plates welded in specified conditions. Images are obtained by composing several low-magnification optical micrographs. Advancing side (AS) and Retreating Side (RS) as well as the distinct regions examined in deep detail are superimposed

The microstructure of the stir zone (Nugget) of all welded samples is very similar to the base material with blocky and Widmanstätten ferrite and pearlite (Fig. 5). This structure is consistent with a steel heated over A_3 temperature, and heavily formed by the tool, undergoing dynamic austenite recrystallization[10]. The rapid cool down rate leads to a slightly higher proportion of acicular ferrite than in base metal.

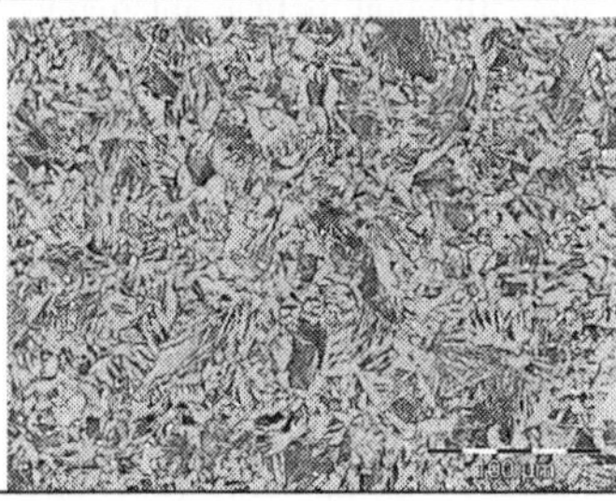

Figure 5.- Optical micrograph of the centre of the stir zone (Nugget-C) of a IA-FSW-150 sample (500x)

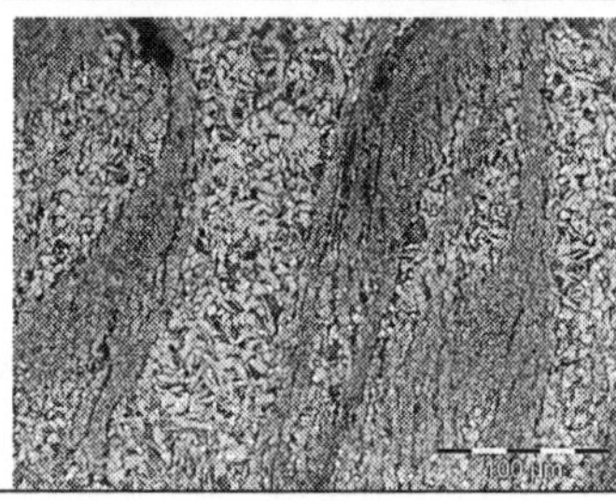

Figure 6.- Microstructure at the Nugget-S zone of a IA-FSW-150 sample (500x)

In all the samples, at the upper surface of the nugget zone (Nugget-S) evidence of an important grain refinement is found, consequence of the higher strain experienced by the metal closest to the shoulder. But in some cases, as the Fig. 6 shows, inclusions like oxides coming from the surface or sulfides from the central band are also present. In the same way, in almost all samples, at the root of the welding (Root, at fig. 4) some inclusions are also detected. EDS analyses show that inclusions mostly are iron oxides from the surface and particles from the talcum powder used on the backing plate to avoid the adherence between both surfaces. In fact, no evidence indicating that Ni from the Inconel plate has been diffused into any of the weldments, was detected. Fig. 7 presents a SEM micrograph of one of the root defects.

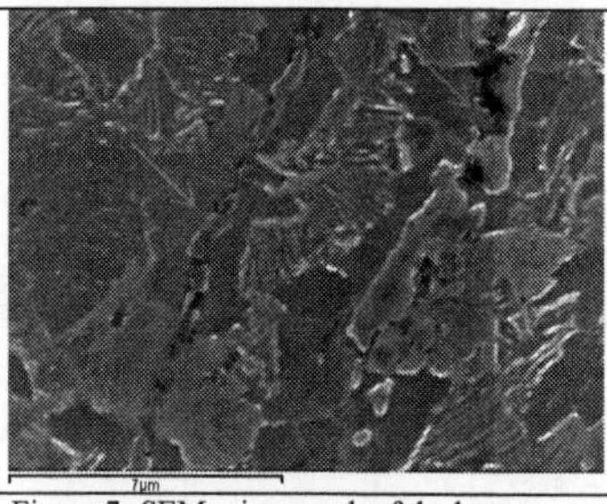

Figure 7.-SEM micrograph of the bottom root of a IA-FSW-150 sample, showing the presence of entrapped oxide inclusions

The stir zone is surrounded by a finer grained region, the HAZ, where two zones can be differentiated: the external part (denoted as HAZ 1 in Fig. 4) consistent in fine and almost equiaxed ferrite with a small amount of perlite, forming a clearly banded structure, and the internal zone (HAZ2) where the grain refinement and homogeneity of the structure are much higher.

This structure can be found both in the retreating and the advancing side of the weld, but size grain differences between HAZ1 and HAZ2 are even more significant in the RS. Fig. 8 shows the transition between both zones in the advancing side of a IA-FSW-150 sample.

Mechanical properties

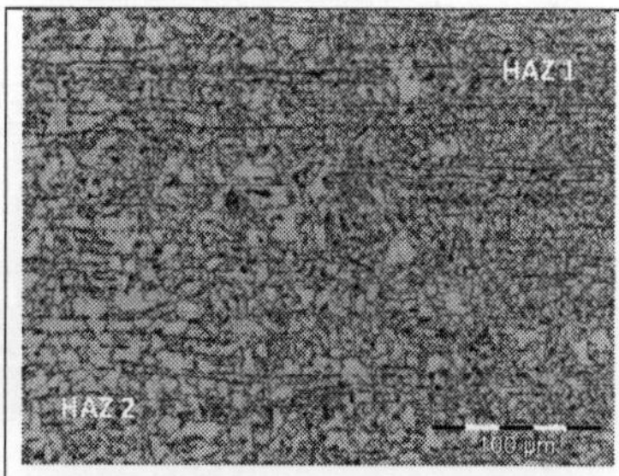

Figure 8.- Light optical micrograph showing the transition between HAZ 1 and HAZ 2 in the advancing side of a IA-FSW-150 sample

Fig. 9 shows a plot of the Vickers microhardness profiles of the carbon steel FSW joints welded at the different welding conditions, as a function of position. The hardness of as received material was approximately 143 HV. Basically the hardness within the weld regions is higher than the base metals and varied with the position depending on the grain size and phases sampled by each indentation. Hardness, in general, increased with increasing the distance from the center of the nugget, although some hardness peaks are measured when indentation is performed on a microstructurally refined region. As it was previously commented, no substantial variations in microstructure are induced by the welding conditions considered in this work, so the welding speed and the preheating treatment used ahead of FWS do not significantly affect the hardness profiles of the joints.

The mechanical resistance of all joint types and base material specimens was addressed using bending tests, in order to see what influence does preheating have on mechanical properties. The obtained results are summarized in figure 10.

As can be seen on the figure, face bending tests were OK in all cases. However root bending failed in all the studied samples. A possible reason that could explain this behavior is that these tests are very sensitive to defects near the surface of the weld bead, such as root flaws. As it was previously commented, transverse metallographic sections of almost all the samples revealed the presence of rolled-in scales and inclusions, coming from the oxidized surface, the primer and also from the talcum powders used on the Inconel backing plates. Those inclusions can strongly limit the ductility of the welds. Moreover, the fact that in all cases the bending test failed, suggests that it is not a matter of travel speed or preheating the plates, but of welding procedure.

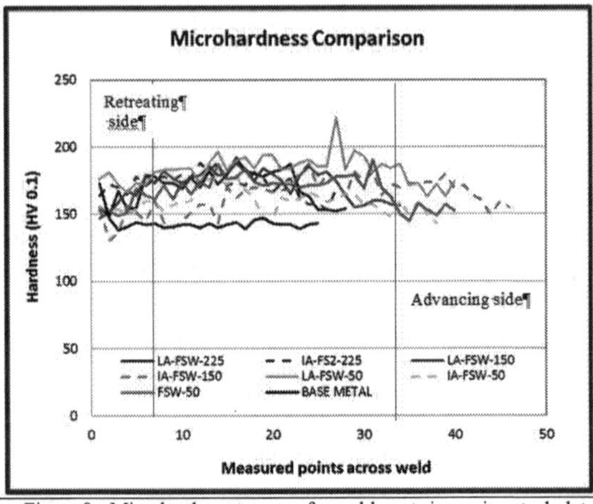

Figure 9.- Microhardness traverse for weldments in marine steel plates grade A

Sample	Face bending	Root bending
FSW-50	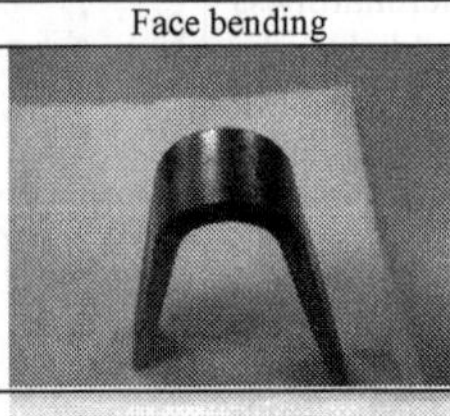	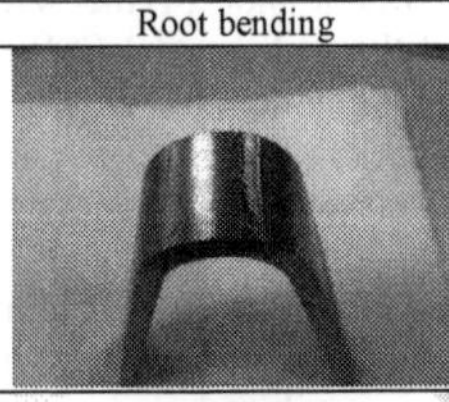
IA-FSW-50		
LA-FSW-50	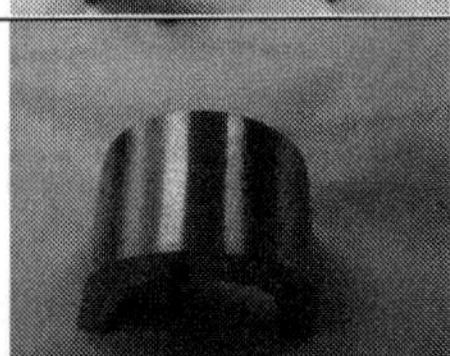	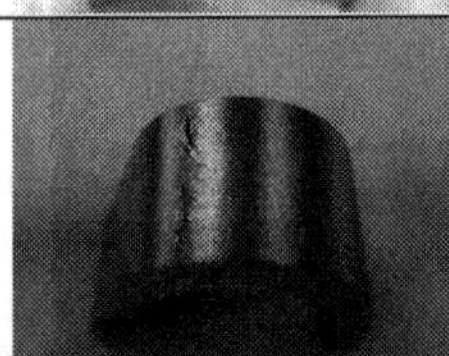
IA-FSW-150	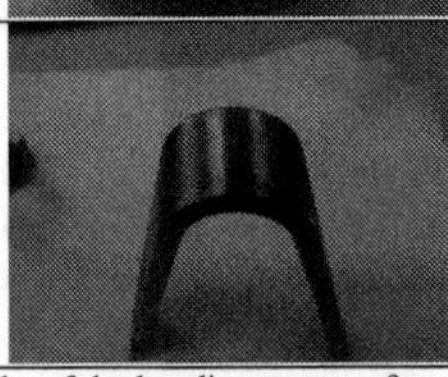	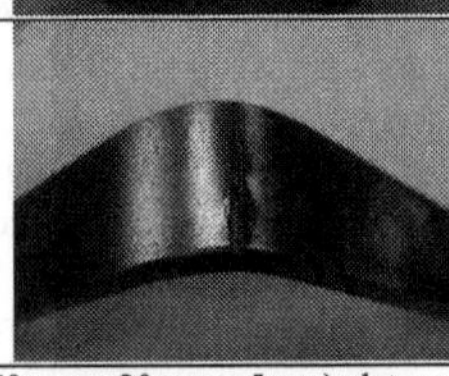

Figure 10: Results of the bending tests performed on (250mm x 20mm x 5mm) plates of welded samples, using a die block of 20mm diameter

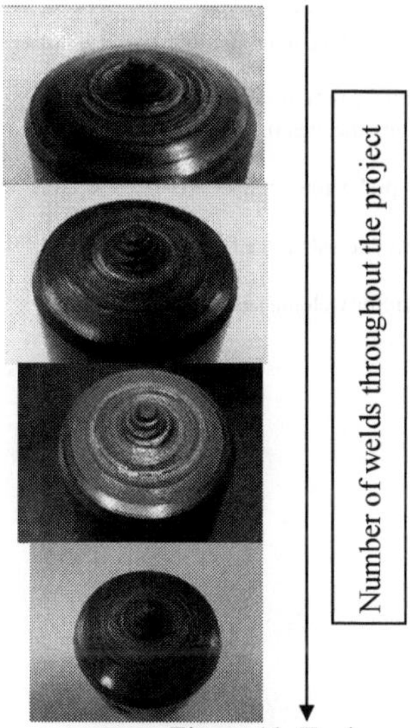

Figure 11: Tool wear

Besides those inclusions, another point that has to be taken into account to analyze these bending fails is tool wear. The initial state of the tool was not new, because it was used for previous tests. However, with the different trials carried out in the project, the wear was increasing. Although this wear has not been quantitatively measured yet (in terms for example of loss of material), it is clear from the pictures shown in Fig.11 that it has affected tool integrity. The scrolls of the shoulder almost disappeared and the pin stepped spiral features were damaged. This was not expected because PCBN tools tend to have brittle fracture rather than excessive wear. A possible overheating of the tool could explain this excessive wear. In any case, a damaged tool (both shoulder and pin) can affect material flow during welding, causing a poor mixturing and even lack of penetration, which may result in these bending tests failures.

Conclusions

In this study marine steel plates grade A were successfully joined by FSW using two preheating techniques: laser and induction. An important decreasing forge force has been observed that can lead to a longer tool life. The preheating treatments increase the hardness of the stir zone, with no evident effect of the travel speed or kind of preheating technique. The failures detected on root bending tests can be attributed to the particles entrapped in the root as a consequence of a welding procedure that can be easily modified.

Acknowledgements

The authors acknowledge financial support from Xunta de Galicia under program INCITE for the development of the project 10DPI008CT: "Investigación do proceso de soldeo de aceiros mediante batimento por fricción asistida por técnicas de prequecemento".

References

[1] H.K.D.H. Bhadeshia and T. DebRoy, "Critical assessment: friction stir welding of steels", *Science and Technology of Welding and Joining*, 14 (3) (2009), 193-196

[2] J. Perret, "Evaluation of tool material performance for the friction stir welding of 6mm AISI 304L stainlees steel" (Report TWI 915/2008)

[3] R. Johnson, "Friction Stir Welding of Steels - A Further evaluation" (Report TWI 884/2007)

[4] S.R. Cater and J. G. Perret "Friction Stir Welding of Steel" (Report TWI 1008/2011)

[5] Kwanghyun Park, "Development and analysis of ultrasonic assisted friction stir welding process" (Ph.D. thesis, University of Michigan, 2009)

[6] Paul Sinclair, "Heated Friction Stir Welding: an investigation into how preheating aluminum 6061 affects process forces" (M. Sc. Thesis, Vanderbilt University, 2009)
[7] G. Kohn at al. "Laser-assisted friction stir welding", Welding Journal, 2002. 81(2): p. 46-48
[8] S. Daftardar "Laser assisted friction stir welding: Finite volume method and metaheuristic optimization" (M. Sc. Thesis, Louisiana State University, 2009)
[9] B. Tweedy et al. "Friction Stir Welding of ferrous alloys using induction preheating" (Paper presented at TMS annual meeting, San Francisco, California, February 2005)
[10] P. J. Konkol and M. F. Mruczek, "Comparision of friction stir weldments and submerged arc weldments in HSLA-65 steel", Welding Research, July 2007
[11] T. Lienert, W. L. Stellwag, Jr, B.B. Grimmet and R.W. Warke, "Friction Stir welding studies on mild Steel" Welding Journal, January 2003, p. 1S-9S.

Friction Stir Welding and Processing VII
Edited by: Rajiv Mishra, Murray W. Mahoney, Yutaka Sato, Yuri Hovanski, and Ravi Verma
TMS (The Minerals, Metals & Materials Society), 2013

INFLUENCE OF HEAT INPUT ON FRICTION STIR WELDING FOR THE ODS STEEL MA956

B.W. Baker[1], L.N. Brewer[1*], E.S.K. Menon[1], T.R. McNelley[1], B. El-Dasher[2], S. Torres[2], J.C. Farmer[2], M.W. Mahoney[3], and S. Sanderson[4]

[1]Naval Postgraduate School, Monterey, CA
[2]Lawrence Livermore National Laboratory, Livermore, CA
[3]Consultant, Salt Lake City, Utah
[4]MegaStir Technologies, Salt Lake City, Utah

Keywords: Friction Stir Welding, Oxide Dispersion Strengthened Steel, MA956

Abstract

The oxide dispersion strengthened steel MA956 was friction stir welded using eight different rotational speed/translational speed combinations using a polycrystalline cubic boron nitride tool. Weld parameter conditions with high thermal input produced defect-free, full penetration welds. Electron backscatter diffraction showed a significant increase in grain size in the stir zone, a body centered cubic torsional texture in the stir zone, and a sharp transition in grain size across the thermo-mechanically affected zone. Micro-indentation results showed an asymmetric reduction in hardness across the transverse section of the weld that was sensitive to the heat input. This change in hardness is explained by the increase in grain size and may be described using a Hall-Petch type relationship.

Introduction

Because of their high temperature strength, radiation damage resistance, creep resistance, and corrosion resistance, oxide dispersion strengthened (ODS) ferritic-martensitic (FM) steels are attractive candidates for high temperature power production applications including proposed fusion reactors[1] and fast breeder reactors[2]. FM steels experience far less radiation swelling than austenitic steels, and the addition of Y_2O_3 and other oxide dispersoids gives these materials exceptional high-temperature strength and creep resistance due to the pinning of grain boundaries and dislocations. The dispersed oxides also mitigate radiation swelling and embrittlement by providing sites for the accumulation of hydrogen and helium atoms. Although the thermo-mechanical and radiation resistance properties of ODS FM steels are promising, traditional fusion joining processes for these alloys such as gas tungsten arc and gas metal arc welding are ineffective due to agglomeration of the oxide particles during melting of the base metal. The resulting inhomogeneous distribution of the oxide particles significantly reduces strength in depleted areas[3, 4]. To preserve the strengthening capability of the dispersed oxides

---

* Correspondence should be addressed to lnbrewer@nps.edu

and high temperature creep resistance, friction stir welding (FSW) is a promising solid state joining method to join ODS steels for fusion or advanced energy production designs. Several authors have demonstrated the capability to join different ODS alloys via FSW with good success[5-10].

FSW was invented at The Welding Institute (TWI) of the United Kingdom in 1991 as a solid state joining technique[11]. A detailed review of process parameters, process modeling, microstructure evolution, material properties, and specific material issues for FSW has been completed by Mishra which focuses primarily although not exclusively on aluminum alloys[12]. Due to early limitations on tool materials, FSW was initially constrained to low melting temperatures alloys such as aluminum, magnesium, and lead. However, improvements in FSW tool materials such as the use of polycrystalline cubic boron nitride (PCBN) and W-Re pins have expanded the use of FSW to higher temperature alloys such as steels. Expanding the use of FSW to steels and polymers, Nandan provides a comprehensive review of heat generation, heat transfer and plastic flow, tool design, defect formation, and material properties for FSW applications[13].

Based on the current literature, increasing the heat input during FSW of MA956 is expected to increase the grain size in the stir zone (SZ) while consequently lowering hardness[5, 6, 8, 9]. The research results reported herein quantifies the heat input during FSW as a function of the tool rotational speed and tool traverse speed. The ratio of these terms has been analyzed as the pseudo-heat index by Mishra or weld pitch by Nandan and has been previously related to the formation of defects in the weld as well as grain growth in the SZ[12, 13]. This paper will demonstrate the effect of this pseudo-heat index upon the microstructure and properties of the friction stir weld microstructure in MA956.

Experimental Procedure

The material used in this study was MA956, a high Cr ferritic ODS steel with a measured material composition (Anamet Inc., inductively couple plasma-mass spectrometry) shown in Table I. MA956 was canned and extruded at 1100°C (2025°F) and subsequently hot-rolled in three passes at 1100°C (2025°F) over 4 hours with reheating to 1100°C (2025°F) for 30 minutes before and after each rolling pass before final machining into 4 mm (0.157 inch) thick plates.

Table I. Chemical Composition of MA956 (wt%)

C	Cr	Al	Ti	Y_2O_3	Mo	Mn	Ni	S	Si	P	Fe
0.023	19.93	4.75	0.39	0.51	0.02	0.09	0.04	0.008	0.08	0.006	Bal.

FSW of MA956 plate was accomplished by MegaStir Technologies using a convex step spiral scrolled shoulder (CS4) tool with an MS80 grade PCBN tip that requires no tool tilt. FSW parameters of tool rotation rate in revolutions per minute (RPM) and tool traverse speed in inches per minute (IPM) were varied to produce welds of differing quality and consolidation. Plunge force was maintained constant at 17.8 kN (4000 lbf).

Samples of friction stir welded MA956 plate were sectioned and analyzed by optical microscopy (OM), scanning electron microscopy (SEM), electron backscatter diffraction (EBSD), and micro-indentation. Cross sections across the weld path were metallographically prepared for each welding condition by standard metallographic preparation techniques with SiC papers, aluminum oxide polishing solutions, and 0.05 mm colloidal silica solution and electropolishing at 20 V using an electrolyte containing 10% perchloric acid in ethanol maintained at 250 K (-23 C). Microstructural examinations were conducted using a Zeiss Neon 40 field emission SEM at 20 keV. All EBSD analyses were carried out using the TSL OIM 6.0 system with a Hikari camera at 20 kV using a 60μm objective aperture with an approximate probe current of 1 nA. EBSD data was de-noised using TSL OIM software cleanup functions of grain dilation and grain confidence index standardization in accordance with recommended settings. To ensure an adequate number of grains were counted for pole figure analysis, data was collected from areas with minimum dimensions of 150x300 μm equating to a minimum of 2000 grains for the most limiting case. For areas of very fine grain size such as the base material (BM), a step size of 0.25 μm was used equating to over 20,000 grains in the analysis area. For other areas with a larger average grain size, a step size of 0.5 μm was used. Microhardness measurements were accomplished using a HVS-1000 microhardness tester with a diamond indenter and settings of 1 kg-force (2.2 lbf) load and a dwell time of 15 seconds. The hardness tester calibration was verified before and after indentation using a National Institute of Standards and Technology (NIST) certified specimen with a hardness of 726 HV and a certified error of 1.9%.

Results

Full penetration, defect free welds were observed for a range of FSW conditions with higher heat input (Table II and Figure 1). Heat input was quantified using the ratio of the rotational speed to the translational speed. Weld quality and penetration were determined by macrographs of the weld nugget as a function of RPM and IPM combinations in order to establish weld parameters to achieve defect free microstructures. Macroscopic metallographic observations are summarized in Table II. A higher heat input, i.e., increasing tool rotation rate or decreasing tool traverse speed, resulted in a defect free weld nugget. Defect free welds were produced when the ratio of rotational speed to translation speed was greater than 100 (Figure 1a). For conditions with heat input ratios less than 100, welds were observed to be defective with tunnel defects occurring at the bottom or edges of the SZ (Figure 1b).

Table II. Friction stir welding parameter summary.

Tool Rotation Rate (RPM)	Tool Traverse Speed (IPM)	RPM/IPM	Weld Quality	Weld Penetration
200	2	100	Lack of Consolidation	Incomplete
300	2	150	Defect Free	Full
300	4	75	Lack of Consolidation	Incomplete
300	6	50	Lack of Consolidation	Incomplete
400	2	200	Defect Free	Full
400	4	100	Defect Free	Full
400	7	57	Lack of Consolidation	Incomplete
500	1	500	Defect Free	Full

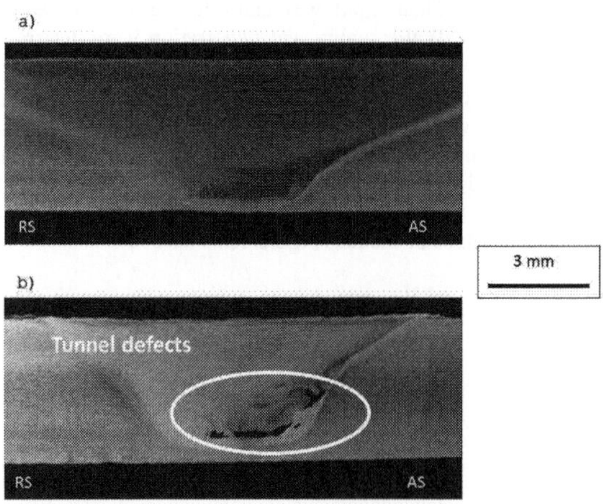

Figure 1. Optical macrographs of transverse metallographic cross-sections for (a) 400 RPM/2 IPM showing a full penetration defect free weld and (b) 400 RPM/7 IPM showing tunnel defect formation at a higher tool traverse rate. (RS=retreating side, AS=advancing side)

The grain size of the SZ increased systematically with increasing heat input (Table III). In all cases, the grain size in the stir zone was larger than the BM (Figure 2). The highest heat input (RPM/IPM=500) resulted in an order of magnitude increase in grain size compared with the BM. This observation is consistent with the previous work on similar ODS steels[5, 8]. The SZ grain size increased with increasing pseudo-heat index, but the relationship was non-linear.

Table III. SZ Hardness and SZ Grain Size Diameter for FSW parameter conditions.

FSW Condition	Mean SZ Grain Size Diameter (microns)	Mean SZ Hardness (HV)
BM	0.89	346 ± 6.6
275/4	1.5	248 ± 4.7
300/2	4.16	225 ± 4.3
400/4	6.94	221 ± 4.2
500/1	12.45	218 ± 4.2

Figure 2. Backscatter SEM image of FSW MA956 with parameters of 400 RPM/4 IPM across SZ, TMAZ, and BM showing large change in grain size and little change in distribution of oxide particles.

In addition, electron microscopy for each FSW parameter combination showed a sharp transition in grain size across the thermo-mechanically affected zone (TMAZ) (Figure 2). This effect was most pronounced for high heat input conditions, particularly on the advancing side (AS) as opposed to the retreating side (RS). The most gradual grain size transition at the AS

TMAZ occurred for the 400 RPM/4 IPM combination, corresponding to the lowest heat input. However, the grain size gradients on the RS were less sharp and extended over the area of several orientation maps. Figure 3 is a composite representation of inverse pole figure (IPF) maps across the entire weld nugget.

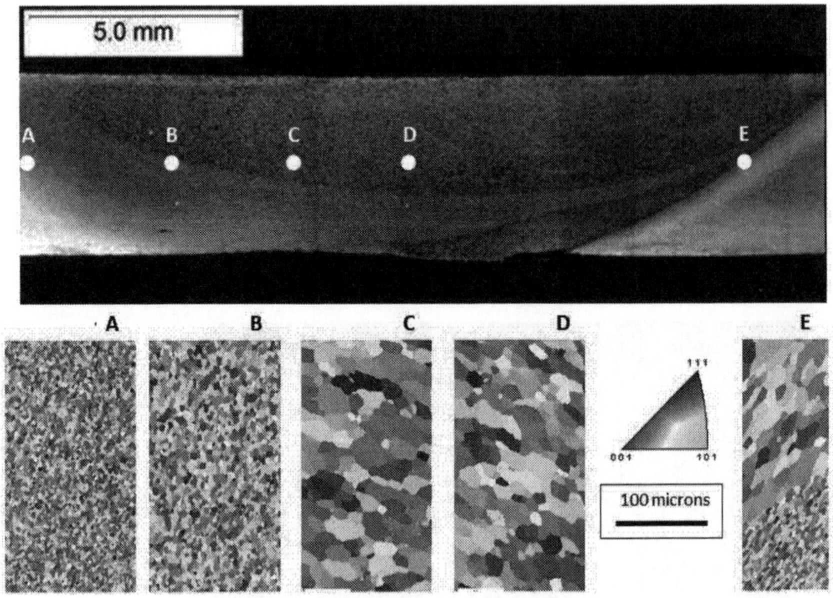

Figure 3. Representation of IPF maps across weld nugget for 500 RPM/1 IPM.

The microhardness in the welded microstructure varied significantly across the weld section and with the level of heat input. In addition, the hardness profiles were asymmetric across the weld nugget. The mean hardness decreased systematically in the SZ as the grain size increased (Table III). For all FSW combinations, the SZ hardness was substantially lower than the BM hardness. An example of the hardness profile is shown for the 500 RPM/1 IPM condition (Figure 4).

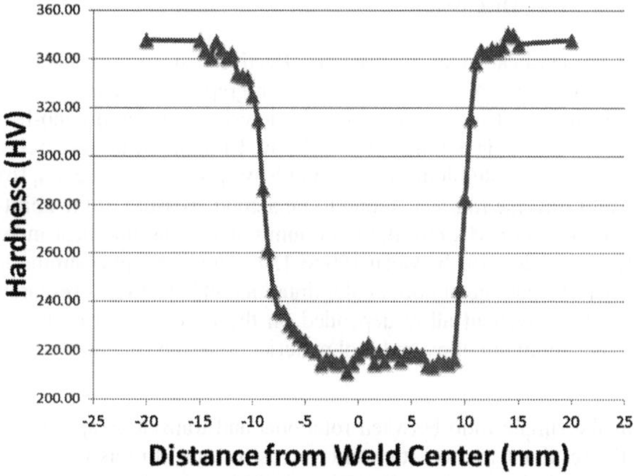

Figure 4. Hardness profile across the SZ for 500 RPM/1 IPM combination.

Pole figures for each FSW condition displayed a consistent torsional texture which is distinctly different from the rolling texture in the base material. The intensity did not change dramatically as heat input increased, but may have decreased slightly. Each FSW combination produced a persistent body centered cubic (BCC) torsional texture compared to the rolled texture of the BM as shown in Figure 5. Similar results have been shown for both torsional steel textures and FSW SZ textures for other ODS steels.[8, 14, 15]

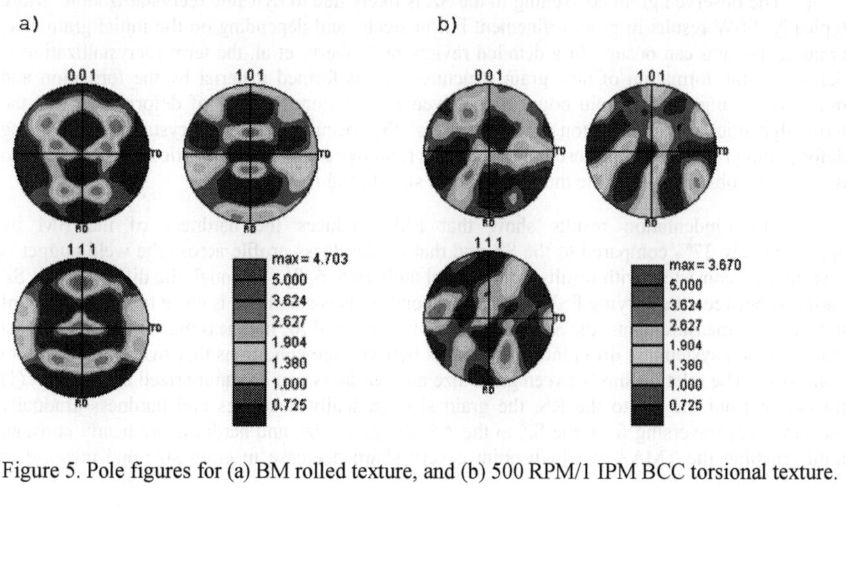

Figure 5. Pole figures for (a) BM rolled texture, and (b) 500 RPM/1 IPM BCC torsional texture.

Discussion

This works suggests that pseudo-heat index may be an effective way to estimate FSW conditions required for successful joining. As shown in the macrographs of Figure 1, weld defects systematically occurred for low heat input conditions, as described by the pseudo-heat index (RPM/IPM). This observed trend is similar to work by Chimbli who researched minimizing lack of consolidation defects in aluminum by varying FSW process parameters[16]. The ratio of tool rotation rate to tool traverse rate was found to be a good measure of the ability to form defect free consolidated welds. For MA956, the minimum range of pseudo-heat index that produced consolidated defect free welds was between 100 to 150. This concept is similar to research by Biswas in a review of FSW parameters on an aluminum alloy[17]. In this research, the overall mechanical response of the welded alloy depended on the ratio of tool rotational speed to tool traverse speed. For the aluminum alloy and tool material used, the suitable value of this ratio was between 380 to 400.

Quantifying heat input as the simple ratio between rotational and translation speed may not be sufficient for predicting the resultant friction stir weld microstructure. In this work, two FSW conditions with the same pseudo-heat index (400 RPM/4 IPM and 200 RPM/2 IPM) produced successful and defective welds, respectively. In very recent research by Wang et. al., MA956 with a coarser starting grain size (hundreds of microns) was friction stir welded with a high heat input (1000 RPM/2 IPM) [10]. Using the pseudo-heat index, this welding condition should apply the same heat input as the 500 RPM/1 IPM condition used in this work. However, the SZ grain size observed by Wang et al. is approximately 1-2μm, compared with the 12μm grain size observed in this work. This difference points out the importance of both the starting microstructure and the heat input model for predicting the SZ grain size. More elaborate heat input models [12, 13, 18] have been developed and should be applied to systematic studies connection FSW conditions with the starting and resultant microstructures.

The observed grain coarsening in the SZ is likely due to dynamic recrystallization. More typically, FSW results in grain refinement but in steels, and depending on the initial grain size, grain coarsening can occur. In a detailed review by Doherty et al. the term recrystallization is defined as the formation of new grain structures in a deformed material by the formation and migration of high angle grain boundaries driven by the stored energy of deformation and the term dynamic recrystallization is defined as the occurrence of recrystallization during deformation[19]. Based on these definitions, the term dynamic recrystallization is appropriate to describe the observed dramatic increase in grain size during FSW.

Micro-indentation results show that FSW reduces the hardness of the BM by approximately 37% compared to the SZ and that the hardness profile across the weld nugget is asymmetric, consistent with results from several authors[5, 6, 8]. Although the differences in SZ hardness between the varying FSW parameters conditions is small and is close to the accuracy of the hardness measurements, the average values of each set of SZ hardness measurements suggest that there is a systematic difference in hardness between the conditions that may be relatable to grain size. The relationship between grain size and hardness can be summarized as follows: (1) traversing from the SZ to the RS, the grain size gradually decreases and hardness gradually increases, (2) traversing from the SZ to the AS, the grain size and hardness are nearly constant until reaching the TMAZ at which point a very sharp decrease in grain size and increase in hardness occurs, and (3) these effects are enhanced as the heat input increases, i.e., the SZ grain size increases and the abruptness of the transitions, both on the RS and most noticeably on the AS, is more distinct for higher heat input conditions. The change in grain size across the weld explains the asymmetric nature of the hardness profiles and at lower heat input values, the asymmetry is less noticeable.

Little change in micro-scale oxide particle distribution was observed for these microstructures for all FSW conditions. For example, no meaningful change in oxide particle size or distribution could be seen in backscattered electron images across the SZ/TMAZ interface for the 400 RPM/4 IPM condition, (Figure 2). It is well known that MA956 has both a micro-scale and a nano-scale oxide particle distribution and the results in this work cannot speak to changes in the nanoscale oxides[9, 20]. Atom probe tomography (APT) measurements by other authors have concluded that the oxide particle size and distribution does not change much during FSW[7].

The coupled increase in SZ grain size and reduction in SZ hardness with increasing heat input during FSW suggests that the Hall-Petch relationship may aptly describe the strengthening mechanism in the SZ. A plot of SZ hardness versus inverse square root of SZ grain size is shown in Figure 6 and shows a nearly linear relationship. The agreement of this hardness data suggests that much of the strength at low temperatures in the stir zone is controlled by the grain size; however, hardness is a function of both yield strength and work hardening. Tensile measurements that can separately measure the yield strength should clarify the strengthening mechanism in stir zone material. It should be noted that the hardness of the BM is much greater than for any of the SZ microstructures, and it does not lie on the Hall-Petch line, suggesting more than one strengthening mechanism is operative.

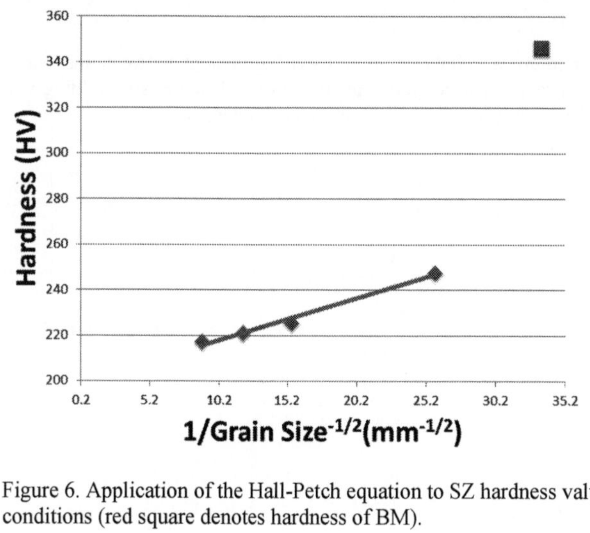

Figure 6. Application of the Hall-Petch equation to SZ hardness values for successful FSW conditions (red square denotes hardness of BM).

Conclusions

This paper examined the correlation between the FSW parameters, tool rotational speed, and tool traverse speed and the resulting welded microstructures and mechanical properties in the ODS steel MA956. Four of eight welding conditions produced defect-free, full penetration welds. Conditions with high thermal input produced defect free welds. The following conclusions are drawn:

(1) Welds with low tool rotational speeds or high tool traverse speeds produce tunnel defects in the weld root or SZ edges. The ratio of tool rotation speed to tool traverse speed can be used as a parameter for predicting defect-free weld consolidation.

(2) Grains in the SZ are substantially coarsened increasing with higher heat input conditions due to dynamic recrystallization.

(3) A persistent, simple torsional texture in the SZ was observed for all FSW conditions.

(4) An abrupt change in grain size and hardness exists across the TMAZ to SZ interface. This change is most pronounced on the advancing side for each welding condition. Higher heat input conditions produce a more abrupt change in both grain size and hardness equating to a smaller but steeper TMAZ.

(5) Hardness decreases from the BM to the SZ by 37% for each welding condition and may be correlated to observed grain growth in the SZ using the Hall-Petch relationship. This suggests that grain coarsening is the dominant, low-temperature strengthening mechanism of the welded material.

Acknowledgements

This work was sponsored by a collaborative research agreement between the Naval Postgraduate School and Lawrence Livermore National Laboratory. Friction stir welding was performed and technical assistance was provided by MegaStir Technologies.

References

[1] E. E. Bloom, S. J. Zinkle, and F. W. Wiffen, "Materials to deliver the promise of fusion power - progress and challenges," *Journal of Nuclear Materials,* vol. 329, pp. 12-19, Aug 1 2004.

[2] S. Ukai, T. Nishida, T. Okuda, and T. Yoshitake, "Development of oxide dispersion strengthened steels for FBR core application, (II) - Morphology improvement by martensite transformation," *Journal of Nuclear Science and Technology,* vol. 35, pp. 294-300, Apr 1998.

[3] M. G. McKimpson and D. Odonnell, "JOINING ODS MATERIALS FOR HIGH-TEMPERATURE APPLICATIONS," *Jom-Journal of the Minerals Metals & Materials Society,* vol. 46, pp. 49-51, Jul 1994.

[4] V. G. Krishnardula, N. I. Sofyan, W. F. Gale, and J. W. Fergus, "Joining of ferritic oxide dispersion strengthened alloys," *Transactions of the Indian Institute of Metals,* vol. 59, pp. 199-203, Apr 2006.

[5] S. Noh, R. Kasada, A. Kimura, S. H. C. Park, and S. Hirano, "Microstructure and mechanical properties of friction stir processed ODS ferritic steels," *Journal of Nuclear Materials,* vol. 417, pp. 245-248, Oct 1 2011.

[6] F. Legendre, S. Poissonnet, P. Bonnaillie, L. Boulanger, and L. Forest, "Some microstructural characterisations in a friction stir welded oxide dispersion strengthened ferritic steel alloy," *Journal of Nuclear Materials,* vol. 386-88, pp. 537-539, Apr 30 2009.

[7] A. Etienne, N. J. Cunningham, Y. Wu, and G. R. Odette, "Effects of friction stir welding and post-weld annealing on nanostructured ferritic alloy," *Materials Science and Technology,* vol. 27, pp. 724-728, Apr 2011.

[8] W. Han, S. Ukai, F. Wan, Y. Sato, B. Leng, H. Numata, N. Oono, S. Hayashi, Q. Tang, and Y. Sugino, "Hardness and Micro-Texture in Friction Stir Welds of a Nanostructured Oxide Dispersion Strengthened Ferritic Steel," *Materials Transactions,* vol. 53, pp. 390-394, Feb 2012.

[9] M. West, B. Jahsthi, P. Hosemann, and V. Sodesetti, "Friction stir welding of oxide dispersion strengthened alloy MA956," in *Friction Stir Welding and Processing VI, TMS,* Warrendale, PA, 2011, pp. 33-40.

[10] J. Wang, W. Yuan, R. S. Mishra, and I. Charit, "Microstructure and mechanical properties of friction stir welded oxide dispersion strengthened alloy," *Journal of Nuclear Materials,* vol. 432, pp. 274-280, 2013.

[11] W. M. Thomas, E. D. Ncholas, J. C. Needham, M. G. Murch, P. Templesmith, and C. J. Dawes, "G.B. Patent Application No. 9125978.8," 1991.

[12] R. S. Mishra and Z. Y. Ma, "Friction stir welding and processing," *Materials Science & Engineering R-Reports,* vol. 50, pp. 1-78, Aug 31 2005.

[13] R. Nandan, T. DebRoy, and H. K. D. H. Bhadeshia, "Recent advances in friction-stir welding - Process, weldment structure and properties," *Progress in Materials Science,* vol. 53, pp. 980-1023, Aug 2008.

[14] U. F. Kocks, C. N. Tome, and H. R. Wenk, *Texture and Anisotropy: Preferred Orientation in Polycrystals and their Effect on Materials Properties*: Cambridge Univesity press, 1998.

[15] W. T. Han, F. R. Wan, B. Leng, S. Ukai, Q. X. Tang, S. Hayashi, J. C. He, and Y. Sugino, "Grain characteristic and texture evolution in friction stir welds of nanostructured oxide dispersion strengthened ferritic steel," *Science and Technology of Welding and Joining,* vol. 16, pp. 690-696, Nov 2011.

[16] S. K. Chimbli, D. J. Medlin, and W. J. Arbegast, *Minimizing lack of consolidation defects in friction stir welds*, 2007.
[17] P. Biswas, D. A. Kumar, and N. R. Mandal, "Friction stir welding of aluminum alloy with varying tool geometry and process parameters," *Proceedings of the Institution of Mechanical Engineers Part B-Journal of Engineering Manufacture,* vol. 226, pp. 641-648, Apr 2012.
[18] R. Nandan, G. G. Roy, T. J. Lienert, and T. DebRoy, "Numerical modelling of 3D plastic flow and heat transfer during friction stir welding of stainless steel," *Science and Technology of Welding and Joining,* vol. 11, pp. 526-537, Sep 2006.
[19] R. D. Doherty, D. A. Hughes, F. J. Humphreys, J. J. Jonas, D. J. Jensen, M. E. Kassner, W. E. King, T. R. McNelley, H. J. McQueen, and A. D. Rollett, "Current issues in recrystallization: a review," *Materials Science and Engineering a-Structural Materials Properties Microstructure and Processing,* vol. 238, pp. 219-274, Nov 15 1997.
[20] M. F. Hupalo, M. Terada, A. M. Kliauga, and A. F. Padilha, "Microstructural characterization of INCOLOY alloy MA 956," *Materialwissenschaft Und Werkstofftechnik,* vol. 34, pp. 505-508, May 2003.

FRICTION STIR WELDING AND PROCESSING VII

Friction Stir Welding: High Temperature Materials – II

Friction Stir Welding and Processing VII
Edited by: Rajiv Mishra, Murray W. Mahoney, Yutaka Sato, Yuri Hovanski, and Ravi Verma
TMS (The Minerals, Metals & Materials Society), 2013

MICROSTRUCTURAL AND MECHANICAL INVESTIGATIONS OF FRICTION STIR WELDED TI/TI- AND TI-ALLOY/TI-ALLOY-JOINTS

N. Buhl[1], G. Wagner[1], D. Eifler[1], M. Gutensohn[2], F. Zillekens[2]

[1]Institute of Materials Science and Engineering (WKK),
University of Kaiserslautern, Gottlieb-Daimler-Strasse, 67667 Kaiserslautern, Germany
[2]PFW Aerospace AG, Am neuen Rheinhafen 10, 67346 Speyer, Germany

Keywords: FSW of cp-Titanium and Ti6Al4V,
Microstructure, Texture, Mechanical Properties

Abstract

Friction Stir Welding (FSW) is an efficient welding technique to join light-weight materials in ductile material condition. Especially for aerospace applications FSW of cp-titanium and Ti-alloys is of high scientific and technological interest. At the Institute of Materials Science and Engineering (WKK) friction stir welds of 1.2 mm thick cp-titanium and Ti6Al4V sheets were produced. To analyze the microstructure of the welding zone in detail light- and scanning electron-microscopic investigations were carried out. By EBSD it was possible to describe the material flow in the welding zone. The mechanical properties were characterized by two dimensional micro-hardness measurements and strain controlled tensile tests. For welds with cp-titanium a yield point of 345 MPa and an ultimate tensile strength of 450 MPa were achieved.

Introduction

The relevance of titanium and titanium alloys for aerospace applications increases continuously. Commercial cp-titanium and Ti6Al4V are still the most used titanium materials. Advantages of these materials are the high specific strength, a good corrosions resistance and an operation temperature of up to 540°C [1]. But to join these materials suitable welding techniques are necessary. In the most cases Tungsten Inert Gas Welding (TIG) is used. But TIG welding leads also to undesired effects as an intensive grain growth which results in a decreasing ductility of the welding seam. In the early 1990's developed friction stir welding (FSW) enables to join materials below the melting temperature in the solid state. Therefore a rotating tool, consisting of a pin and a shoulder area, is pressed in the contact area of two abutting metal sheets. In combination with a perpendicular to the sheets acting welding force, frictional heat is generated and the material around the tool is softened. After the temperature in the weld zone is high enough, the tool starts the movement to realize a seam weld. The advantages of friction stir welding are high strength of the joints on the level of the base materials, low-level energy consumption, low process temperatures and as a result of this the possibility to avoid changes of phases during the welding process. In comparison to ductile aluminum alloys the friction stir welding of cp-titanium and Ti-alloys is more difficult and especially in the aircraft industry the requirements on the quality of Ti-welds are very high. Hence the character of the developing microstructure in the FSW zone is of special interest. For example for cp-titanium a grain refinement in the welding zone and an intensive influence on the texture of the material could be observed [2-4]. To realize the Ti-welds expensive high strength welding tool materials based on W or Mo are necessary [4], because the welding temperatures are in a range of 800°C up to 1200°C.

Materials and Experimental Setup

The thicknesses of the used materials amounted 1.2 mm for cp-titanium and 1.25 mm for Ti6Al4V. Both investigated materials were in rolled and annealed condition. The microstructure of the base material of pure titanium possesses a homogenous α-titanium structure (Fig. 1a). The grain size category (ASTM E112) of the base material is 10. Ti6Al4V is characterized by α-titanium grains with β-titanium phase on the α-grain boundaries (Fig. 1b).

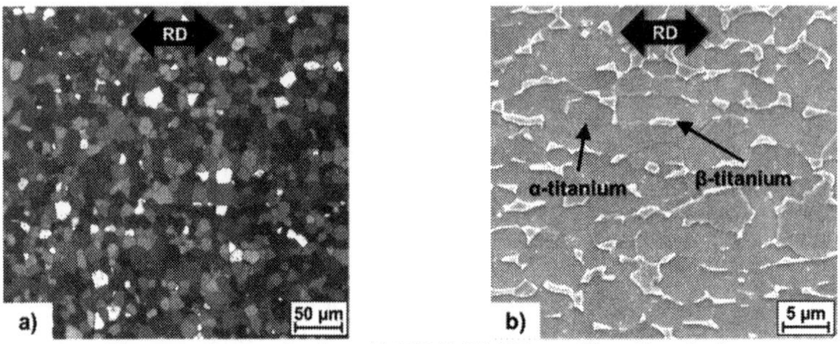

Figure 1. a) pure titanium, b) Ti6Al4V (RD = rolling direction).

All sheets were machining along the abutting edges before welding. To ensure reproducible welds the sheets were positioned in a pneumatic clamping developed at the WKK. All welds were carried out perpendicular to the rolling direction of the sheets. To prevent adhesion of the cp-titanium joining parts on the anvil side a 0.075 mm thin rolled stainless steel foil was positioned. For Ti6Al4V welds a 0.1 mm thick rolled tungsten foil has proved as suitable. All welds were carried out with a tungsten-lanthanum based tool with a conical pin and a shoulder diameter of 11 mm. To prevent oxidation of the titanium sheets the tool was fixed in an oil cooled tool holder with a shielding gas shroud. The tool holder was also equipped with an infrared pyrometer to measure the surface temperature of the tool and a video camera to record the whole welding process. Furthermore the conventional milling machine was upgraded for the special requirements of FSW with a 3-axis force measurement system and a force controller. The used welding parameters for welding cp-titanium and Ti6Al4V are mentioned in Table I.

Table I. Welding parameters friction stir welding of cp-titanium and Ti6Al4V

cp-titanium	
Welding force F_Z	5.5 kN
Welding speed v	55 mm/min
Rotation speed n	500 1/min
Angle of attack φ	0.5°
Ti6Al4V	
Welding force F_Z	3.9 kN
Welding speed v	80 mm/min
Rotation speed n	700 1/min
Angle of attack φ	1.0°

Results cp-titanium

In subject-specific literature a load controlled FSW process of cp-titanium was described as very difficult [5, 6]. But at the WKK investigations have shown, that a stable force controlled process with high quality welding seams can be realized at a process temperature close below the beta transus temperature of 910°C for cp-titanium [7]. In this case the welds were carried out with an average process temperature of 830°C [1]. The developed welding zone is pictured in Fig. 2.

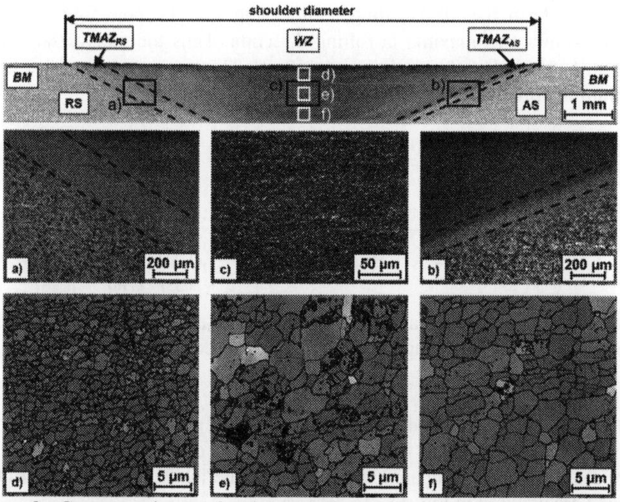

Figure 2. Cross section and microstructure of a cp-titanium FSW-joint; light microscopic pictures (a-c) and EBSD mappings (d-f).

The welding zone is trapezoidal with a 220 µm thick thermo-mechanical affected zone ($TMAZ_{RS}$) on the retreading side (Fig. 2 a) and a 110 µm thick $TMAZ_{AZ}$ on the advancing side (Fig. 2 b). The center of the welding zone is characterized by an extreme grain refinement (Fig. 2 c). Using EBSD mappings the microstructure of the nugget can be visualized in detail (Fig. 2 d-f). In the shoulder contact area the microstructure possesses a homogenous structure with equiaxed globular α-titanium grains (Fig. 2 d). The grain size category is 16. In the middle and in the lower area a microstructure with globular α-titanium grains is developed, too (Fig. 2 e, f). In these areas the grain size increases to 14.

2D-microhardness measurements were carried out (Fig. 3) to detect the microhardness distribution in the welding zone.

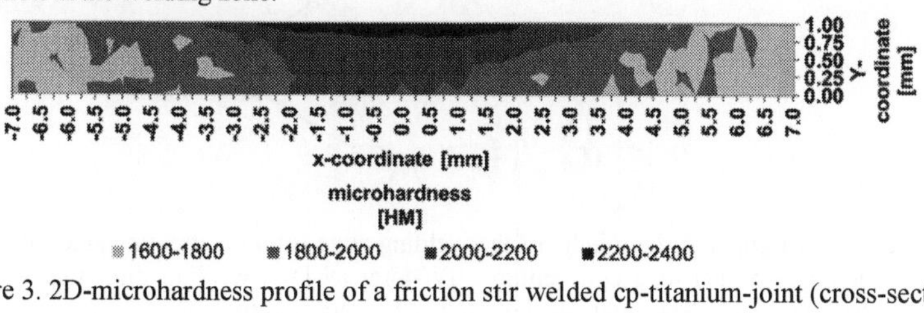

Figure 3. 2D-microhardness profile of a friction stir welded cp-titanium-joint (cross-section).

The hardness has increased from 1600 HM in the base material to more the 2000 HM in the welding zone. The maximum hardness values with up to 2400 HM have been reached in the shoulder contact area.

The material texture of the base material and the welding zone were also investigated to describe the changes in the microstructure. The pole figures of the basal planes $\{0001\}$, the transversal planes $\{10\bar{1}0\}$ and the main slip direction $\langle 11\bar{2}0\rangle$ of the base material are shown in Fig. 4. The pole figure of the base material in the basal planes $\{0001\}$ is characterized by a deviation of 45° of the two maxima from the ideal rolling texture (Fig. 4). The pole figure of the main slip direction $\langle 11\bar{2}0\rangle$ shows two maxima in rolling direction. Thus the transversal texture with the main slip system $\{10\bar{1}0\}$, $\langle 11\bar{2}0\rangle$ predominates [8, 9]. This texture character was observed for rolled cp-titanium in previous investigations, too [10, 11].

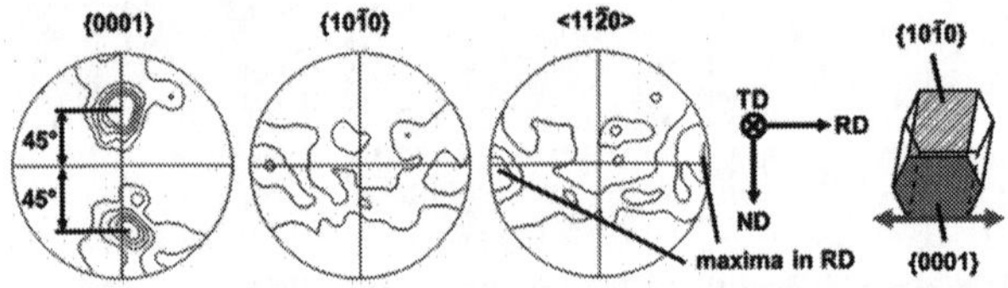

Figure 4. Pole figures of the basal planes, transversal planes and the main slip direction of the cp-titanium base material (RD = rolling direction, ND = sheet normal direction, TD = tangential direction).

During the FSW process the welding temperature was permanent below the beta transus temperature, so that phase changes in the material can be excluded. Accordingly the resulting microstructure can be directly compared with the base material. The pole figures of three areas of the welding zone are pictured in Fig. 5.

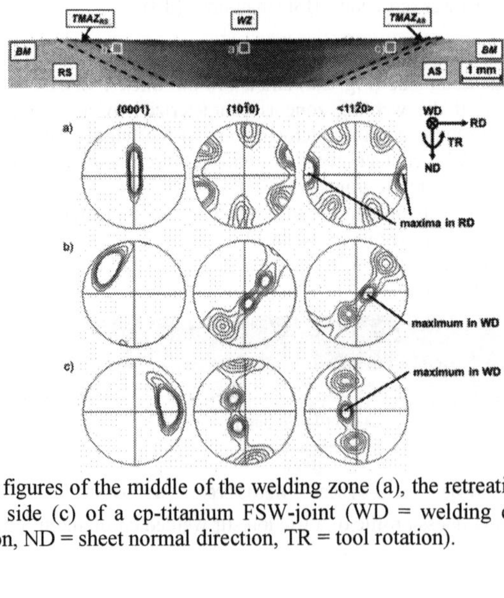

Figure 5. Pole figures of the middle of the welding zone (a), the retreating side (b), and the advancing side (c) of a cp-titanium FSW-joint (WD = welding direction, RD = rolling direction, ND = sheet normal direction, TR = tool rotation).

The pole figures in the middle of the welding zone (Fig. 5 a) are characterized by a distinct rolling texture as a result of the FSW process described by a sharp maximum in the middle of the basal plane pole figure {0001}. In the $\{10\bar{1}0\}$ pole figure the six transversal planes can be clearly identified at 90°. The pole figure of the main slip direction $\langle 11\bar{2}0 \rangle$ shows two maxima in rolling direction (RD) at 90°. At the retreating side (Fig. 5 b) and the advancing side (Fig. 5 c), the pole figures of the basal planes are turned to the left and right side. The maxima of the main slip directions are in welding direction (WD) (Fig. 5 b, c). A possible approach for the development of this material texture will be explained in Fig. 6. Based on the hexagonal crystal structure, α-titanium shows anisotropic mechanical properties [8, 9]. The hexagonal elementary cells were plastically deformed by the welding tool during the FSW process. Thereby the deformation of the hexagonal α-titanium preferential occurs in the transversal slip system. Hence the elementary cells slip with their main slip direction $\langle 11\bar{2}0 \rangle$ in rotation direction of the welding tool. As a result of this the elementary cells are orientated along a circular path in the welding zone.

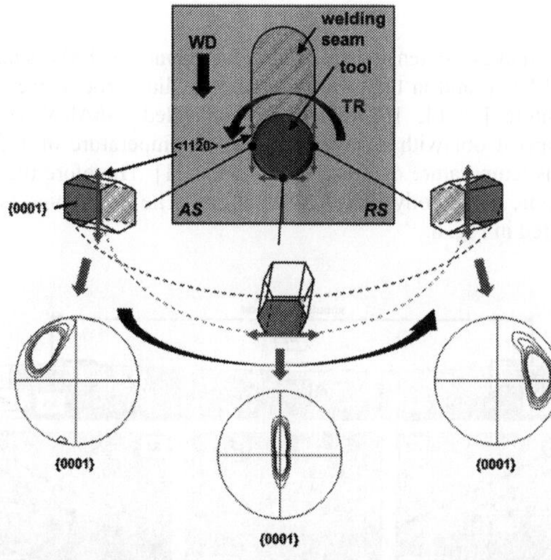

Figure 6. Schematically illustration of the relationship between welding tool rotation and created material texture of the welding zone (top view with WD = welding direction, TR = tool rotation, AS = advancing side, RS = retreating side, WZ = welding zone).

To describe the mechanical properties of the cp-titanium FSW-joints strain-controlled tensile tests were carried out. The stress-strain-diagrams of the cp-titanium base material and the cp-titanium FSW welding zone are pictured in Fig. 7. All FSW specimens failed in the middle of the welding zone. The base material shows an upper and lower yield point. In contrast, the FSW zone possesses only a technical elastic limit which increases in the welding zone up to 388 MPa in comparison to an upper yield point with 348 MPa of the base material. At the same time the ultimate tensile strength decreases from 463 MPa for the base material to 449 MPa for the FSW-joint. Also the fracture strain has been reduced from 32% to 17% in the FSW welding zone.

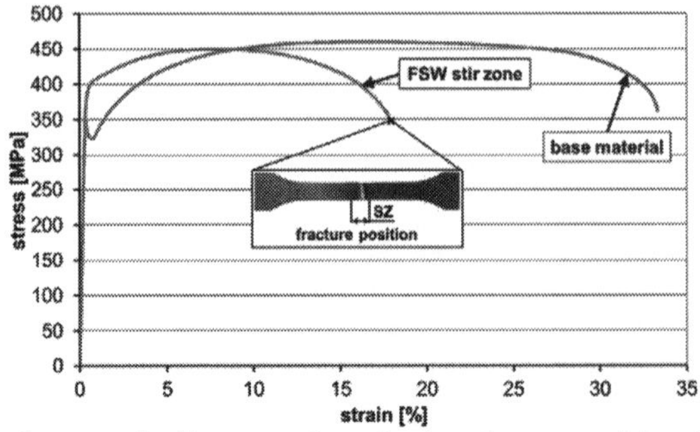

Figure 7. Stress-strain-diagrams of cp-titanium base material and cp-titanium FSW welding zone.

Results Ti6Al4V

In comparison to cp-titanium extensive investigations about the FSW suitability of Ti6Al4V were carried out [6, 12-15], and in the most cases the welding process were realized above the beta transus temperature [4, 14, 16]. The force-controlled Ti6Al4V welds in the present investigation were carried out with an average process temperature of 1120°C which is also above the beta transus temperature of Ti6Al4V at 995°C [1]. Therefore the plastic deformation has occurred in isotropic cubic body centered β-titanium. The developing microstructure of the welding zone is pictured in Fig. 8.

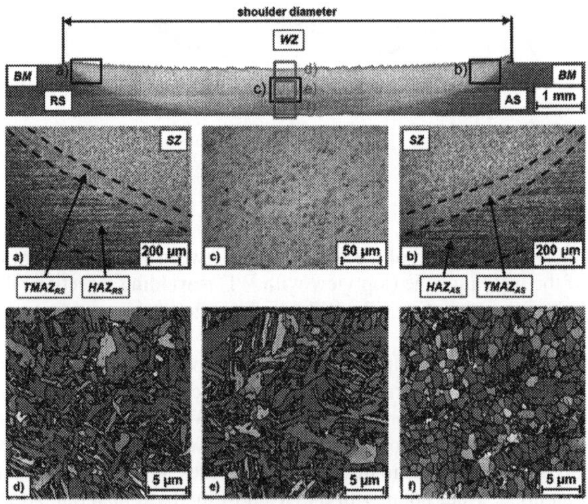

Figure 8. Cross sections and microstructure of a Ti6Al4V FSW-joint; light microscopic pictures (a-c) and EBSD mappings (d-f).

The welding zone possesses the geometry of a half-shell with small symmetrical TMAZ and HAZ on the retreating side (Fig. 8 a) and advancing side (Fig. 8 b). The HAZ is characterized by

a banding of the microstructure. This is a result of the partial transformation of the α-titanium part of the base material in β-titanium during the welding process and the retransformation to α-titanium during the cooling. The HAZ was not plastically deformed during the welding process. In comparison to the base material the microstructure in the welding zone has completely changed. The nugget is extremely fine grained (Fig. 8 c). By EBSD mappings the character and orientation of the grains can be describe in detail (Fig. 8 d-f). In the shoulder contact area (Fig. 8 d) and the middle of the welding zone (Fig. 8 e) an acicular martensitic microstructure can be detected, hence the welding root shows a fine equiaxed globular microstructure (Fig. 8 f). The acicular microstructure results from the plastic deformation above the beta transus temperature followed by a fast quenching [8, 9] because the transformation from β-titanium to acicular martensite is not diffusion controlled and follows the burgers relationship [8]. Thus the martensitic structure possesses particular orientations limited by the burgers relationship. The globular microstructure results from a plastic deformation in the α+β-area followed by a recrystallization process. Therefore a process temperature below the beta transus temperature and a high plastic deformation is necessary. The reasons for the lower temperature of the welding root is the distance from the tool shoulder in combination with the low thermal conductivity of Ti6Al4V [4],[4, 17]. With longer annealing time the grain size increases [9].
Moreover microhardness measurements were carried out. In Fig. 8 a 2D-microhardness profile of a friction stir welded Ti6Al4V-joint is shown. The microhardness increases from 3000 HM in the base material to 3400 HM in the welding zone. The maximum hardness occurs in the middle of the welding zone.

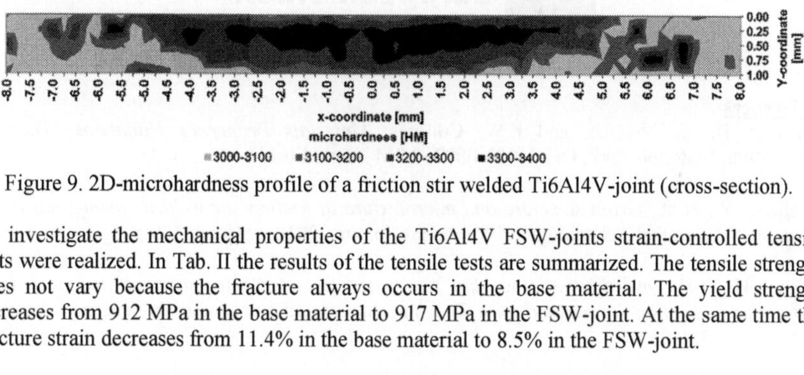

Figure 9. 2D-microhardness profile of a friction stir welded Ti6Al4V-joint (cross-section).

To investigate the mechanical properties of the Ti6Al4V FSW-joints strain-controlled tensile tests were realized. In Tab. II the results of the tensile tests are summarized. The tensile strength does not vary because the fracture always occurs in the base material. The yield strength increases from 912 MPa in the base material to 917 MPa in the FSW-joint. At the same time the fracture strain decreases from 11.4% in the base material to 8.5% in the FSW-joint.

Table II. Mechanical properties of Ti6Al4V FSW-joints

		Base material	FSW joint
Yield point	[MPa]	912	917
Ultimate tensile strength	[MPa]	1020	1020
Fracture strain	[%]	11.4	8.5

Summary

Investigations at the WKK have shown that for force controlled cp-titanium welds below the beta transus temperatures in the shoulder contact area a distinctive grain refinement from category 10 in the base material to 16 can be observed. In the middle and the root area of the welding zone the grain size category increases as well but only up to 14. The pole figures of the welding zone texture have shown that as a result of the FSW process the elementary cells have been orientated with their main slip direction along the circular path around the welding tool. The microhardness

increases from 1600 HM in the base material to 2400 HM in the welding zone. The maximum microhardness values have been reached in the shoulder contact area. With strain controlled tensile tests of the base material and the FSW welding zone it could be proved that for the welds the tensile strength decreases 3.1%, but the yield point increases 11.5% simultaneously.
Investigations of force controlled Ti6Al4V friction stir welds above the beta transus temperature showed different microstructures in the welding zone. The shoulder contact area and the middle of the welding zone possess an acicular martensitic microstructure. The process temperature in these areas was during the FSW process above the beta transus temperature and the high deformed microstructure has quenched fast. In the welding root an equiaxed microstructure is distinguishable which has been developed by high deformation below the beta transus temperature and a following recrystallization process. Moreover an increase of the microhardness from 3000 HM in the base material to 3400 HM in the welding zone could be observed, which results in a reduction of the tensile strain while the yield strength and the tensile strength is nearly unchanged.

Acknowledgment
The investigation of friction stir welding of cp-titanium and titanium alloys is supported by the German Federal Land Rhineland-Palatinate, the European Union with funds of the European Regional Development Fund (ERDF) and the PFW Aerospace AG (Speyer, Germany).

European Union
European Regional Development Fund

References

1. Boyer, R., G. Welsch, and E.W. Collings, *Materials Properties Handbook: Titanium Alloys*1998, Materials Park, OH 44073-0002: ASM International.

2. Zhang, Y., et al., *Grain structure and microtexture in friction stir welded commercial purity titanium.* Science and Technology of Welding & Joining, 2010. 15(6): p. 500-505.

3. Fujii, H., et al., *Investigation of welding parameter dependent microstructure and mechanical properties in friction stir welded pure Ti joints.* Materials Science and Engineering: A, 2010. 527(15): p. 3386-3391.

4. Mishra, R.S. and Z.Y. Ma, *Friction stir welding and processing.* Materials Science and Engineering: R: Reports, 2005. 50(1-2): p. 1-78.

5. Jata, K.V. and A.P. Reynolds, *Microstructure And Mechanical Behavior Of Friction Stir Welded Titanium Alloys.* Metallic Materials With High Structural Effiency, 2004: p. 10.

6. Edwards, P. and M. Ramulu, *Identification of Process Parameters for Friction Stir Welding Ti-6Al-4V.* Journal of Engineering Materials and Technology, 2010. 132(3): p. 031006-10.

7. Buhl, N., et al., *Friction Stir Welding of Ti/Ti- and Ti-Alloy/Ti-Alloy-Joints.* 9th International Symposium on Friction Stir Welding, USA, 2012.

8. Lütjering, G. and J.C. Williams, *Titanium*2007, Berlin: Springer.

9. Peters, M. and C. Leyens, *Titan und Titanlegierungen*2002, Weinheim: WILEY-VCH.

10. Singh, A.K. and R.A. Schwarzer, *Texture and Anisotropy of Mechanical Properties in Titanium and Its Alloys.* Zeitschrift für Metallkunde, 2000. 91: p. 15.

11. Mironov, S., Y.S. Sato, and H. Kokawa, *Development of grain structure during friction stir welding of pure titanium.* Acta Materialia, 2009. 57(15): p. 4519-4528.

12. Zhang, Y., et al., *Microstructural characteristics and mechanical properties of Ti-6Al-4V friction stir welds.* Materials Science and Engineering: A, 2008. 485(1-2): p. 448-455.

13. Ramirez, A.J. and M.C. Juhas, *Microstructural Evolution In Ti-6Al-4V Friction Stir Welds.* Materials Science Forum, 2003. 426-432: p. 2999.

14. Edwards, P. and M. Ramulu, *Peak temperatures during friction stir welding of Ti6Al4V.* Science and Technology of Welding and Joining, 2010. 15: p. 468-472.

15. Zhou, L., et al., *The stir zone microstructure and its formation mechanism in Ti-6Al-4V friction stir welds.* Scripta Materialia, 2009. 61(6): p. 596-599.

16. Pilchak, A., et al., *Microstructure Evolution during Friction Stir Welding of Mill-Annealed Ti-6Al-4V.* Metallurgical and Materials Transactions A, 2011. 42(3): p. 745-762.

17. Ahmed, M.M.Z., et al., *Through-thickness crystallographic texture of stationary shoulder friction stir welded aluminium.* Scripta Materialia, 2011. 64(1): p. 45-48.

Friction Stir Welding and Processing VII
Edited by: Rajiv Mishra, Murray W. Mahoney, Yutaka Sato, Yuri Hovanski, and Ravi Verma
TMS (The Minerals, Metals & Materials Society), 2013

Microstructure and Mechanical Properties of FSW Lap Joint between Pure Copper and 1018 Mild Steel Using Refractory Metal Pin Tools

Md Shamsujjoha[1, 2], Bharat K. Jasthi[1], Michael West[2] and Christian Widener[1]

[1.] Arbegast Advanced Materials Processing Laboratory
South Dakota School of Mines and Technology
501 East Saint Joseph Street, Rapid City, SD 57701, USA

[2] Department of Materials and Metallurgical Engineering
South Dakota School of Mines and Technology
501 E. St. Joseph St., Rapid City, SD 57701 USA

Keywords: Lap welding; Friction stir welding; Copper; Steel; Microstructure

Abstract

Friction stir welding was conducted on a copper/steel lap joint configuration. Two different W-25%Re-4% HfC pin tools, having two different pin diameters and pin lengths, were investigated to examine the combined effect of plunge depth and bonding area on joint properties. The effects of travel speeds and alternating the top and bottom sheet materials on the microstructure and mechanical properties were also investigated. Structures of the joining interface were analyzed by optical and scanning electron microscopy. Joint strength and fatigue strength were also evaluated. The results showed a "hook-like" feature on both the advancing and retreating sides, but more severe on the retreating side. The interface was bonded both by mechanical mixing between copper and steel and metallurgically as evidenced by a diffusion zone formed between Fe-Cu.

Introduction

Joining of dissimilar metals has been identified as a top priority in materials joining technology because of the increased application of dissimilar combinations in the automotive, aerospace, and power generation industries [1]. Recently, the solid state nature of friction stir welding (FSW) has shown promise in welding dissimilar metal combinations. FSW has become a very important welding technique for welding aluminum and its alloys and other soft alloys like magnesium and copper. Advancements in tool materials have allowed FSW to be successfully applied to weld high temperature, hard materials like steel, titanium, and nickel-based alloys.

FSW has been employed to make lap joints between dissimilar soft-soft metal combinations. Research that has been reported has shown that FSW is capable of producing good quality welds with bond strengths close to the base metals. However, limited success has been achieved with soft/hard metal combinations because of the tool degradation during welding. Recent work with friction stir lap welding (FSLW) has been conducted with soft/hard combinations with low-temperature pin tools where the pin is plunged into the top soft metal and the probe tip is standing off or adjacent to the bottom hard plate in order to prevent tool wear. These studies reported that bonding strength increases with depth of the pin tools into the hard metals. Consequently, the results reported fairly low bonding strength [2]. The research work presented here investigates the potential to improve mechanical properties of lap joints between the soft/hard metal combinations of Cu/steel using tungsten rhenium hafnium carbide (W-25%Re-4%HfC) pin tools, which give the freedom to plunge into the bottom hard metals.

Copper-steel combinations are widely used in the fields of power generation and transmission, cryogenics, electrical and electronics because of the combination of properties like high electrical conductivity and stiffness. However, differences in chemical and thermomechanical properties between copper and steel make it very difficult to fusion weld these two materials [3]. Because of the high thermal conductivity of copper, heat dissipates rapidly from the joint interface and thus makes it very difficult to reach the melting point of copper during fusion welding. Cu also has very low solubility in Fe. These factors make fusion joining of copper to steel with acceptable mechanical properties very difficult [4]. Typically, explosive welding has also been used for welding copper to steel [5].

Despite FSW's success in joining dissimilar metals, it is surprising that very limited work has been reported for this combination. Imani et al. [6] studied butt welding between copper and stainless steel. The effect of pin offset for producing sound welds was investigated and the researchers concluded that a 30% pin offset toward the copper produce the best quality weld. The reported joint strength was 30% less than that of the copper base material.

In this study, the feasibility of joining Cu to steel with FSWL was investigated. To get the best quality joint properties, the effect of process parameters, plunge depth, bonding area, and top sheets position on joint properties were investigated.

Experimental

The materials used in this study are readily available 0.185" thick 1018 mild steel (ASTM A108) and 0.125" thick pure copper (Alloy 110) for Cu/steel lap weld configurations. Copper was placed on the top and mild steel was on the bottom for copper/steel lap joints. The sheets overlapped 2" and the total weld length was 10.5". The joint configuration is shown schematically in Figure 1. Both right handed and left handed lap welds were made during the course of this study.

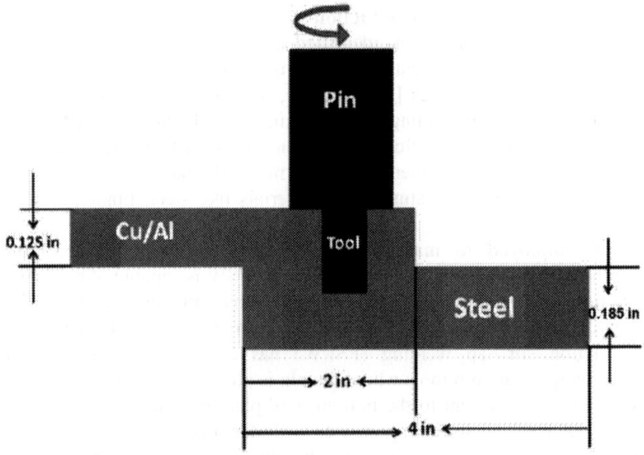

Figure 1: Schematic view of joint configuration.

Before welding, all of the plates were cleaned using Scotch-Brite pads followed by alcohol. All the welds were made in this work using the MTS ISTIR 10 system and water-cooled Megastir Head welding machine in the Arbegast Advanced Materials Processing and Joining (AMP) Lab at the South Dakota School of Mines and Technology.

Friction stir lap welds were made using W-25% Re-4% HfC tools having featureless tapered geometries with two different pin lengths and shoulder diameters to investigate the combined effects of plunge depth and bonding area on joint properties. Pin tool A was a tool with 0.23" pin length and 0.4" shoulder diameter; pin tool B was a tool with 0.15" pin length and 0.352" shoulder diameter. Photographs of the pin tools are shown in Figure 2.

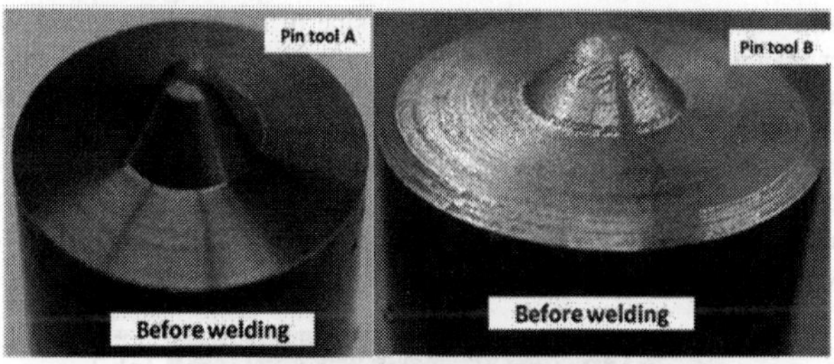

Figure 2: Photographs of the pin tools.

A wide range of parameters were investigated to produce welds with acceptable weld qualities using both the pin tools. After welding was conducted, samples were cross sectioned using a water jet machine perpendicular to the welding direction for metallurgical examination using standard metallographic techniques. Polished specimens were etched using solution of 5g FeCl3, 10ml nitric acid, 20ml hydrochloric acid and 50ml water by swabbing for approximately 30 seconds. Grain size measurements were conducted using Buehler Omnimet software according to ASTM E-112. A ZEISS Gemini Supra 40VP scanning electron microscope (SEM) was used to acquire high magnification images and to perform energy dispersive spectrometry (EDS) on points of interest and also for elemental mapping.

Vickers microhardness measurements were made at a load of 500 g and a loading time of 15 seconds accordance with ASTM E92-82. Lap shear strength and fatigue tests were conducted on an MTS 810 hydraulic mechanical test machine using a 55 KN load cell. A schematic of lap welded samples used for tensile and fatigue testing is given below in Figure 3. A stress ratio ($R = (\sigma_{min}/\sigma_{max})$) was chosen to be 0.1 for all of the fatigue tests. Tests were conducted at 15 Hz frequency at room temperature. Two samples were tested in each load cycle condition.

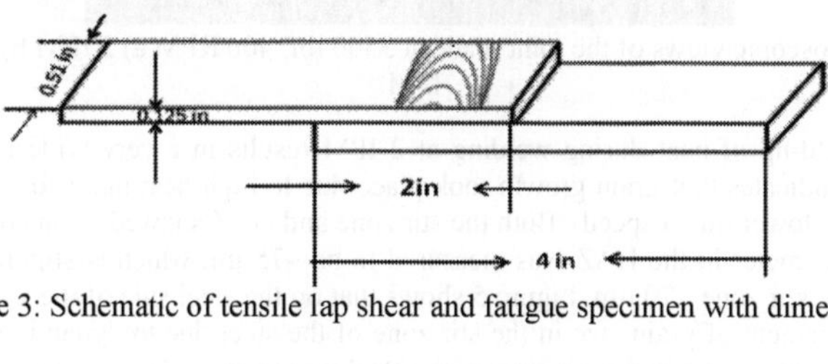

Figure 3: Schematic of tensile lap shear and fatigue specimen with dimension.

Results and Discussion

Characteristic Features of the Joints:

Figure 4 shows the macroscopic view of the cross-section of the welds made using longer pin tools. It can be seen from the macrograph that very few microscopic voids at the interface were observed for all of the travel speeds. Both copper and steel underwent plastic deformation during lap welding as the pin tools plunged deep into the bottom steel plates. Frictional heat generated between the pin tools and the hard steel and plastic deformation was sufficient to recrystallize the steels. Two distinct stir zones were observed and a typical friction stir weld zone (stir zone, HAZ on advancing and retreating side) on both sides was observed, as shown in Figure 6. The microstructure of the stir zone corresponding to copper was characterized by relatively fine equiaxed grains. On the copper side of the joint, the grain size near the top surface is finer than grains at the middle of the stir zone and close to joint interface. Grain size near the top surface is approximately 18 μm during weld made at 3 IPM while at the bottom of the stir zone it is close to 35 μm. A wider HAZ was observed on the copper side of the joint. Grain size in the HAZ was ~30μm on both advancing and retreating side and is smaller than the parent metal. Similar characteristics were observed during welding at 4 IPM and the first part of the weld made at 2 IPM. However, the grain size in stir zone and HAZ slightly decreased with increasing travel speed, as heat input decreased with increasing travel speed.

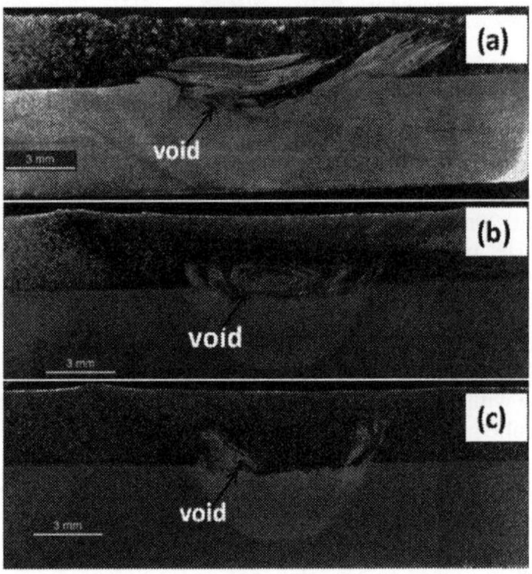

Figure 4: Macroscopic views of the joint made at 3500 lbf, 400 RPM a) 2 IPM b) 3 IPM and c) 4 IPM.

The build-up of heat during welding at 2 IPM results in a very wide HAZ and coarse grains, which indicates that grain growth took place due to high heat input for relatively longer period of time (slower travel speed). Both the stir zone and HAZ showed grain coarsening effects and the grain diameter in the HAZ was measured to be ~75 μm which is slightly coarser when compared to the stir zone ~70 μm. Figure 5 shows that on the steel side of the Cu/Steel lap joint, significant refinement of grain size in the stir zone of the steel due to dynamic recrystallization occurred and the grain size in the stir zone of the steel was measured to be ~4 μm. The grain size in the HAZ is slightly coarser than the stir zone and slightly deformed.

Similar weld zones were observed during welding using pin tool B. However, since the pin tools did not plunge as deep into the steels as with welding conducted with pin tool A, the stir zone in the steels formed just below the pin tip. As a result, steel undergoes very high plastic deformation and the microstructure of the stir zone consists of slightly deformed grains.

The interface of Cu/Steel lap joints can be viewed as four distinctive zones in each case of welds made using different pin tools, as shown in Figure 6. Zone A and Zone D are identified by the "hook-like" feature at the retreating and advancing side of the welds made with both short and long pin tools. Both zone A and B were the result of vertical motion of the processed steels. As during welding both short and long pin tools pin is plunged fairly deep into the bottom plates and frictional heat between probe and materials and plastic deformation of the hard materials made it very soft and rotational speed of the pin tools induced vertical motion into the materials and this pulled up steel into the copper stir zone. Besides the "hook-like" feature in the advancing side (zone D), presence of a "tooth- like" feature was also observed. The stirring force of FSLW extruded the copper into steel and may have resulted into micro interlocking [7].

Zone B is characterized by iron rich particles at the top of the interface into the stir zone of copper. Combined action of pin tool force and rotational speed stripped up the processed steels into the copper and resulted into complex mechanical mixing of copper and steel. This zone showed different morphology for different parameters and also for different pin tools as it can be seen from the macrograph in Figure 5.

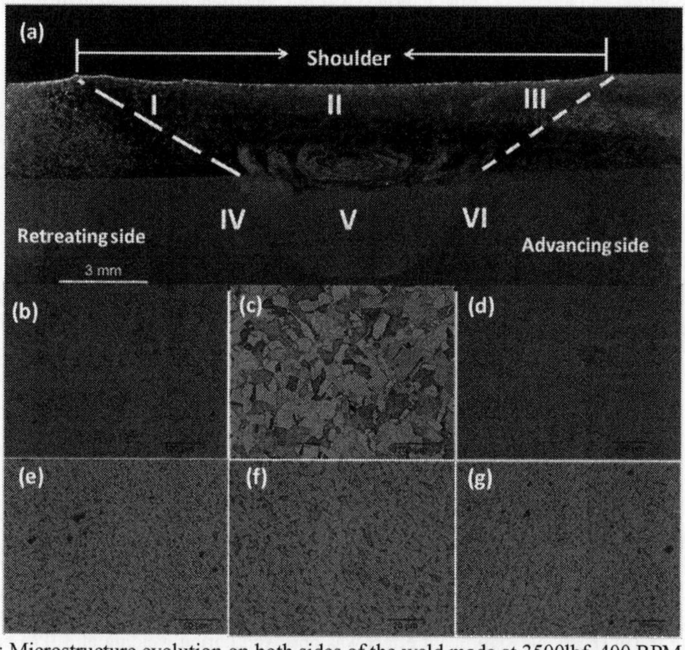

Figure 5: Microstructure evolution on both sides of the weld made at 3500lbf, 400 RPM, 3 IPM: a) Macrograph of the joint b) HAZ on the retreating side of copper c) fine grained stir zone of copper d) HAZ on the advancing side of copper e) relatively coarser grain HAZ on the advancing side of steel f) fine recrystallized stir zone g) HAZ on the retreating side of steel.

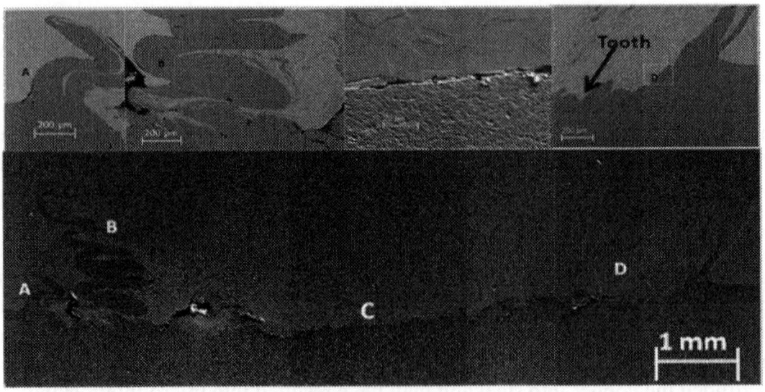

Figure 6: Different significant interfacial zones (bottom) and their close view (top).

Zone C is the interface between copper and steel. The interface for Zone C enlarged in Figure 7 shows an area of extensive diffusion bonding between the Cu and steel. The EDS spectrum analysis suggests this may be a non-equilibrium solid solution of Cu in steel, however the composition and thickness of the interface was not consistent and varied in different areas.

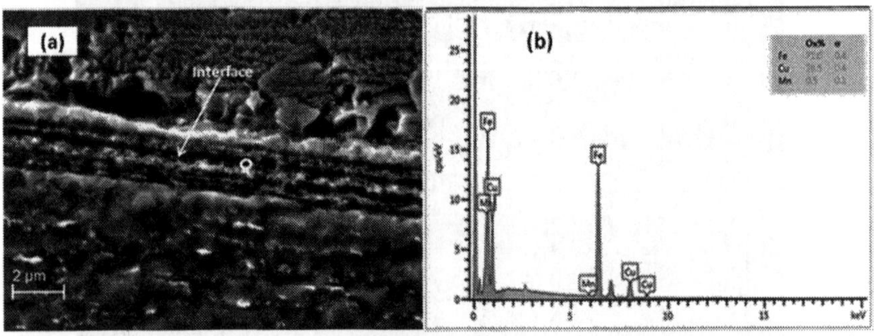

Figure 7: a) Location of EDS spectrum taken at the interface between copper and steel b) EDS spectrum of the interface.

Mechanical Properties:

Tensile and fatigue properties of the joint were evaluated and discussed in terms of effect of travel speed, plunge depth and position of the top plates.

Effect of Travel Speed

Figure 8 shows the effect of travel speed on the lap shear strength of the copper/steel lap joints. All the welds were made at 3500 lbf force and 400 RPM using pin tool A and top plates were on the retreating side. As can be seen from the Figure 8, shear load increases with increasing travel speed. No significant effect was observed during welding at 3 IPM and 4 IPM but the load was significantly lower at 2 IPM. Joint efficiency was calculated using a simple relationship based on the weaker parent sheet. Results showed that joint efficiency at 4 IPM is 88%, at 3 IPM is 87.5 % and at 2 IPM is 78 %.

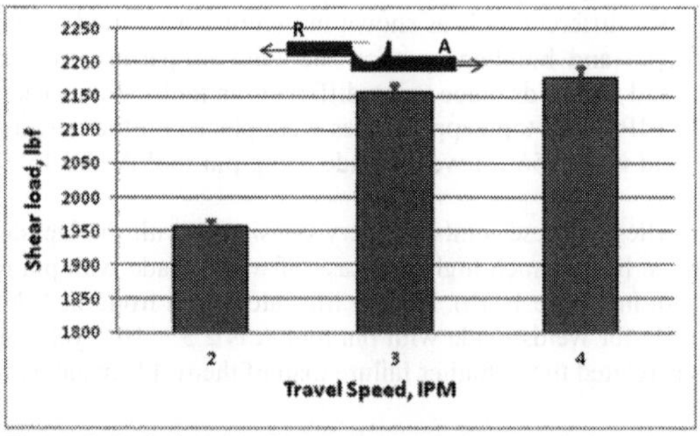

Figure 8: Relationship between travel speed and shear load.

All the welds made using pin tool A at 2, 3 and 4 IPM failed in the HAZ of welds. Figure 10 shows the failure location of welds made at 4 IPM (Figure 9(a)) and 2 IPM (Figure 9(b)). Lap shear samples from 3 and 4 IPM failed outside the shoulder area which is supposed to be parent metal, yet 100 % joint efficiency was not achieved. Because of the high thermal conductivity of copper, the welds had very large HAZ and softening effect was observed for a wide area. Figure 9(c) shows the hardness value in the HAZ is roughly 58 HV. Microhardness value is lower compared to hardness values of the parent metal (98 HV). Failures may have occurred because of the softening of a large area during welding. It should be noted that though the grain size is smaller than parent metal in both stir zone and HAZ, the hardness value still decreased, which indicated that the hardness value does not depend solely on grain size of the copper; rather, it depends on the dislocation density. This result was consistent with the findings of Lee et al. [8]. On the other hand, welds made at 2 IPM failed inside the shoulder area. It has been reported that [9] softening of the HAZ increases with the heat input. Slower welds such as the 2 IPM had higher heat input resulting in relatively lower hardness values observed in the HAZ. Hardness values were found to be roughly 53 HV in this zone. The high degree of softening during welding at 2 IPM caused a significant drop in lap shear strength.

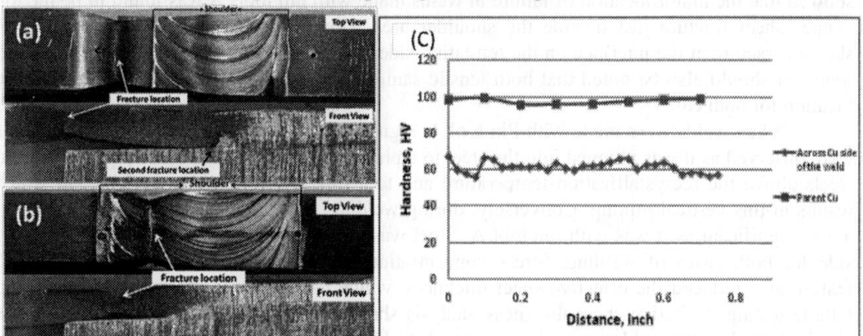

Figure 9: Top view and front view of the fracture parts of the welds made at 3500 lbf, 400 RPM and a) 3 IPM and b) 2 IPM and C) Hardness value across the parent copper and copper side of the weld for welds made at 3500 lbf, 400 RPM and 3 IPM.

Combined Effects of Plunge Depth and Bonding Area

In this study two different tools as shown in Figure 2 were used to examine the combined effects of plunge depth and bonding area on the joint properties. Figure 10 (a) shows the differences in shear load for welds made using different pin tools. Both these welds were made at 3500 lbf, 400 RPM, 3 IPM and top copper plates were placed on the retreating side of the joint. The average shear load was 2156 lbf welds made using pin tool A and 1580 lbf in welds made using pin tool B.

Fatigue properties of these joints are very consistent with the lap shear strength results. At a given load fatigue life is much higher in case of welds made with pin tools A, as it can be seen from the S-N plot in Figure 10 (b). Fatigue life calculation from the S-N plot showed that at 900 lbf, the fatigue life for welds made with pin tool A is 2.3×10^5 cycles and 2.5×10^4 cycles for pin tool B. This is related to the higher failure load of the welds made with pin tool A.

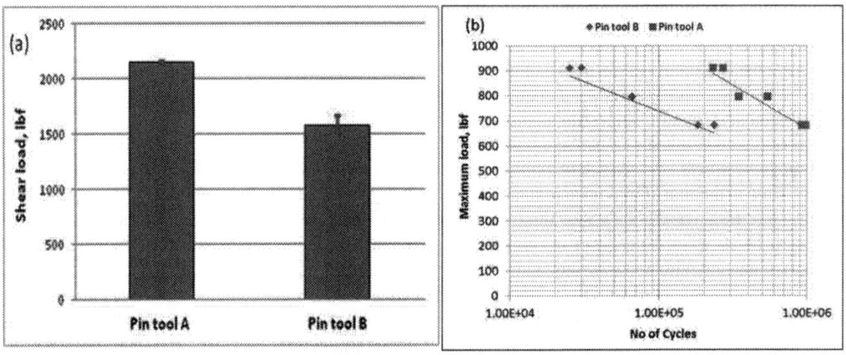

Figure 10: a) Lap shear strength and b) S-N plot for Cu/Steel friction stir lap welded specimens using different pin tools.

Fracture mode analysis was conducted by examining the fracture surfaces. The results showed that the major location of failure in welds made with pin tools A was found to be the top copper sheet fracture just outside the shoulder area. However, welds made with pin tool B showed fracture at the interface on the retreating side of weld where the "hook-like" feature was found. It should also be noted that both tensile samples and fatigue samples failed in the same location for both cases.

When welds were made with Pin tool A, significant vertical mixing of copper and steels were observed as it was plunged into the steel to a plunge depth of 0.105". Stirring of the bottom steels above the recrystallization temperature and tool rotation induced vertical motion which results in this vertical mixing. Conversely, during welding with pin tool B, vertical mixing was not as significant as it was with pin tool A. Steel was observed to be pulled up in the retreating side for both cases of welding. Stress concentration occurred in this area. This "hook-like" feature also reduced the effective sheet thickness, which carries the load during lap shear and fatigue testing. Effective sheet thickness analysis shows that the effective sheet thickness was also higher when the weld was made by pin tool A. Also, a large bonding interface was produced because of the larger diameter of the pin tool A, which also gave additional strength of the joints. It was observed that the larger bonding area, higher plunge depth, and higher mechanical mixing between materials, created by welding using pin tool A, gave joints higher strength and fatigue life compared to the weld strength and fatigue life obtained from welds made using pin tool B.

Effects of Top Sheet Position

It has been observed by several authors [10, 11, 12] that the position of top sheets has a significant impact on properties of friction stir lap joints. The effect of top sheet position depends mainly on interface microstructure characteristics. In this section, effects of top sheet position on the lap shear strength and fatigue strength of the joints will be discussed. Welds for this investigation were made using pin tool B at 3500 lbf, 400 RPM and 3 IPM. Figure 11 (a) shows the effects of top sheet position on lap shear strength of the joints. Lap shear strength increased significantly when the top sheet was placed advancing side of the joints. The lap shear strength was observed to be 1580 lbf when the top sheet was placed on the retreating side of the welds and was increased to 2011 lbf when the top sheet was placed on the advancing side of the joints. Fatigue test measurements in Figure 11 (b) shows that the properties of the joints followed the same trends. From the S-N curve it is clear that welds have higher fatigue life when the top sheet was placed on the advancing side of joints compare to when it was placed on the retreating side of the joints because of the higher load carrying capability of the joint in that configuration. At 740 lbf loads, fatigue life is approximately 1×10^5 cycles for joint made by placing the top sheet on the retreating side and at the same load fatigue life is 2.45×10^5 cycles when top sheet was placed on the advancing side.

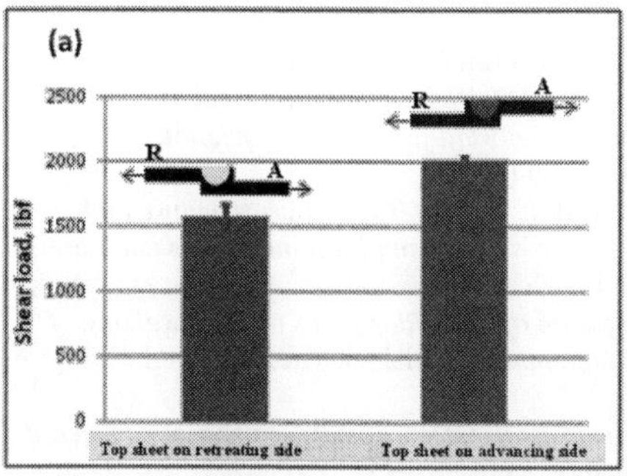

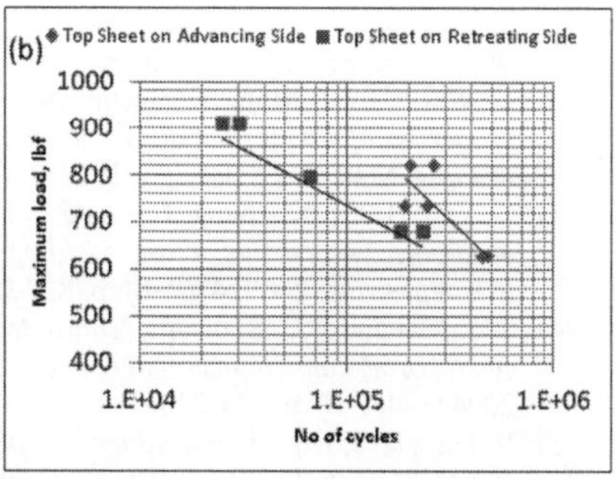

Figure 11: Effect of top sheet position on a) lap shear strength and b) fatigue properties.

The load distribution varies depending on the position of the top sheet which usually causes the variation in properties during welding with different top sheet positions. Retreating side becomes the loaded side when top sheet is placed in the retreating side and vice versa. From the interface characteristics of the welds it was observed that most of the "hook-like" features occurred in the retreating side of the joint. Few or no "hook-like" features were observed on the advancing side of the joint. As a result, when the top copper sheet was placed on the retreating side it becomes the loaded side and the presence of this 'hook-like" feature reduced its effective sheet thickness. When the copper top sheet was placed on the advancing side, this retreating side becomes the unloaded side and this "hook-like" feature will not have any significant effect on joint strength and effective sheet thickness will increase. Thus, a significant increase in lap shear strength and fatigue life was observed for this configuration. Fracture occurred at the interface of the retreating side when the top sheet was placed on the retreating side and at the interface of the advancing side when top sheet was placed on advancing side.

Conclusions

W-Re-HfC pin tools were applied for lap welding of soft/hard metal combinations: copper/steel, so that the pin can be penetrated into the bottom hard sheets without any significant degradation of the pin tools. . Both the pin tools produced successful welds with or without small microvoid defects. Weld joint properties were very consistent with the microstructural evolution on either side of the joints and the interface morphology. Based on interfacial morphologies and mechanical properties, the following conclusions can be drawn:

a) Bottom steel material was pulled up into the top soft sheets on the retreating side of the joints for each welding combinations and for both pin tools.
b) Positioning of the top sheets on the advancing side produced better joint strength for both weld combinations and pin tools, since the "hook-like" feature on the retreating side is on the unloaded side of the joint. . This is opposite to the weld design that conventionally gives better joint strength in FSLW.
c) Pin tool A produced better joints than pin tool B, since it provided better bonding area and deeper plunge depth, which resulted in better mechanical and metallurgical bonding for copper/steel and metallurgical bonding for aluminum/steel.
d) Maximum joint efficiency of 88% was achieved for copper/steel lap joints for welds made using pin tool A. This joint efficiency falls around the shear strength of butt joints between friction stir welded copper.

References

1) V. Firouzdor and S. Kou, *Al-to-Mg Friction Stir Welding: Effect of Positions of Al and Mg with Respect to the Welding Tool,* Welding Journal, 2009. 88(11).
2) Y. Wei, J. Li, J. Xiong, F. Huang and F. Zhang, *Microstructures and mechanical properties of magnesium alloy and stainless steel weld-joint made by friction stir lap welding,* Materials & Design, 2012. 33(0): p. 111-114.
3) C. Yao, B. Xu, X. Zhang, J. Huang, J. Fu and Y. Wu, *Interface microstructure and mechanical properties of laser welding copper–steel dissimilar joint,* Optics and Lasers in Engineering, 2009. 47(7–8): p. 807-814.
4) T.A. Mai and A.C. Spowage, *Characterisation of dissimilar joints in laser welding of steel–kovar, copper–steel and copper–aluminium,* Materials Science and Engineering: A, 2004. 374(1–2): p. 224-233.
5) V. Ghizdavu, *Explosive welding of copper to steel, International Conference of Scientific Paper,* May 26-28, 2011. (Brasov, Germany).
6) Y. Imani, M.K. Givi and M. Guillot, *Improving Friction Stir Welding between Copper and 304L Stainless Steel,* Advanced Materials Research, 2012. 409: p. 263-268.
7) Y. Wei, J. Li, J. Xiong, F. Huang and F. Zhang, *Microstructures and mechanical properties of magnesium alloy and stainless steel weld-joint made by friction stir lap welding,* Materials & design, 2011.
8) W.B. Lee and S.B. Jung, *The joint properties of copper by friction stir welding,* Materials Letters, 2004. 58(6): p. 1041-1046.
9) P. Xue, B.L. Xiao, Q. Zhang and Z.Y. Ma, *Achieving friction stir welded pure copper joints with nearly equal strength to the parent metal via additional rapid cooling.* Scripta Materialia, 2011.
10) H. Badarinarayan, Y. Shi, X. Li and K. Okamoto, *Effect of tool geometry on hook formation and static strength of friction stir spot welded aluminum 5754-O sheets.* International Journal of Machine Tools and Manufacture, 2009. 49(11): p. 814-823.
11) M.K. Yadava, R.S. Mishra, Y.L. Chen, B. Carlson and G.J. Grant, *Study of friction stir joining of thin aluminium sheets in lap joint configuration,* Science and Technology of Welding & Joining, 2010. 15(1): p. 70-75.
12) L. Cederqvist and A. Reynolds, *Factors affecting the properties of friction stir welded aluminum lap joints,* Welding Journal, 2001. 80(12): p. 281.

FRICTION STIR WELDING AND PROCESSING VII

Friction Stir Welding: Light Materials – I

Friction Stir Welding and Processing VII
Edited by: Rajiv Mishra, Murray W. Mahoney, Yutaka Sato, Yuri Hovanski, and Ravi Verma
TMS (The Minerals, Metals & Materials Society), 2013

EFFECT OF TOOL PIN FEATURES AND GEOMETRIES ON QUALITY OF WELD DURING FRICTION STIR WELDING

Md. Reza-E-Rabby, Wei Tang, Anthony Reynolds

Department of Mechanical Engineering, University of South Carolina, 300 Main Street, Columbia, SC 29208, USA

Keywords: Friction Stir Welding, tool geometries, pin features, flats, flutes

Abstract

Tool pin geometric features are important parameters in friction stir welding (FSW). They influence weld quality through their effects on material flow and consolidation during FSW while also affecting the in-plane process forces and the necessary level of forge force. In this paper, volumetric defect content as a function of pin features on mildly tapered conical, threaded, pins was examined in 6061-T6. Features included flats, co-flow flutes and counter-flow flutes. Feature effects on the magnitude and direction of the resultant in-plane force was also examined. Variation of applied forge force, other process response variables (torque, power, temperature) were also extracted to provide information regarding their dependence on tool pin features for a given set of welding parameters.

Introduction

The friction stir welding tool is a crucial part that is essentially responsible for frictional heating and plastic dissipation in weld material. The flow of weld material around the tool is highly influenced by the geometric shape and features of the tool pin during FSW which eventually affects the mechanical and microstructural properties to obtain sound welded material [1-4]. The local plastic deformation and mixing of the work piece along the weld seam is primarily due to the pin. Therefore, to achieve an efficient friction stir process it is important to design the tool pin with proper geometric shape and features. A good process results in defect free joints with superior mechanical and microstructural properties as well as lessens in-plane forces on the tool. Furthermore, minimization of the in-plane reaction force is an important consideration for tool life and production of long welds. Within a process window (a set of rpm and weld traverse speed), the design of tool geometries and features may vary depending upon the weld material flow stress behavior (depend upon the alloy type and alloy properties) and dimensions. Hence the optimum tool design may be very specific for a given situation and not readily obtained by perusing the available literature.

The effects of tool geometries on welding force and mechanical properties were investigated by Hattingh et al [1] on AA5083 to obtain an optimum friction stir welding tool. Variation in number of flutes, flute depth and angles, pin diameter and taper angle, thread pitch were reported with the reaction forces on the pin and comparing the tensile strength, a tapered pin with three taper flutes and pitch around 10% of pin diameter was suggested to be a successful tool design. Elangovan et al. [2] and D'Urso et al. [3], at about the same time studied separately the arbitrarily chosen pin geometric shape to study the performance of the geometric shape in terms of tensile strength of FSW. A square cross section pin which is also considered as cylindrical pin with 4 flats was reported to produce quality welds in AA6061 by Elangovan et al [2]. On the other hand D'Urso et al. [3] observed the effectiveness of cylindrical threaded pin to perform better during FSW. Lammlein et al. [5] used a shoulderless, conical, probe and reported a dramatic reduction of process forces, particularly forge force.

Material transportation around the tool also plays a significant role and obviously the flow of material depends on the geometric features of the tool. To monitor the flow behavior studies have been conducted using marker insertion in similar metal welds [6, 7]. Researchers also investigated material flow in dissimilar material FSW in butt weld arrangement for different tool geometries. Shoulder geometries were investigated by Leal et al. [8] to study the material flow during bi-material FSW of AA 5182 and AA 6016 with AA5182 on the advancing side. Comparing the conical cavity shoulder and scrolled shoulder, they reported that intense flow of material was observed around the scrolled shoulder tool. However, pin driven flow was reported to be predominant in the case of the conical cavity shoulder. Aval and co-workers [9] studied the tool geometric effect on the mechanical and microstructural properties for dissimilar friction stir butt welding of AA5086-O and AA6061-T6 with 5086 on advancing side. They reported that a tool with concave conical shoulder and tapered, threadless pin with three grooves provided more homogenous mixture at higher heat input compared to a tool with cylindrical smooth or threaded pin with grooves. Silva et al. [10] studied the material flow in the friction stir butt weld of dissimilar alloy AA2024-T3 and AA7075-T6 with 7075 on the advancing side using a single threaded pin and reported a non-stable rotational flow around the thread with a formation of a cavity behind the pin because of transportation of bulk material underneath the shoulder. It is noted that alloy placement on advancing and retreating side in dissimilar friction stir butt welding has a very important role in the flow behavior and production of sound welds.

In the present paper the effect of various tool pin features (flats, flutes) were investigated with constant shoulder and pin dimensions to observe macrostructures (detection of defects), variation of applied forge forces, response variables (in-plane reaction forces, torque, temperature) for a set of process parameters.

Experimental Procedure

Bead on plate welds were performed on 25.4 mm thick aluminum alloy AA6061-T651. All the welds were performed at partial penetration of tool pin depth (12.7 mm). The reason of selecting this depth is that the process window is much larger at this depth and also the area of influence underneath the pin during friction stir welding is initially unknown.

The tool used in this current study consists of two parts made from H13 tool steel with (a) 25.4 mm diameter single scrolled shoulder and (b) 8° taper, conical, right hand threaded (2.12 mm pitch) pin with shank diameter of 15.88 mm. The tool pins were featured with thread only, thread + 3 flats, thread + 3 co-flow flutes and thread + 3 counter-flow flutes. The depth and width of cut for flats and flutes were 1.35 mm and 6.35 mm respectively. The pitch for both co-flow and counter-flow flutes was 50.8 mm. The tool materials were heat treated at 980° C for 20 minutes followed by oil quench.

All welds were performed on an MTS friction stir welding Process Development System (PDS) in Z axis force control. The forge force in each weld was adjusted to maintain similar plunge depth for different tool pin features and variable process control parameters (rpm and traverse speed). For bead on plate welds in AA6061, the following sets of process parameters (rotational speed/feed rate) were employed:

Table 1: Process response parameters

Rotation rate (rpm)	160	200	240	240	320	400
Feed Rate (mm/min)	101.6	101.6	101.6	203.2	203.2	203.2

Metallographic specimens from the welds were cut using a water jet machine and subsequently ground, polished, and etched with Keller's reagent (5 ml HNO3, 3 ml HCl, 2 ml HF and 190 ml distilled water) to observe the macro and micro structures. The process response variables (in-plane reaction forces) were also recorded by the MTS FSW PDS. The temperature during FSW was recorded using K-type thermo-couple spot welded into the pin at mid depth on the axis of rotation.

Results and Discussions

Examination of AA6061

Tool feature effects on the macro and microstructural features in the nugget zone within a process control window are examined as discussed in the experimental procedure. Figure 1 shows the weld cross sectional macrostructures for different pin features at different rotational and traverse speed where the advancing side is on the left in each macrograph. Except for two welds made using thread +3 co-flow flutes (240rpm & 101.6mm/min) and thread+3 counter-flow flutes (240rpm & 203.2 mm/min) all the welds were defect free (see figure 1). For the two defective welds, the one made with thread+3 co-flow flutes had wormholes near the crown while the other made with thread+3 counter-flow flow flutes had a wormhole near root (see figure 1 & 2).

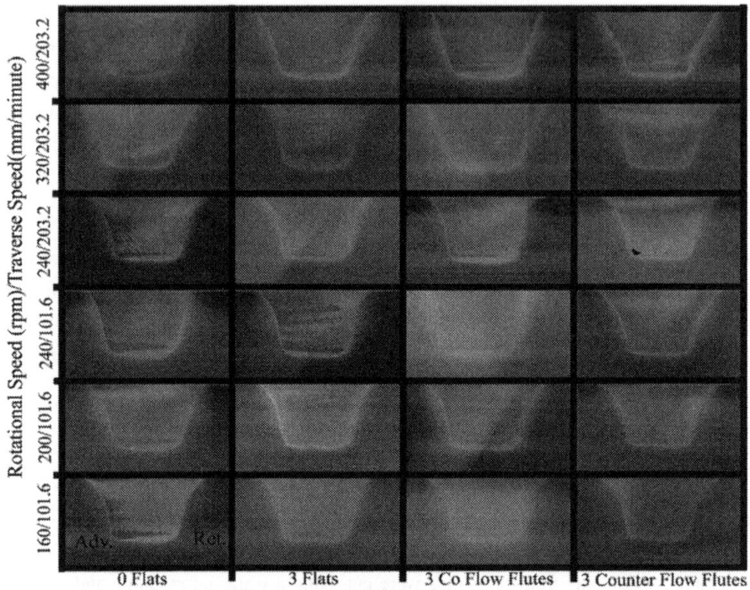

Figure 1: Macrostructure for 8° Taper threaded Pin with different pin features at different rotational and traverse speed on AA6061

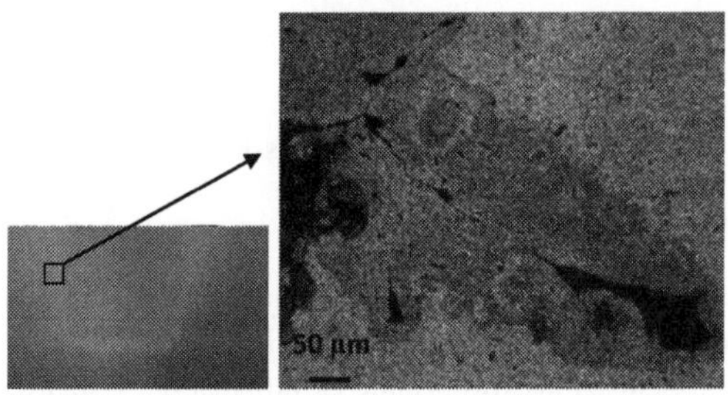

Figure2: Microscopic defects in the nugget zone near crown for weld with 3 Co-flow flutes at 240rpm & 101.6mm/min

<u>Process response variables for weld on AA6061</u>

The average in-plane reaction forces on the pin for different welds with all pin features are plotted in a polar coordinate system (Figure 3 (a)). The sign convention of these reaction forces is shown in Figure 3 (b) where the positive X forces act in the opposite direction of tool traverse and the positive Y force act perpendicular to the X force towards the advancing side.

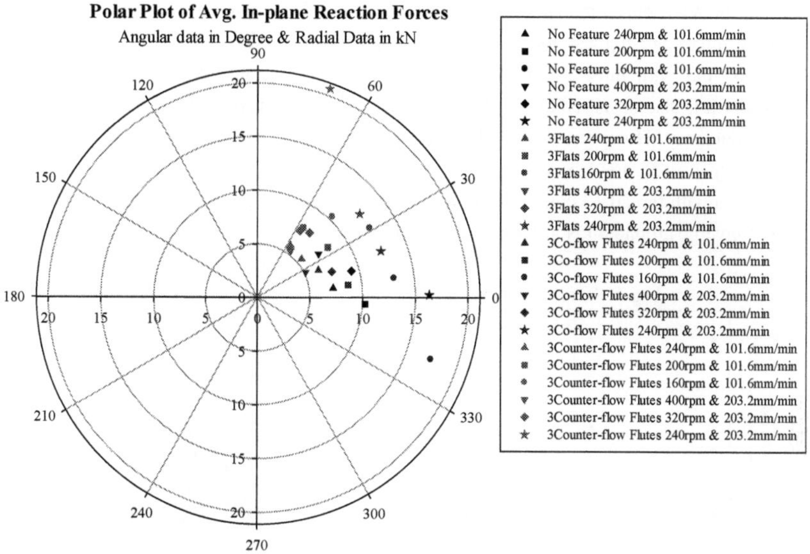

(a)

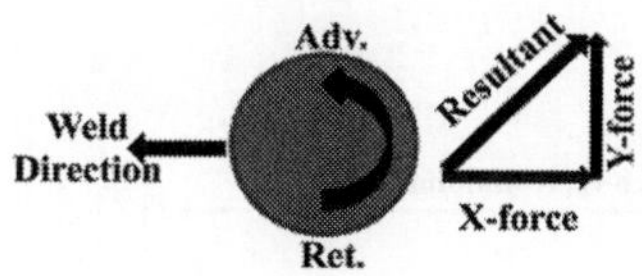

(b)

Figure 3: (a) Average In-plane reaction forces in Polar Plot for Pin features with different process control parameters & (b) Sign convention for reaction forces on pin

Figure 3 (a) is the vector sum of the average X and Y forces, however these forces are not necessarily in phase, therefore at any instant, the resultant force on the tool is not necessarily equal to the reported resultant force. In this study the average X and Y forces are measured over a weld distance of 20 mm. This position was selected at approximately where steady state is obtained during friction stir welding and near the location from where the specimens were cut for the microscopic investigation. Figure 3 (a) illustrates how the orientation of resultant reaction changes with changing pin features. It is noted in most cases that, for a set of rotational and traverse speed, the arc-tangent (Y-force/X-force) follows the incremental sequence in ascending order: pin with thread+ 3 counter-flow flutes, pin with thread+ 3 flats, pin with thread+3 co-flow flutes and threaded only pin. This phenomenon is also explained in figure 4 (a) & (b), where average X forces are highest for threaded only pin and then pin with thread+ co-flow flute followed by pin with thread+3 flats and pin with thread+3 counter-flow flutes Figure 4(a). Additionally it is also observed that with the increase of rotational speed the X force decreases as a general tendency. However, the trend for average Y force is opposite to that observed in X forces as shown in figure 4(b) where, at lower rotation rate of 160 to 320 rpm, the average Y forces for pin with thread +3 counter-flow flutes are highest. However, Y forces for different pin features converge around 4 kN at rotation rate of 400 rpm.

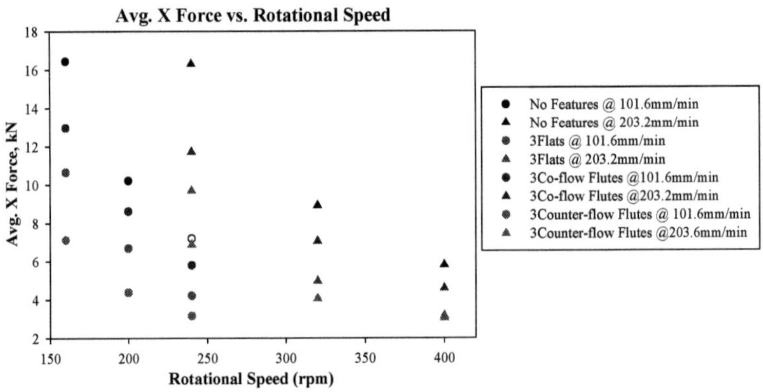

(a)

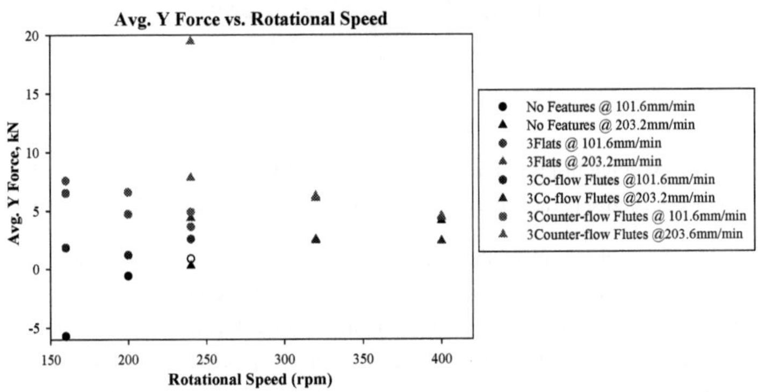

(b)

Figure 4: (a) Average X forces & (b) Average Y Forces- as a function of rotational speed for different pin features

It is interesting to note here that thread interruptions lead to decreases in the X force on the pin. As illustrated in Figure 4 (a), the threaded only pin encounters the highest X force while the pin with thread+3 flats and pin with thread+3 counter-flow flutes exhibit lower X force. Hence it can be noted here that thread interruption features have significant effect on the observed reaction forces on the tool pin.

Figure 5 shows the trend of torque as a function of forge forces for different tool pin features within process window. It is noted here that for a given set of rotational and traverse speed, measured torque does not vary significantly (less than 10%). This is likely due to dominance of shoulder effects in torque. However, the applied forge forces along the Z axis direction vary a lot due to the effect of pin features (see figure 5). During friction stir welding, right hand threaded pins rotate in the counter clockwise direction and therefore, pins which push material down require a definite additional forge force to maintain similar depth of penetration. Because of this mechanism the threaded only pin requires the highest Forge force and the pin with co-flow flutes required higher forge forces than those for pins with flats and counter-flow flutes. It is noted here that thread interruption features have significant effect on controlling the required forge force during friction stir welding.

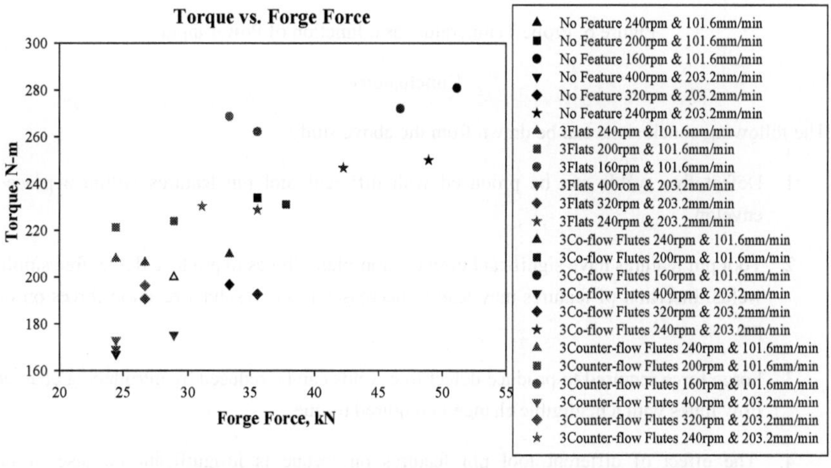

Figure 5: Measured Torque for different tool features vs. the applied force

Temperature as a function of power input is presented in figure 6 for different rotational and traverse speed with various pin features. Temperature was found to increase with increasing power input which is a commonly observed phenomenon. It is also noted here that for a constant feed rate, the threaded only pin requires higher power input resulting in higher pin temperature than other features. Interestingly for two cases (240 rpm & 400 rpm) the temperature for the threaded only pin was much higher with lower power input.

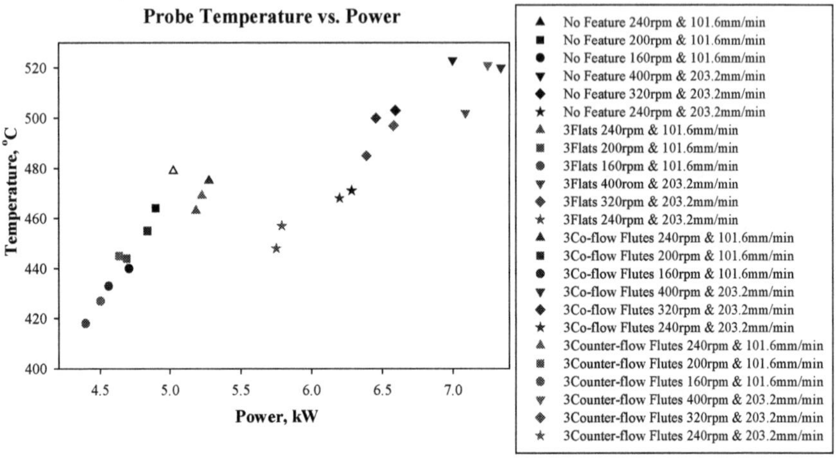

Figure 6: Probe Temperature as a function of Power input

Conclusions

The following conclusions can be drawn from the above study:

1. Defect free welds can be produced with different tool pin features within a process envelop.
2. Tool pin features have significant effects on in-plane forces to produce defect free similar welds. Insertion of features may lead to decreases in the resultant reaction forces on the pin.
3. Forge force required to produce defect free welds can be reduced by introducing counter-flow flutes with a negligible change in required torque.
4. The effect of different tool pin features on torque is insignificant because of the dominant effect of the shoulder.

Acknowledgements

This work was supported by the Center for Friction Stir Processing, a NSF-I/UCRC, grant number EEC-0437341.

References

1. D.G. Hattingh, C. Blignault, T.I. van Niekerk, M.N. James, "Characterization of the influences of FSW tool geometry on welding forces and weld tensile strength using an instrumented tool", *Journal of Materials Processing Technology* 203, 2008, no. 1: 46-57.

2. K. Elangovan, V. Balasubramanian, and M. Valliappan, "Influences of tool pin profile and axial force on the formation of friction stir processing zone in AA6061 aluminum alloy," *The International Journal of Advanced Manufacturing Technology*, 38 (2008), 285-295.

3. G. D'Urso, E. Ceretti, C. Giardini, G. Maccarini, "The effect of process parameters and tool geometry on mechanical properties of friction stir welded aluminum butt joint", *The International Journal of Material Forming*, Vol. 2 Suppl. 1 (2009), 303-306.

4. Yan-hua Zhao, San-bao Lin, Lin Wu, Fu-xing Qu, "The influence of pin geometry on bonding and mechanical properties in friction stir weld 2014 Al alloy", *Materials Letters*, 59 (2005), 2948-2952.

5. D.H. Lammlein, D.R. DeLapp, P.A. Fleming, A.M. Strauss, G.E. Cook, "The application of shoulderless conical tool in friction stir welding: An experimental and theoretical study", *Materials and Design*, 30 (2009), 4012-4022.

6. A.P. Reynolds, "Flow visualization and simulation in FSW," *Scripta Materialia*, 58 (2008), no. 5: 338-342.

7. C.R. Tolle, T.A. White, K.S. Miller, D.E. Clark, H.B. Smartt, "Investigation of material flows within FSWs using 3D tomography", *Trends in Welding Research: Proceedings of the 8th International Conference*, June 2008, 126-132.

8. R.M. Leal, C. Leitão, A. Loureiro, D.M. Rodrigues, P. Vilaça, "Material flow in heterogeneous friction stir welding of thin aluminum sheets: Effect of shoulder geometry", *Material Science and Engineering A*, 498 (2008), 384-391.

9. H. Jamshidi Aval, S. Serajzahed, A.H. Kokabi, A. Loureiro, "Effect of tool geometry on mechanical and microstructural behaviors in dissimilar friction stir welding of AA5086-AA6061", *Science and Technology of Welding and Joining*, Vol 16 No. 7 (2011), 597-604.

10. A.A.M. da Silva, E. Arruti, G. Janeiro, E. Aldanondo, P. Alvarez, A. Echeverria, "Material flow and mechanical behavior of dissimilar AA2024-T3 and AA7075-T6 aluminum alloys friction stir welds", *Materials and Design*, 32 (2011), 2021-2027.

Friction Stir Welding and Processing VII
Edited by: Rajiv Mishra, Murray W. Mahoney, Yutaka Sato, Yuri Hovanski, and Ravi Verma
TMS (The Minerals, Metals & Materials Society), 2013

Effect of Friction Stir Processing on Armor Grade Materials

Timothy Johnson[1,2], Todd Curtis[1,2], Bharat Jasthi[1], Eric East[1,2], Christian Widener[1], Michael West[1]

[1]Arbegast Materials Processing Laboratory
South Dakota School of Mines and Technology
501 E. St. Joseph St Rapid City SD, 57701
[2]Zone Four Engineering, LLC
2241 S. Plaza Dr. Unit 7 Rapid City SD, 57702

Abstract

The objective of this paper is to investigate the effect of friction stir processing (FSP) on the ballistic performance of armor grade aluminum and steel plates. FSP is a solid state microstructural modification process that was adapted from the concepts of friction stir welding (FSW). FSP has shown to locally refine the microstructures and eliminate many casting defects for improved mechanical and corrosion properties. Armor grade aluminum (6061 and 7039) and high-strength low-alloy (HSLA- MIL-A-12560) steel were prepared in various thicknesses. The prepared plates were then FSP in partial penetration overlapping passes to provide for the largest area of processed material possible. The plates were then tested in accordance with MIL-STD-662E to determine the V_{50} ballistic limit of the armor. Results from the ballistic testing of processed panels were analyzed and compared with the parent material. Microstructural differences observed between the processed and unprocessed panels were characterized and correlated with the ballistic performance.

Keywords: Friction Stir Processing, Armor, Ballistic Limit

1. Introduction

Friction stir processing (FSP) is an innovative microstructural modification process developed initially for achieving superplasticity in commercial Al alloys The intense plastic deformation in this process creates a fully recrystallized, equiaxed, and fine grained microstructure in the weld nugget. This novel solid state joining process has also been used to join or process mild steels [1-3], titanium alloys [4, 5], oxide dispersion strengthened (ODS) alloys [6], metal matrix composites [7], dissimilar alloys [8], and nickel alloys [9-12].

The main parameters that are used to control the process are forge force, spindle rotation speed, and travel speed. The forge force is the downward pressure that creates a forging action that forces the material to consolidate behind the pin tool. By varying the travel speed, forge force, and spindle speed the overall energy input can be controlled.

There are a great variety of pin tool designs to choose from. The main components of a pin tool are the pin and shoulder. The length of the pin can be set to match the thickness of the material or the desired depth of processing. The type of material being processed and desired effect in the weld determines the size and shape of the pin and shoulder.

FSP significantly reduces the grain size of the material in the weld zone, thereby increasing its toughness and impact properties. In a recent study by the authors, FSP was shown to increase the ballistic performance in thin (2mm) aluminum alloys. The FSP material showed signs of adiabatic shear bands after ballistic impact. It was observed that the material softened during impact allowing more energy to be absorbed. The parent material was observed to strain harden during impact which led to a more brittle failure. The parent material had significant deformation and cracking after impact of a .45 caliber projectile at approximately 195 m/s. The FSP material showed significant plastic deformation but only a small crack was observed [13].

This paper is a continuation of that work to show the effect of FSP on thicker armor grade materials. For this paper 25.4 mm thick 7039 aluminum, 12.7 mm thick 6061 T6, and 12.7 mm thick HSLA steel (MIL-A-12560) were friction stir processed and tested for ballistic resistance.

2. Friction Stir Processing

2.1 Aluminum Plates

The 6061 and 7039 aluminum plates were friction stir processed using an H13 tool steel pin tool. Parameters were developed with linear processing to optimize travel speed, processing overlap, and forces. A spiral pattern was then developed in order to make a continuous processing path on the plates. This was done to minimize processing time, exit holes, and to push the heat input toward the outer edges of the plates. The pin tool was moved over 7.62 mm for each time around. The tool processed 6.35 mm deep in the material. For the plates that were processed on both sides this results in a layer of parent material in the center of the plate 12.7 mm thick. Figure 1 shows the tool and processing parameters used to process the aluminum plates.

Control Mode	Forge and Position
Rotational Speed, RPM	600
Travel Speed, IPM	1 2.0
Tool Tilt, Degree	0
Plunge Depth, Inches	0.250

Figure 1: Aluminum tool with process parameters.

2.2 HSLA Steel Plates

The steel plates were processed using a W-25%Re-4%HfC pin tool. Linear processing was done in the steel, processing 4.32 mm deep in the material. The processing paths were overlapped by 12.7 mm. Figure 2 shows a micrograph of a HSLA steel plate with FSP on both sides. Figure 3 shows a high temperature tool and general process parameters. The HSLA parent material had an average hardness of 35 HRC. After FSP the hardness increased to an average of 46 HRC. The minimum hardness for MIL-A-12560H armor is 35 HRC [14].

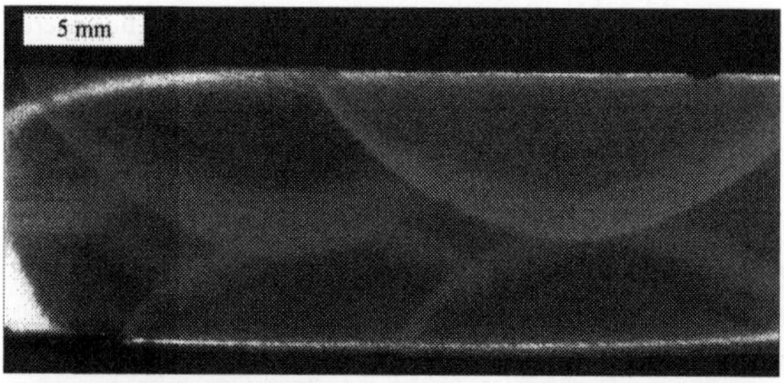

Figure 2: Micrograph of HSLA steel plate with FSP on both sides.

Control Mode	Forge and Position
Rotational Speed, RPM	200 - 400
Travel Speed, IPM	1 - 2
Tool Tilt, Degree	0 - 1
Plunge Depth, Inches	0.13 – 0.15

Figure 3: High temperature tool with parameters.

During FSP a small amount of flash was developed on the surface of the plates. This is the case with both aluminum and steel plates. This flash was removed prior to ballistic testing to ensure a uniform plate thickness. FSP produced a grain size on the order of 5 μm in the processed plates, which was approximately an order of magnitude reduction in grain size for these alloys. This reduction of grain size contributes to both strength and toughness in these materials; however, the treatment also affects the distribution and size of precipitates which can also have a strong effect on material properties.

3. V_{50} Ballistic Testing

3.1 Procedure

All V_{50} testing was done in accordance with MIL-STD-622F [15]. The armor plates were clamped to a test frame with a 0.51 mm thick 0.61 m x 0.61m 2024 T3 witness plate 152 mm behind the test plates. A chronograph was placed 122 cm from the plate to measure the projectile velocity. The plates were then tested using a 7.62 mm AP M2 projectile, weighing 165 grains. The projectile velocity was adjusted by varying the propellant charge until velocities for complete and partial penetrations were found for each material type and thickness.

3.2 Testing

Each material type and thickness was tested as parent material, FSP on impact face, FSP on back face, and FSP both faces. This was done in order to test the effectiveness of FSP. The

12.7 mm 6061 aluminum was tested with a 30° obliquity angle, this increases the effective thickness of the material. The 12.7 mm and 25.4 mm HSLA and 7039 aluminum were tested with a 0° obliquity.

4. Results

4.1 Aluminum Plates

The 7039 aluminum plates were tested one day after the plates were friction stir processed. The plates were in a very soft condition, near a W-temper, and had a V_{50} of 561 m/s, which was lower than expected. The required V_{50} for 25.4 mm thick plate is 624 m/s[16]. The parent material was tested at 678 m/s. The remaining 7039 aluminum plates were then artificially aged to a stable condition before testing. One plate was artificially aged at 107°C for 76 hours[17]. The V_{50} for this plate was 625 m/s. Another plate was naturally aged for 7 days at room temperature. This plate had the same V_{50} of 625 m/s. While this represented a substantial improvement, it still did not meet the minimum requirements for the V_{50} and requires further investigation in order to achieve substantial improvements over parent material. Positive changes in the failure mode were observed, however, which will be discussed later. Figure 4 shows the V_{50} values for various material conditions.

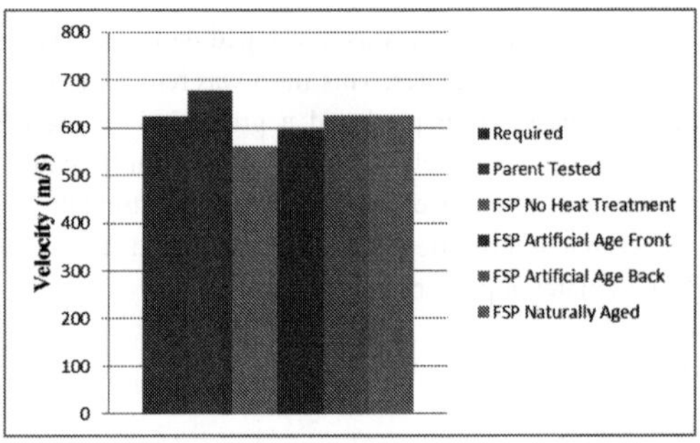

Figure 4: 7039 Aluminum V_{50} values for various material conditions.

The friction stir processed 6061 aluminum plates were aged to the T6 condition after processing by holding the plates at 160°C for 18 hours[17]. This was done before any 6061 plates were tested because of the 7039 results. The required V_{50} for 12.7 mm is 412 m/s[18]. The V_{50} for the FSP material was 418 m/s. This also meets the required velocity but did not demonstrate a substantial improvement in penetration velocity. However, again, a positive change was observed in the failure mode which is described in the next section. Figure 5 shows the V_{50} values for 6061 aluminum. The FSP material in both cases showed very different characteristics around the impact site verses the parent material. Figure 6 shows a typical impact site for the parent material (a) and the FSP material (b). In the parent material, the copper jacket, lead core, and surface of the aluminum plate are fragmented from the plate during impact as the hardened steel penetrator passed through or was stopped. This is consistent with parent material impacts. In the FSP material the copper jacket and lead core were stripped off the penetrator as it passed through or was stopped. The copper jacket, lead core, and surrounding aluminum did not fragment. These materials were retained in the plate, leading to a much smaller entrance hole as compared to parent material. This is due to the increased ductility of the FSP material and represents a significant improvement in the failure mode because the shrapnel from an impact on armor could also pose a threat to anyone standing nearby.

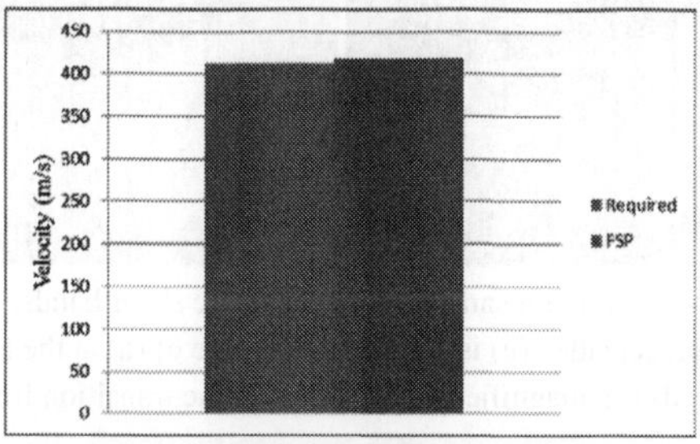

Figure 5: 6061 aluminum V_{50} values.

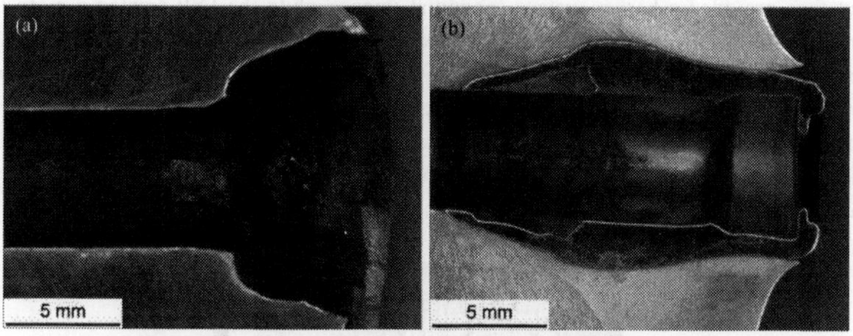

Figure 6: Impact site of (a) parent material, (b) FSP material.

Figure 7 show the adiabatic shear bands that are formed during high strain rate and high plastic deformation events[19]. Figure 7 (a) shows that there is an absence of adiabatic shear bands in the FSP material and as the projectile advances through the FSP material to the unprocessed material shear bands develop. This is opposite when the FSP material is on the back side of the plate. Figure 7 (b) show the shear bands stop as the projectile enters the FSP material. Figure 7(c) and Figure 7(d) are magnified images of (a) and (b) at the transition from FSP to unprocessed material.

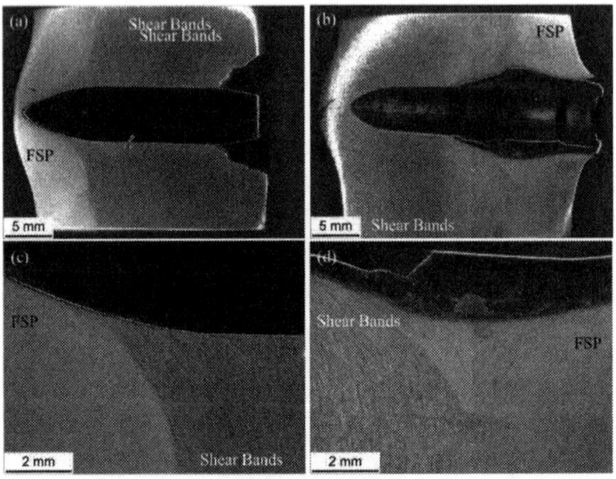

Figure 7: Macrograph of Aluminum samples with adiabatic shear bands, (a) FSP opposite side of impact, (b) FSP on impact side. (c) is a magnified image of (a) at the transition from FSP to unprocessed material. (d) is a magnified image of (b) at the transition from FSP to unprocessed.

FSP on the opposite side of impact showed the best results. The parent material on the impact face provided a harder surface for initial impact while the FSP created a more ductile material on the back side, resulting in more extensive plastic deformation. This is due to the increased ductility and softening of the FSP material during impact. Figure 8 shows the difference between the exit side of the parent material and the FSP material. The parent material has cracked as the penetrator breaks through the back side, while the FSP material continues to plastically deform. This is also because the increased ductility and toughness in the FSP material due to smaller grain size of the FSP material.

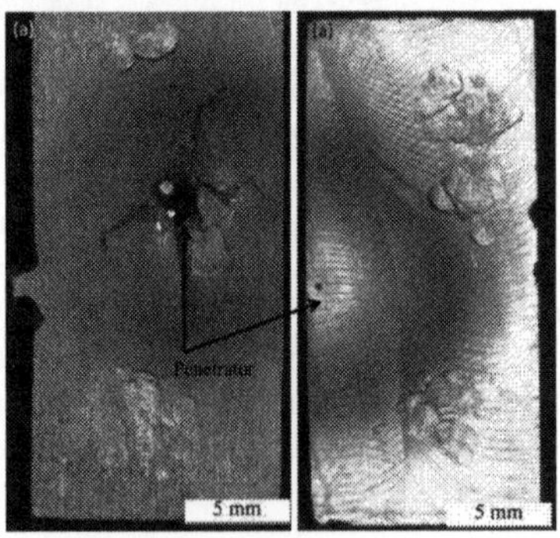

Figure 8: Exit side of (a) parent material, (b) FSP material.

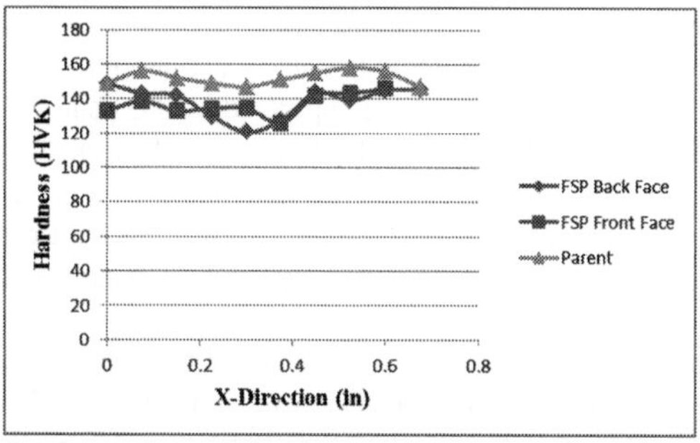

Figure 9: Microhardness of tested aluminum samples along the void left by penetrator.

Figure 9 shows the variation in microhardness for the tested samples. The values were measured along the void left by the penetrator. The FSP material showed softening due to over aging from excessive heat generated during the 45 minute processing time. Conductivity measurements showed that the plates were overaged. The strengthening phases were solutionized during FSP due to the extensive time of processing the plate remained at an elevated temperature for an extended period of time.

4.3 HSLA Steel

HSLA steel has a required V_{50} of 744 m/s[14]. The reported V_{50} value for the material tested was 755 m/s. The tested parent material had a V_{50} of 797 m/s. The FSP material with impacts on the opposite side of the plate had a V_{50} of 811 m/s. The full penetration processing plate had a V_{50} of 762 m/s. The plate with the FSP on the impact side had a V_{50} of 827 m/s, with a peak partial penetration of 844 m/s. This is an increase of 30 m/s over the tested parent material and an increase of 83 m/s over the required $V_{50.}$ The V_{50} of 827 m/s meets the requirement for a 15.24 mm thick plate of parent HSLA steel, which is 2.54 mm thicker. Figure 10 shows the V_{50} values for various material conditions.

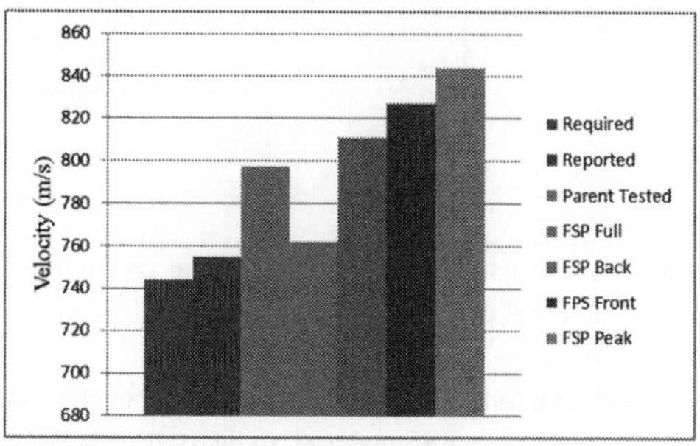

Figure 10: HSLA steel V_{50} values for various material conditions.

Adiabatic shear bands are also seen in the HSLA steel samples. Figure 11 shows that in HSLA steel shear bands are formed in the FSP material and not the parent material. The best results were seen with the FSP material on the impact side of the plate. The plate with FSP on the opposite side of the plate showed that the FSP material cracked as the softer material deformed into it. Smaller grain size allows for a tougher material at higher hardness. With the FSP material on the impact side the penetrator interacted with the harder, tougher material first then the softer parent material was able to deform in a ductile manner and absorb energy from the impact.

Figure 12 shows the microhardness values measured along the void left by the penetrator. The tested parent material showed strain hardening caused by impact. The FSP material started at a higher hardness value and showed the characteristic peak and valley associated with adiabatic shear bands[19]. After the transition to unprocessed material the microhardness shows the same trend as the parent material.

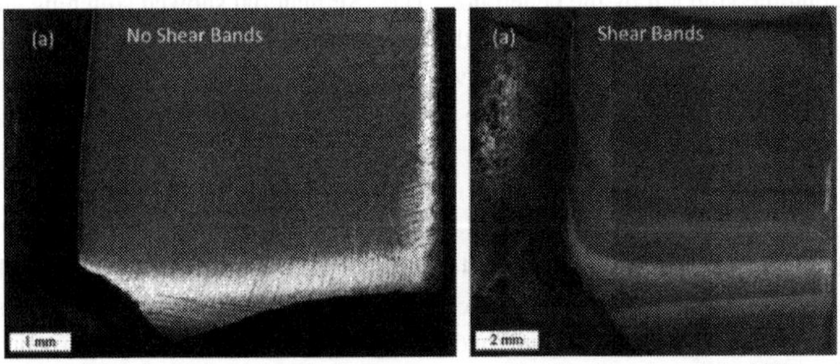

Figure 11: (a) Lack of shear bands in parent material, (b) shear bands in FSP material.

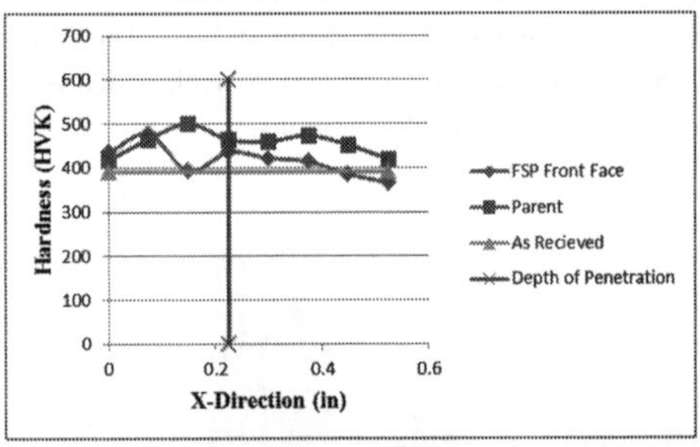

Figure 12: Microhardness of the tested HSLA steel along the void left by the penetrator. The vertical line is the depth of penetration for the FSP material.

Figure 13(a) shows the microhardness of the FSP partial penetration HSLA steel. The material to the left of the vertical line is FSP material, to the right is unprocessed material. The partial penetration showed a harder surface and softer unprocessed material than the parent sample. This is due to the increased hardening from FSP and the softening of the unprocessed material because of the heat input. Figure 13(b) shows the microhardness of the full penetration material. The full penetration processing showed an overall softer condition than the parent material due to the heat generated during processing on both side of the plate. The material to the left of the vertical is FSP that was processed first. The plate was then turned over and processed on the other side, this is the material to the right of the vertical line.

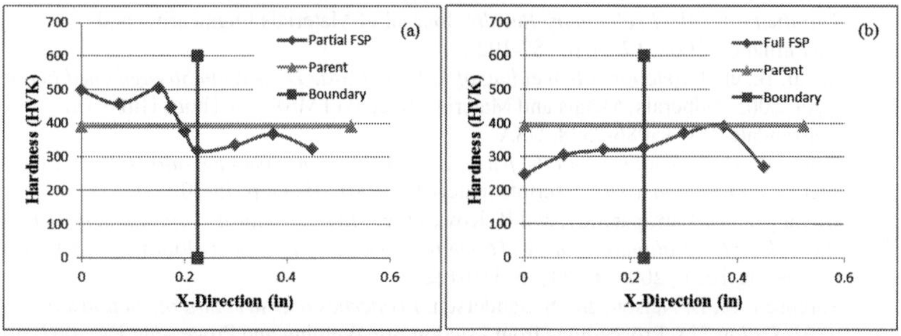

Figure 13: Microhardness of (a) partial and (b) full penetration processing and parent material. The vertical lines are boundaries between material types.

6. Conclusions

Friction stir processing has been shown to increase the ballistic performance of HSLA steel and change the interaction of the aluminum alloys with the penetrator. Adiabatic shear bands formed in the aluminum and HSLA FSP material. FSP HSLA steel showed an increase in V_{50} ballistic limit of 83 m/s over the required minimum and 30 m/s over the parent material. Thick plate aluminum alloys showed softening and a lower V_{50} parent material, so additional development is needed for thick plate FSP aluminum. However, there was a marked improvement in the failure mode as the entrance sites of the FSP aluminum material contained the fragments of the copper jacket and lead core, rather than ejecting them as shrapnel from the front of the plate in the parent material. It has been observed that to get the greatest improvement in V_{50} values for HSLA steel the FSP material should be on the impact face because of increase hardness of the material, whereas for aluminum alloys the FSP material should be on the face opposite of impact. This allows the FSP material to absorb energy during plastic deformation during impact. Overall, friction stir processing is a viable option for the improvement of ballistic performance in armor materials and additional development is warranted.

References

1. Thomas, W., P. Threadgill, and E. Nicholas, *Feasibility of friction stir welding steel.* Science and Technology of Welding &# 38; Joining, 1999. **4**(6): p. 365-372.

2. Lienert, T., et al., *Friction stir welding studies on mild steel.* WELDING JOURNAL-NEW YORK-, 2003. **82**(1): p. 1.
3. Tweedy, B., W. Arbegast, and C. Allen. *Friction Stir Welding of Ferrous Alloys using Induction Preheating.* 2005: Minerals, Metals and Materials Society(TMS), 184 Thorn Hill Road, Warrendale, PA, 15086-7528, USA.
4. Lee, W., et al., *Microstructural investigation of friction stir welded pure titanium.* Materials Letters, 2005. **59**(26): p. 3315-3318.
5. Sanders, D., et al., *Characterization of Superplastically Formed Friction Stir Weld in Titanium 6AL-4V: Preliminary Results.* Journal of Materials Engineering and Performance, 2008. **17**(2): p. 187-192.
6. Jasthi, B., et al. *Friction Stir Welding of MA 957 Oxide Dispersion Strengthened Ferritic Steel.* 2005: Minerals, Metals and Materials Society(TMS), 184 Thorn Hill Road, Warrendale, PA, 15086-7528, USA.
7. Prado, R., et al., *Tool wear in the friction-stir welding of aluminum alloy 6061+ 20% Al2O3: a preliminary study.* scripta materialia, 2001. **45**(1): p. 75-80.
8. Ouyang, J., E. Yarrapareddy, and R. Kovacevic, *Microstructural evolution in the friction stir welded 6061 aluminum alloy (T6-temper condition) to copper.* Journal of Materials Processing Tech., 2006. **172**(1): p. 110-122.
9. Sorensen, C., B. Nelson, and S. Sanderson, *Properties and structure of friction stir welded alloy 718.* Proceedings from Friction Stir Welding and Processing IV, 2007: p. 285-293.
10. Jasthi, B., W. Arbegast, and S. Howard, *Thermal Expansion Coefficient and Mechanical Properties of Friction Stir Welded Invar (Fe-36% Ni).* Journal of Materials Engineering and Performance, 2009. **18**(7): p. 925-934.
11. Jasthi, B.K., et al., *Friction Stir Processing of Cast Inconel 718*, in *Friction Stir Welding and Processing VI.* 2011, Minerals, Metals and Materials Society: San Diego p. 25-32.
12. Jasthi, B.K., W.J. Arbegast, and S.M. Howard, *Friction Stir Welding of Alloy 22*, in *Friction Stir Welding and Processing VI.* 2011, The Minerals, Metals and Materials Society: San Diego. p. 11-18.
13. Johnson, T.D., Hinz, B.J.,West M.K.,Jashti, B.K.,Ellingsen, M.D.,Widener, C.A.,Muci, K.H., *EVALUATION OF THE RESPONSE OF FRICTION STIR PROCESSED PANELS UNDER BALLISTIC LOADING*, in *26th International Symposium on Ballistics.* 2011, DEStech Publications, Inc.: Miami, FL.
14. Defense, U.S.D.o., *Armor Plate, Steel, Wrought, Homogeneous.* 1990, U.S. Department of Defense.
15. Defense, U.S.D.o., *V_{50} Ballistic Test For Armor.* 1997, U.S. Department of Defense. p. 1-23.
16. Defence, U.s.D.o., *Aluminum Alloy Armor, Forged.* 1998.
17. Chandler, H., ed. *Heat Treater's Guide- Practices and Procedures for Nonferrous Alloys.* 1996, ASM International.
18. Defence, U.S.D.o., *Armor Plate, Aluminum Allooy, Unweldable Applique 6061.* 2007.
19. Murr, L., et al., *Adiabatic shear bands and examples of their role in severe plastic deformation.* Journal of materials science, 2002. **37**(16): p. 3337-3360.

Friction Stir Welding and Processing VII
Edited by: Rajiv Mishra, Murray W. Mahoney, Yutaka Sato, Yuri Hovanski, and Ravi Verma
TMS (The Minerals, Metals & Materials Society), 2013

ANALYSIS OF MECHANICAL AND METALLURGICAL PROPERTIES OF FRICTION STIR BUTT WELDED AA2024

Sarah F. Jurak[1], Dwight Burford[2] and Michael McCoy[1]

[1]Advanced Joining and Processing Lab, College of Engineering, Wichita State University
1845 Fairmount Street Wichita, KS 67260
[2]Joining Innovations LLC, 6321 N. Ulysses Street, Wichita, KS 67219

Keywords: Friction Stir Welding, AA2024-T3, Microhardness Testing, Conductivity Testing

Abstract

The data from a round robin study of friction stir welded 6.35 mm (0.25-in) and 3.175 mm (0.125-in) AA2024 was statistically evaluated. Friction stir welded butt welds were produced by four participant sites and tested in two independent test labs. Each participant site selected a weld tool of their choice and then developed a process window. The welds were produced with the weld parameters from the four corners of their weld process windows using their own welding machinery. The weld properties were then characterized by tensile testing, full field microhardness testing, and conductivity testing of a transverse line across the welds. The transverse tensile strength of the welds was found to be at least 88% of the base material strength. Statistical analysis of the data showed that site-to-site variability was significant with respect to mechanical properties. Overall, variability in the mechanical properties between welds produced in the round robin study was found not to be significant suggesting that the data from all sites was statistically able to be pooled for developing design allowables.

Introduction

The friction stir welding (FSW) process was developed in 1991 by The Welding Institute (TWI) a British research facility [1]. FSW is a sub-solidus thermo-mechanical process. At least two work-pieces are joined by mechanically stirring the material which is softened as heat flows into the area from friction, stirring and forging forces produced by the weld tool. A rotating probe moves through the material forging material from the leading edge of the probe into a void which forms behind the probe [2]. A shoulder at the base of the probe rotates on the surface of the material and produces frictional forces.

A depiction of the FSW process is shown in Figure 1(a). The material is heated just enough to soften it but not to the solidus temperature. There is no change in the bulk chemistry of the material during the FSW process because no fillers are required. The material does not reach the solidus temperature so no bulk solid-to-liquid transformation occurs. This results in less residual stress and distortion as well as better mechanical properties in the friction stir weld than with other welding methods, particularly fusion welding. FSW is an environmentally friendly process, which does not produce toxic by-products, does not require cover gases, and does not require large amounts of energy.

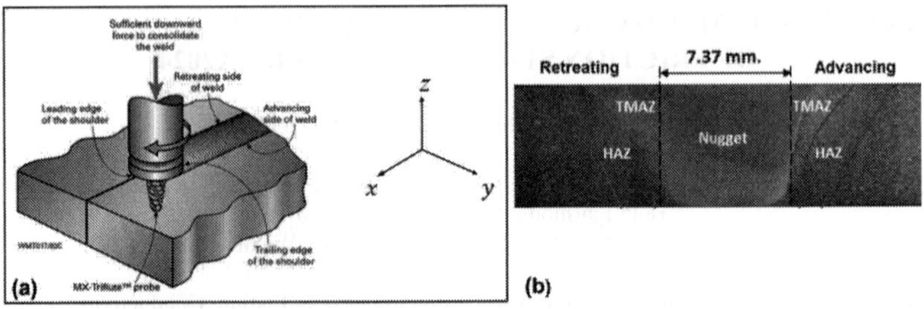

Figure 1: (a) depiction of the friction stir welding process as presented by TWI [3]with the axis convention added (b) metallographic image of cross section of friction stir weld with microstructural areas labeled.

Several unique microstructural zones are formed within the joint as a result of the FSW process. The complex flow of material around the probe forms the nugget or stir zone. The thermo-mechanical affected zone (TMAZ) is the result of mechanical forces experienced at the edge of the stir zone as the probe moves through the material and the heat that flows into the area from the nugget. In the nugget, temperatures are high enough to cause dissolution of precipitates. In precipitate-hardened metals, such as AA2024, the microstructure in the nugget is fine equiaxed grains with few dislocations and S′(S) (Al_2CuMg) particulates well dispersed and coherent in the matrix. This is due to dynamic recrystallization which occurs as the probe moves through the area. The TMAZ has rotation of the grains with a high concentration of dislocations due to the deformation field at the edge of the stir zone. The temperatures reached in the TMAZ result in S′(S) precipitation and coarsening of the grains. The Heat Affected Zone (HAZ) is formed as heat flows into the area through the TMAZ and from frictional energy produced by the rotation of the shoulder on the surface of the material. When higher temperatures occur, or lower temperatures for extended periods, coarse incoherent phases form, which lead to a loss of strength as is seen in the HAZ. Due to rotation of the grains and coarse grain structure in the TMAZ/HAZ region, this region is the primary area of failure in the friction stir weld when mechanically tested transverse to the joint line weld [4] [5] [6]. The metallographic cross-section in Figure 1(b) is of an FSW butt-weld which shows the approximate location of the microstructural areas of the weld.

Since FSW was first developed, research has investigated the flow of the material around the probe and the effect of tool design on material flow [5] [7]. In addition, research has looked at the heat input into the weld [8] [9] [10] and the effect of weld parameters and tool design on the quality of the weld [10] [11], the microstructural changes in the weld and flow of weld material [2] [9] [12], as well as other aspects of the FSW process. Arbegast presented a flow-partitioned deformation zone model for understanding the formation of weld defects [13]. The research results have contributed to the understanding of the production of defect-free welds, and the data has shown that the process is reproducible.

A study at the National Institute for Aviation Research (NIAR), at Wichita State University (WSU) indicated that the friction stir welding process is path independent [14]. In other words, a defect-free weld is not dependent on the tool used or machinery used, but rather on the weld parameters required to make a defect-free weld with the particular weld tool used. Each tool design has a range of spindle speed (rpm), travel speed (mm/s) and forge load which can be used to produce a defect free butt-weld. These weld parameters of spindle speed versus the travel speed define a process window for that tool design.

As part of the Path Independence study, a round robin was conducted at WSU to evaluate variability between sites when welds are produced by different suppliers. In this paper, the round robin portion of the Path Independence Study conducted at WSU will be discussed and the data analyzed.

Experimental Procedure

Welds were produced at four different participating sites with the equipment and processes of each site. Each participant site was asked to produce four strong butt-welds using the four corner weld parameter combinations from their process windows. A depiction of the areas of the weld process window where the weld parameters used were found is shown in Figure 2(a). The designation given to the welds with parameters from each weld process window area is shown in Figure 2(b). The participating sites were Wichita State University (WSU), Lockheed Martin, Alcan (now Constellium), and Airbus. Four welds in 6.35 mm (0.25-in) AA2024-T3 material and four welds in 3.175 mm (0.125-in) AA2024-T3 material were to be produced by each site. The material welded was supplied by Aluminum Company of America (ALCOA), and Kaiser Aluminum. The participant sites are referred to as SITE A, SITE B, etc. in this paper.

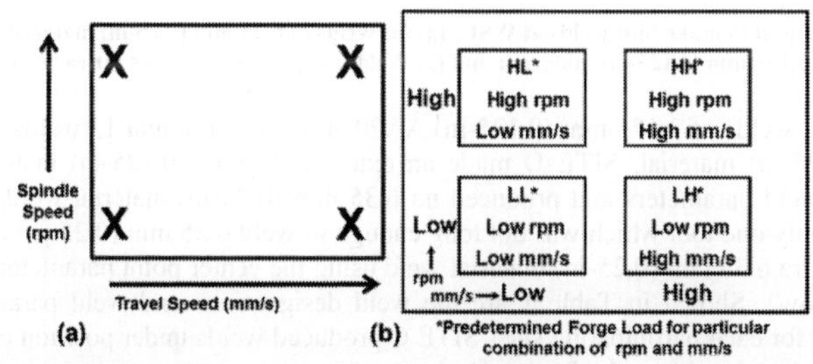

Figure 2: Location in weld process window of weld parameters used to make welds for the study.

The welds produced at WSU were welded on an MTS® ISTIR PDS 5-axis motion, 7-axis force monitoring FSW gantry machine with an anvil made of 4130 steel. The panels to be welded were secured by placing 12.7 mm (0.5-in) by 25.4 mm (1-in) steel bars positioned 50.8 mm (2-in) from the joint-line on each side with top-clamps positioned normally to the joint-line on the steel bars and tightened to approximately 5.3 kN (1,200 lbf per foot). The clamps were placed 101.6 mm (4-in) apart. The material was welded with the direction of the grain. The weld tools chosen to make the WSU welds are shown in Figure 3(a) and Figure 3(b). Each WSU weld tool has a flat shoulder with Wiper™ scrolls [15]. Each participating site produced welds with their own welding tools and the process windows used by each participant site is provided in Figure 3(c). These process windows indicate that for each tool the process window can be very different.

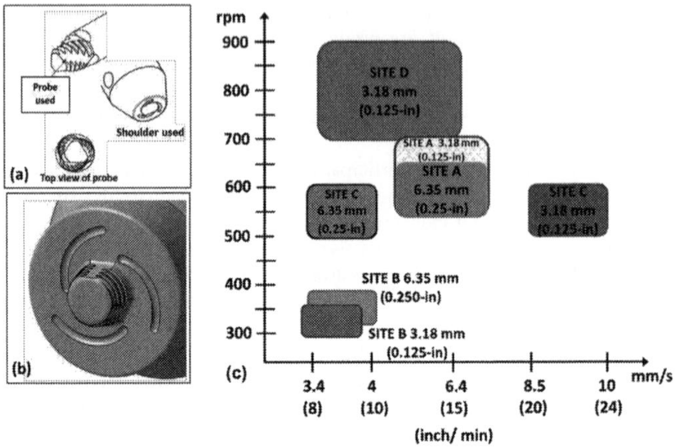

Figure 3: Tools used to make butt welds at WSU; (a) for welds of 6.35 mm (0.25-in) material; (b) for welds of 3.175 mm (0.125-in) material; and (c) Weld tool process windows for each site.

There were 17 welds of 3.175 mm (0.125-in) AA2024-T3 material and 13 welds of AA2024-T3 6.35 mm (0.25-in) material. SITE D made an extra 3.175 mm (0.125-in) material weld with center point weld parameters and produced no 6.35 mm (0.25-in) material welds. In addition, SITE D had only one tool which was not long enough to weld 6.35 mm (0.25-in) material. SITE A made an extra 6.35 mm (0.25-in) material weld using the center point parameters of their weld process window. Shown in Table 1 are the weld designations and weld parameters used to produce them for each participating site. SITE C produced welds under position control, and the other sites produced welds under load control.

Table 1: The welds produced at the participant sites with the weld parameters used.

Round Robin Specimen	Spindle Speed (RPM)	Travel Speed (mm/s)	Forge Load (kN)	Tensile Strength (MPa)	Round Robin Specimen	Spindle Speed (RPM)	Travel Speed (mm/s)	Forge Load (kN)	Tensile Strength (MPa)
SITE A_3_HH	650	7.6	21.7	464.29	SITE C_133_HL	600	8.5	25	472.08
SITE A_4_LL	550	5	21.7	456.98	SITE C_134_LL	500	8.5	23	464.02
SITE A_5_LH	550	7.6	21.7	464.71	SITE C_135_LH	500	10	26	471.6
SITE A_6_CP	600	6.4	21.7	458.78	SITE C_136_HH	600	10	25	475.74
SITE A_7_HL	650	5	21.7	456.71	SITE C_153_LH	500	4	35	434.65
SITE A_27_LL	550	5	21.7	462.91	SITE C_154_HH	600	4	34	439
SITE A_28_LH	550	7.6	21.7	464.71	SITE C_155_HL	600	3.4	31	440.78
SITE A_29_HH	700	7.6	21.7	456.43	SITE C_156_LL	500	3.4	32	436.02
SITE A_30_HL	700	5	21.7	469.67					
SITE B_1_LL	300	3	15.3	430.23	SITE D_1_LL	700	3	9	437.4
SITE B_2_LH	300	3.8	15.3	436.02	SITE D_2_HL	900	3	9	441.4
SITE B_3_HL	350	3	15.3	440.16	SITE D_3_HH	900	7	9	461.95
SITE B_4_HH	350	3.8	15.3	442.78	SITE D_4_LM	700	5	9	447.19
SITE B_11_LL	325	3.4	15.3	428.85	SITE D_5_CP	800	4	9	446.78
SITE B_12_LH	325	4	15.3	440.16					
SITE B_13_HL	375	3.4	15.3	424.92					
SITE B_14_HH	375	4	15.3	436.51					

The welds produced at SITE B and SITE C were tested at WSU. The welds produced at SITE A and SITE D were sent to Pennsylvania for testing at Westmoreland Mechanical Testing and Research, Inc. All testing was performed after at least 100 hours of natural aging, allowing time for the AA2024 weld material to stabilize with natural aging before testing. The microhardness testing conforming to ASTM E384 and conductivity testing conforming to ASTM E1004-09 of selected welds from all participant sites was done at WSU. The welded panels were cut to provide 2 metallographic specimens to be prepared for metallography and microhardness testing and 5 coupons transverse to the weld to meet ASTM E8 specifications for tensile testing.

Tensile testing of the coupons at WSU was carried out on a 22-kip MTS® 810 load frame. A clip-on MTS® axial extensometer with a 50.8 mm (2-in) gauge length was used to obtain strain data during the tensile test to calculate stress. Microhardness testing was performed on selected specimens on a LECO AMH43 automatic microhardness tester, using a Vickers scale with a 500 gf load, 13-second dwell, and 555 micron spacing. Electrical conductivity testing was performed on the specimens that were also microhardness tested. This was accomplished using a Hocking AutoSigma 3000DL Electrical Conductivity Meter with a 500 kHz, 8 mm diameter GE Inspection Technologies model 47P002 probe. Each conductivity chart is the result of the average of three sets of readings every 1.59 mm (1/16 inch) in the transverse direction across the weld, with readings taken in the direction from the advancing side to the retreating side

Results and Discussion

The average ultimate tensile strength (UTS) for the 30 welds produced in the round robin can be found in Figure 4. The UTS values are the average of five tensile coupons (designated as T1 to T5) from the joint-line of each weld. The tensile strength of the welds was 88% to 98% of the base material tested tensile strength of 484 MPa. The failure locations were evaluated for the welds from each participant site. All of the SITE A tensile coupons failed in the TMAZ on the retreating side, except for two, which were from the same 3.175 mm (0.125-in) weld. The T1 tensile from the SITE A_29HH weld failed in the nugget, and the T4 tensile from that weld failed in the nugget at the nugget/TMAZ border. The UTS values for the two tensile specimens were 447.59 MPa and 437.24 MPa respectively, while the rest of the tensile specimens for this weld were 462.76 to 472.41 MPa. This would indicate that this weld may have defects, although none were seen when visually inspected. The T1 and T4 tensile values of the SITE A_29HH weld, although low values for that weld, were comparable to the average UTS of 424.92 to 447.19 MPa for most of the welds produced at SITES B and D. All SITE B weld coupons failed in the boundary of the HAZ and TMAZ. There were no defects noted with visual inspection. The SITE D weld tensile coupons failed at the TMAZ on the advancing side. The average UTS values for the SITE C welds were 434.65 to 475.74 MPa. The SITE C weld numbers 133HL and 136LH each had one coupon fail in the nugget, which could indicate weld defects, so the micrographs were inspected to confirm the absence or presence of voids. There were no visible voids or oxide lines in the micrographs of any of the SITE C welds. A small amount of flash was present at the sides of the weld track on all of the coupons. The rest of the SITE C welds had failures on the nugget / TMAZ boundary.

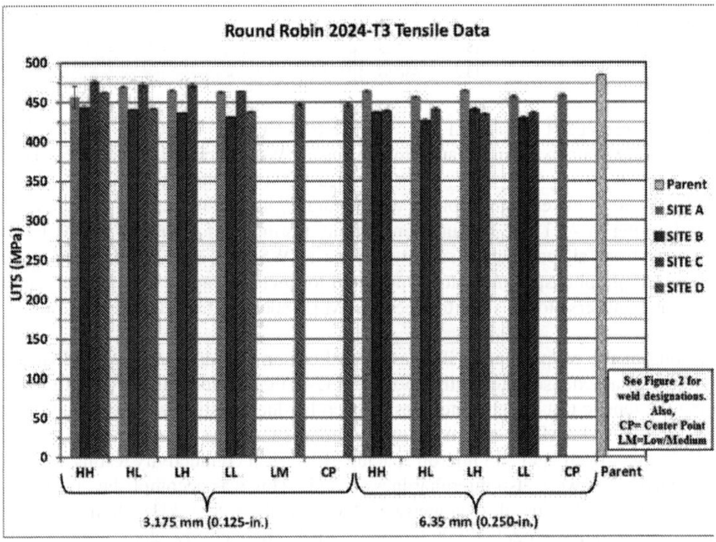

Figure 4: The results of tensile testing of all the welds produced by the participating sites with data grouped by weld parameters; shown with ± one standard deviation.

Figure 5 shows the full-field microhardness map of the SITE D_2HL weld. This weld was produced with a weld tool which had a shoulder diameter of 13 mm and a probe diameter of 5 mm. There are microhardness minimums in the TMAZ/HAZ areas with a hardness increase in the nugget, but the nugget microhardness values are much lower than the base material. An outer minimum is seen outside the shoulder diameter at approximately 11.66 and 33.3 mm.

Jones et al. [5] discussed the outer minimum area. TEM images of the area during that study showed that between the HAZ and outer minimum there was a microstructure similar to the base material but with very fine S-phase precipitates. This was believed to have contributed to the increase in hardness in that area. The outer minimum is too far removed to be affected by the forging forces of the probe. It is only affected by the shoulder on the surface of the material. TEM images indicated that no S-phases were present in the outer minimum area. Jones et al. concluded that the outer minimum hardness area was due to dissolution of the fine S-phase particulates with the small elevation in temperature during the weld cycle. The area closer to the HAZ is affected by the work of the shoulder and heat flowing into the area from the HAZ, which results in an optimal environment for growth of fine grains and retention of the S-phase particulates.

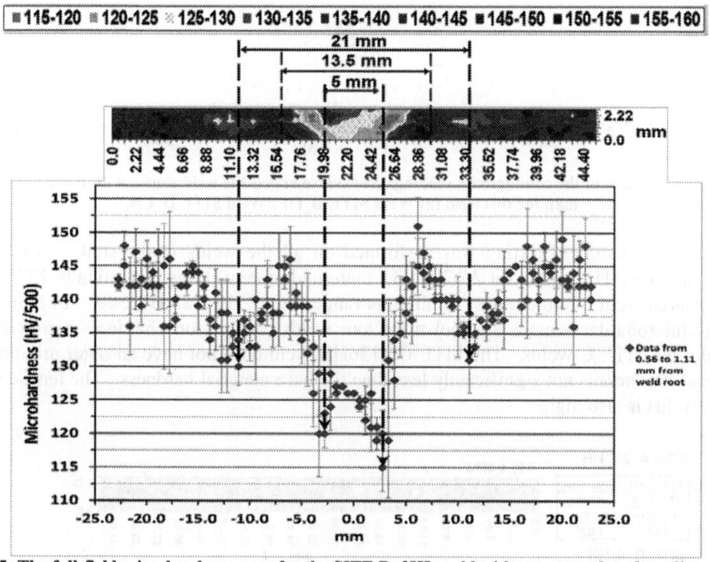

Figure 5: The full-field microhardness map for the SITE D_2HL weld with a scatter plot of two lines of data transverse across the weld.

Conductivity testing can be used to characterize the microstructure of a weld. An increase in the size of precipitates and coarse, incoherent grains occur in the area of the HAZ and TMAZ in precipitate-strengthened metals [6]. A "hot" weld (high spindle speed, low travel speed) is shown in Figure 6(a). This results in an increase in conductivity in this area that can be confirmed by comparing the conductivity curve to the corresponding microhardness map. In the "cold" weld, a peak in conductivity is observed in the area of the nugget, which reflects an over-aged microstructure in this zone of the joint. This could explain the area of minimum hardness in the nugget of the Site D_2HL weld. Conductivity testing was done every 1.59 mm (1/16-in) to obtain the conductivity transverse across the weld. Since the probe used was 8 mm in diameter, each reading represents an average of the conductivity in that area, but the resulting graph sufficiently represents conductivity trends across the weld. Figure 6(b) shows the microhardness map and the graph of conductivity across the weld of the SITE D_2HL weld. The elevation in the nugget represents the increase in conductivity due to the "over-aged" condition of the weld. This is a "cold" weld having a relatively low spindle speed and high travel speed.

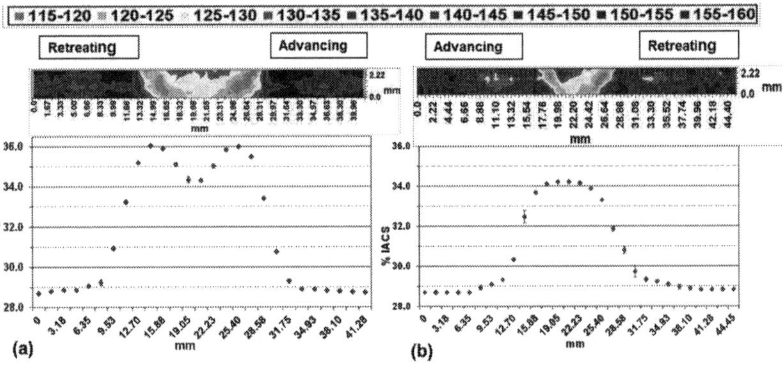

Figure 6: The graph of the conductivity readings transverse across the weld shown with the microhardness map for this weld (a) Weld SITE B_HL; Weld SITE D_LH.

Full field microhardness testing was performed on all the welds in the study. The nominal Vickers hardness (HV) value for AA2024-T3 material before welding is nominally 137 HV [16]. The measured weld microhardness minimums ranged from 110–132 HV/500. In Figure 7, the full-field microhardness maps are shown for four welds with uniform hardness across the weld, especially the SITE C welds. The SITE C _136HH weld does not have an outer minimum, and the nugget hardness is not significantly less than the base material hardness. The tensile strength of these welds is also high.

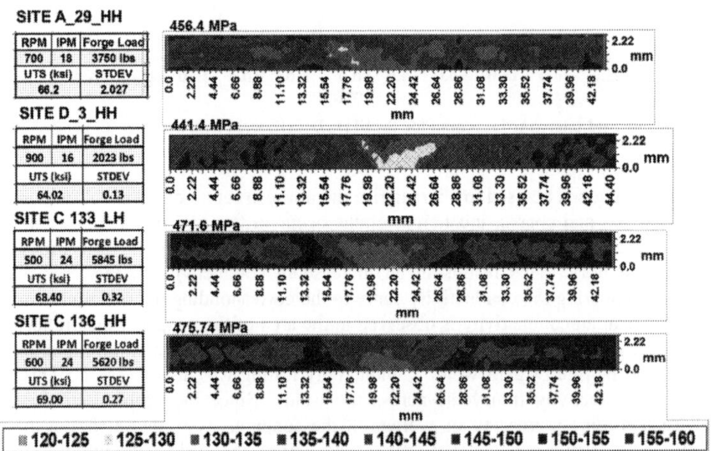

Figure 7: The full-field microhardness maps of welds with fairly uniform microhardness and high tensile strength.

The weld UTS results were entered into Statgraphics®, a computer-based statistics package, to evaluate the variability of the data. A Weibull analysis indicated the data could have come from a normal distribution; therefore, the data was grouped according to the location of the weld parameters in the weld process window and analyzed. A multiple sample comparison showed no significant difference between the means of the data groups, as shown in the box-and-whisker plot from the analysis in Figure 8(a). The variance check indicated no significant difference in the standard deviations between the groups of data with a 95% confidence level.

In addition, the Kruskal-Wallis test of the medians indicated that there is not a statistically significant difference between the medians of the groups of data at a 95% confidence level. The data was grouped and analyzed according to the participant site that produced the welds. According to the results shown in Figure 8(b) there was a statistically significant difference between the standard deviation of SITE C to SITE A and of SITE C to SITE B. The variability between sites is significant with a 95% confidence level, but it is low at 6.2 to 14.2 MPa. Assuming that the tensile coupons came from independent locations within the weld joint-line, the data was grouped according to T1 to T5 and analyzed. The results shown in Figure 8(c) indicate that there is not a significant difference between the means of each group of data. A Kruskal-Wallis test of the medians, which assumes all medians are the same, shows no statistical difference between the medians of the groups with a 95% confidence level. Therefore, there is evidence that the welds are uniform in tensile strength from beginning to end. Finally, the results show that the participant sites were able to produce welds of equivalent quality using their own equipment and processes.

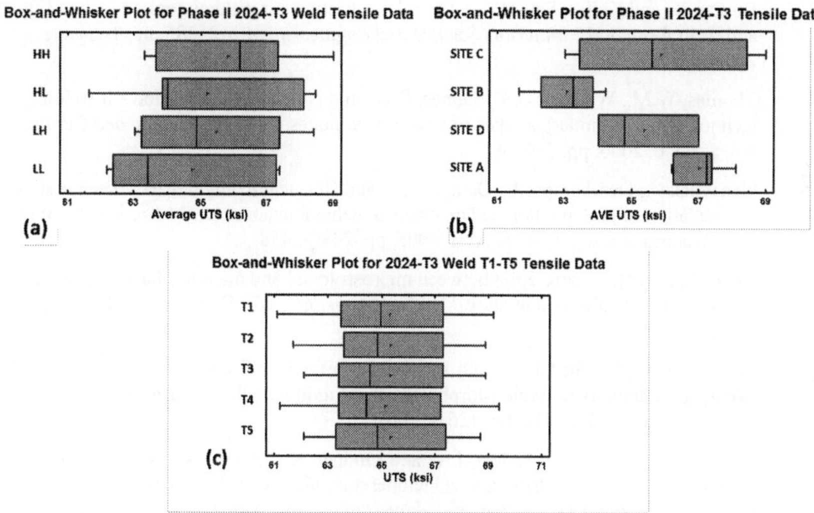

Figure 8: The results of Statgraphics® analysis of the tensile data.

Conclusions

Welds were produced by four participant sites, using their own equipment and processes. The welds were shown to be of equivalent quality with at least 88% and up to 98% of the base material tensile strength. Four of the welds had uniform microhardness profiles with high tensile strengths. Statistical analysis of the data indicated that the data could have come from a normal distribution. When grouped according to location of the weld parameters in the weld process window, no significant difference between the data sets was detected. When grouped according to the site that produced it, a significant difference between the standard deviations of the data from the different sites was observed. Although significant variation was demonstrated between sites, the variation was low at 6.2 to 14.2 MPa. There was evidence that the welds were of uniform tensile strength from beginning to end.

Acknowledgments

We would like to thank the FAA and Kansas N-I-S for funding this study. In addition, we would like to thank the industry participants who provided material, welds, and testing. We would like to recognize and thank those who work in the Advanced Joining Lab, which was part of NIAR at the time of the study, for their invaluable assistance.

Works Cited

[1] Thomas, W.M. et al., "Friction Welding," 5,460,317, October 24, 1995.

[2] Kumar, K. and Kailas, Satish.V., "The role of friction stir welding tool on material flow and weld formation," *Materials Science and Engineering A*, vol. 485, no. 1-2, June 2008, pp. 367-374.

[3] Thomas, W.M., Weisner, C.S., Staines, D.G., and Norris, I.M., "Friction stir welding technology - Preliminary studies of variant techniques - Part 1," *Welding and Cutting*, vol. 5, no. 6, 2006, pp. 339-344.

[4] Genevois, C., Deschamps, A., Denquin, A., and Boisneau-cottignies, B., "Quantitative investigation of precipitation and mechanical behaviour for AA2024 friction stir welds," *Acta Materiala*, vol. 53, no. 8, April 2005, pp. 2447-2458.

[5] Jones, M.J. et al., "Correlation between microstructure and microhardness in a friction stir welded 2024 aluminium alloy," *Scripta Materialia*, vol. 52, no. 8, April 2005, pp. 693-697.

[6] Widener, C.A., Burford, D.A., and Jurak, S.F., "Effects of Tool Design and Friction Stir Welding Parameters on Weld Morphology in Aluminum Alloys," *Materials Science Forum*, vol. 638-642, 2010, pp. 1261-1266.

[7] Reynolds, A.P., Tang, W., Khan, J.A., and Lindner, K., "Relationships between weld parameters, hardness distribution and temperature history in alloy 7050 friction stir welds," *Science and Technology of Welding and Joining*, vol. 2, no. 9, 2005, pp. 190-199.

[8] Burford, D.A., Tweedy, B.M., and Widener, C.A., "Development of Design Data for FSW and FSSW," in *7th International Friction Stir Welding Symposium*, Awaji Island, Japan, 20-22 May 2008.

[9] Gallais, C. et al., "Modelling the Relationship between Process Parameters , Microstructural Evolutions and Mechanical Behaviour in a Friction Stir Welded 6XXX Aluminum Alloys," in *5th International Symposium on Friction Stir Welding*, Metz, France, 14-16 September 2004.

[10] Reynolds, A.P. and Tang, Wei., "Alloy, tool geometry and processing parameter effects on friction stir weld energies and resultant FSW joint properties," in *Friction stir welding and processing*, Indianapolis, IN, United States, 2001, Nov 4-8, pp. 15-23.

[11] Cavaliere, P., Cabibbo, M., Panella, F., and Squillace, A., "2198 Al–Li plates joined by Friction Stir Welding: Mechanical and Microstructural Behavior," *Materials and Design*, vol. 30, no. 9, October 2009, pp. 3622–3631.

[12] Fonda, R.W., Bingert, J.F., and Colligan, K.J., "Development of grain structure during friction stir welding," *Scripta Materialia*, vol. 51, no. 3, August 2004, pp. 243-248.

[13] Arbegast, W.J., "A flow-partitioned deformation zone model for defect formation during friction stir welding," *Scripta Materialia*, vol. 58, no. 5, March 2008, pp. 372-376.

[14] Burford, D., Jurak, S., Gimenez-Britos, P., and Boldsaikhan, E., "Evaluation of Friction Stir Weld Process and Properties for Aircraft Application," Advanced Joining and Processing Lab, National Institute for Aviation Research, Wichita State University, Wichita, FAA Report to be published (submitted June 2011).

[15] Burford, D., "Friction stir welding tool having scroll-free concentric region," Friction stir welding tool 8,016,179, September 13, 2011.

[16] ASM International, "Vol.2 - Properties and Selection: Nonferrous Alloys and Special-Purpose Materials," in *Metals Handbook*, ASM International 10th Ed., 1990.

[17] Burford, D.A., Tweedy, B.M., and Widener, C.A., "Development of Design Data for FSW and FSSW," in *7th International Friction Stir Welding Symposium*, Awaji Island, Japan, 20-22 May 2008, p. Paper 56.

[18] Reynolds, A.P. and Tang, Wei., "Alloy, tool geometry and processing parameter effects on friction stir weld energies and resultant FSW joint properties," in *Friction stir welding and processing*, Indianapolis, IN, United states, 2001, Nov 4-8, pp. 15-23.

Friction Stir Welding and Processing VII
Edited by: Rajiv Mishra, Murray W. Mahoney, Yutaka Sato, Yuri Hovanski, and Ravi Verma
TMS (The Minerals, Metals & Materials Society), 2013

MECHANICAL AND MICROSTRUCTURAL PROPERTIES OF FSW LAP JOINTS

Egoitz Aldanondo[1], Ekaitz Arruti[1], Pedro Alvarez[1], Alberto Echeverria[1]

[1] IK4-LORTEK Centro de Investigación en Tecnologías de Unión
Arranomendia kalea 4A, Ordizia, 20240, Spain

Keywords: Aluminum, hook, un-welded interface, sheet thinning

Abstract

The properties of lap joints performed by FSW have been studied in this investigation. Similar and dissimilar combinations have been studied using AA6082-T6 and AA5754-H22 aluminum alloys. The effect of several joining variables on the mechanical and microstructural properties of the joints have been analyzed using different tool designs and welding parameters such as rotational speed and weld speed. The influence of welding parameters and probe design have been found to produce a very strong impact on effects such as hook feature formation and effective sheet thickness of the joints which greatly influence the properties of the joints. It has been concluded that using appropriate tool designs and welding conditions, the flow of the plasticized material can be driven to avoid the formation of undesired joint features and maximize the strength of the joints.

Introduction

Since it was invented [1], FSW has experienced a large research activity and many applications have been developed especially in products produced by FSW of aluminum alloys [2]. This research and development activity has been conducted primarily to butt joint configurations. However lap joint configurations are highly interesting too, as they are widely used for the manufacturing of structures in different transportation industries. The great potential of FSW to produce lap joints with aluminum alloy materials makes it a very interesting technology to replace other similar joining technologies that are applied currently for the manufacturing of components such as structural panels strengthened by stringers. Therefore it is very important to understand the basics of the joint formation in FSW lap joints in order to realize a high quality manufacturing of FSW lap joints.

There have been a number of studies conducted on FSW lap joints [3-10]. These studies have described the joint formation mechanisms, mechanical properties, as well as the importance of the tool design in the properties of the joints. According to some authors [3, 4, 5, 9], the pull-up or pull-down effects produced by the FSW tool have been identified as the main factors governing the joint formation and subsequent properties of the joints. The study of the mechanical properties of the joints [3-7] revealed that presence of defects such as hook features or un-welded interfaces are largely detrimental for the joint strength. It has been shown that conventional tool designs used for FSW of butt joints are not the best option for lap configurations and some attempts have been made in order to improve the tool design for FSW of lap joints [8-11]. The main goal is to avoid pull-up or pull-down effects on the advancing and retreating sides of the joints resulting in hook features or un-welded interfaces, while generating sufficient welded area to allow a good transmission of the stresses from the top sheet to the bottom sheet of a lap joint.

The study reported in this document was focused on the investigation of the influence of different welding conditions on the microstructural and mechanical properties of FSW lap joints using 1,5 mm thick AA6082-T6 aluminum alloys. Different probe designs (length and geometry) were used under different welding conditions focusing on the microstructural features produced for each tool/parameter combination. Mechanical properties of the joints were analyzed by tensile shear testing paying attention to the fracture type produced in each case. From these data relationships have been established among tool design, welding parameters, microstructural features, mechanical properties and fracture types of the FSW lap joints.

Experimental Procedure

Materials

FSW lap joints were performed at IK4-LORTEK Research Centre (Ordizia, Spain) using similar and dissimilar aluminum alloy combinations. 1,5 mm thick AA6082-T6 aluminum alloy sheets were used to produce similar FSW lap joints. These joints were used to study the mechanical strength as well as the microstructural properties of the joints. In addition to these, dissimilar joints were produced using 1,5 mm thick AA6082-T6 and AA5754-H22 aluminum alloy sheets in order to analyze the material flow behavior and joint formation mechanisms. For this material combination AA6082-T6 alloy was placed on top of the overlap configuration. In all tests, both top and bottom sheets were 180 mm long and 100 mm wide while the overlap distance was 40 mm.

Welding Procedure

FSW experiments were conducted in a MTS ISTIR-PDS FSW machine in position control. A constant penetration depth, which was set to be the same as the pin length used in each case, was used for all welding tests. FSW lap joints were produced under different welding conditions as listed in Table I. Two different welding conditions were studied from a welding heat input viewpoint, considering that high rotational speed (RS) and low welding speed (WS) combinations correspond to high welding heat input in comparison with low RS and high WS. Therefore, high heat input joints were performed using a RS of 1500 rpm and a WS of 150 mm/min. On the other hand low heat input joints were produced at values of RS and WS of 750 rpm and 300 mm/min respectively. These two different welding conditions are referred to as “hot welds” (welds produced at 1500 rpm – 150 mm/min) or “cold welds” (welds produced at 750 rpm – 300 mm/min) hereafter.
An adjustable probe concept tool was used for the welding experiments, allowing the change of the probe or the adjustment of the probe length while using the same shoulder. Thus two different probe designs were used for the tests, one comprising a cylindrical threaded probe and the other one featuring a cylindrical non-threaded probe. FSW lap joints were produced by setting both probe designs to three different probe lengths with values of 1,5 mm, 2 mm and 2,5 mm. This means that at a pin length of 1,5 mm the tip of the probe travelled just in the interface of the bottom sheet while at pin lengths of 2 mm and 2,5 mm the probe penetrated the bottom sheet by 0,5 and 1 mm respectively. The tool had a shoulder diameter of 12 mm for all probe design and probe length combinations and the tilt angle was fixed at 1,5 deg.

Table I: Welding Parameters

Weld Nº	Probe Design	Probe Length	Welding Parameters (RS-WS)
1	Threaded	1,5 mm	"Cold Weld" 750 rpm – 300 mm/min
2	Threaded	1,5 mm	"Hot Weld" 1500 rpm – 150 mm/min
3	Threaded	2 mm	"Cold Weld" 750 rpm – 300 mm/min
4	Threaded	2 mm	"Hot Weld" 1500 rpm – 150 mm/min
5	Threaded	2,5 mm	"Cold Weld" 750 rpm – 300 mm/min
6	Threaded	2,5 mm	"Hot Weld" 1500 rpm – 150 mm/min
7	Non-threaded	1,5 mm	"Cold Weld" 750 rpm – 300 mm/min
8	Non-threaded	1,5 mm	"Hot Weld" 1500 rpm – 150 mm/min
9	Non-threaded	2 mm	"Cold Weld" 750 rpm – 300 mm/min
10	Non-threaded	2 mm	"Hot Weld" 1500 rpm – 150 mm/min
11	Non-threaded	2,5 mm	"Cold Weld" 750 rpm – 300 mm/min
12	Non-threaded	2,5 mm	"Hot Weld" 1500 rpm – 150 mm/min

Mechanical Testing and Macro/Microstructural Characterization

Tensile shear tests were performed for evaluating the mechanical strength of the FSW lap joints. 25 mm wide test specimens were machined perpendicular to the welding direction and tested using a tensile testing machine Zwick Roell Z100 (100 kN load capacity). Three specimens were tested for each welding condition. Shims of the same material and thickness as the test specimens were used when clamping the samples in order to induce pure shear and avoid initial realignments. Fracture strength (R) defined as the maximum failure load divided by the width of the specimens (N/mm) has been used as the term for evaluating the strength of the FSW lap joints. In addition to this, fracture type of each condition has been studied too. Loading condition in lap joints can be different as either the advancing side or the retreating side of the joint can be loaded on the top sheet. In this study the condition of loading on the advancing side on the top sheet was used for all welded specimens.

For microstructural characterization, the specimens were carefully prepared following standard metallographic procedures comprising cross-sectioning, mounting and polishing. Similar FSW lap joint specimens produced with AA6082-T6 alloy were etched using a reagent based on a NaOH solution while Kroll's reagent was used for dissimilar AA6082-T6/AA5754-H22 FSW lap joint specimens. An Olympus GX51 light optical microscope was used to investigate the microstructural properties and the defect formation of the FSW lap joints.

Results and Discussion

Mechanical testing

The R of FSW lap joints produced with a threaded probe tool is shown in Figure 1. It is shown that for a probe length of 1,5 mm, where the probe does not penetrate the bottom sheet, "cold welds" and "hot welds" present similar R values of approximately ≈278 N/mm. However a very different R value can be observed for "cold welds" and "hot welds" when there is penetration of the probe in the bottom sheet (probe lengths of 2 and 2,5 mm). In this situation "cold welds" show a higher R compared to "hot welds", which present a very pronounced decrease in the R values. Generally, the use of a larger probe length reduces the R value of the FSW lap joints, but this effect is much more prominent in "hot welds" than in "cold welds".

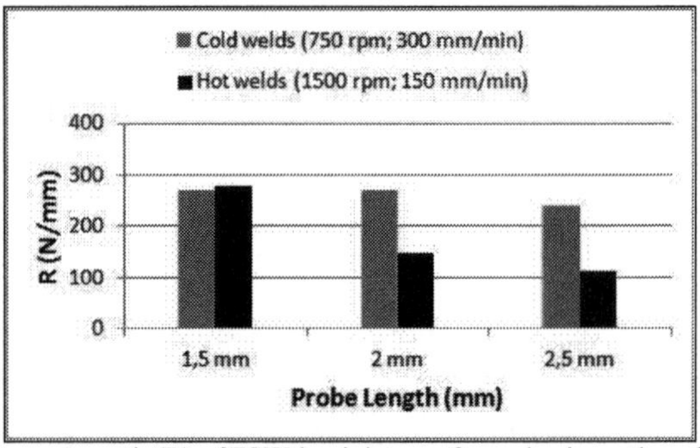

Figure 1: Fracture strength (R) of FSW lap joints performed using a threaded probe tool.

In Figure 2, the R values of the FSW lap joints performed by using a non-threaded probe tool are plotted. Similar R values have been observed for "cold welds" and "hot welds" corresponding to equal probe lengths although "hot welds" present slightly higher R values. Regarding the effects produced by the probe length, it can be clearly seen that the use of larger probe lengths produces an increment of the R values for both "cold welds" and "hot welds". Therefore the use of a probe that penetrates the bottom sheet produces FSW lap joints that present higher R values, for the non-threaded probe tool case.

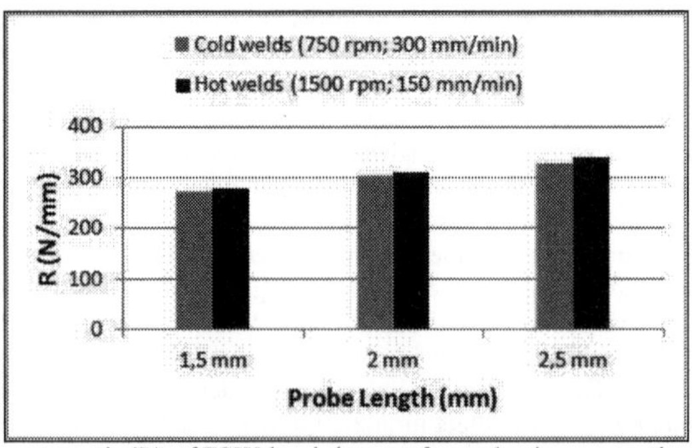

Figure 2: Fracture strength (R) of FSW lap joints performed using a non-threaded probe tool.

From comparison of data that are shown in Figure 1 and Figure 2, it is clear that the use of different tool designs produce radically different results in the mechanical properties of FSW lap joints. Large probes that penetrate the bottom sheet operated at "hot weld" conditions produce higher R values when using a non-threaded probe tool, while the exactly opposite effect is achieved when using a threaded tool at those same conditions. Thus maximum R values of ≈338 N/mm have been achieved with the non-threaded 2,5 mm long probe tool operated at "hot weld" conditions, while minimum R values of ≈114 N/mm have been observed for the threaded probe at the same welding conditions.

Fracture types

The fracture type of all tested FSW lap joints has been studied. Three different fracture types have been observed as shown in Figure 3. Figure 3-A shows a type 1 fracture, which is a shear fracture through the interface between the top and the bottom sheet. This is the most common fracture type that has been observed as all samples welded at "cold weld" conditions have shown this fracture type. Figure 3-B shows a type 2 fracture, which has been observed in FSW lap joints produced by the threaded probe operated at "hot weld" conditions and penetrating the bottom sheet significantly. Type 2 fractures are characterized by a fracture initiation on a hook feature that is present in the advancing side of the joint as it will be explained in detail later. FSW lap joints showing this type of fracture have presented low R values. Figure 3-C shows a type 3 fracture, which is a tensile fracture through the HAZ of the advancing side. Type 3 fractures occur outside the joint due to the softening of the HAZ in the advancing side, which is the loaded side of the FSW lap joints on the top sheet. Type 3 fracture has been observed for the weld N° 12 only, which was produced by a non-threaded 2,5 mm long probe tool operated at "hot weld" conditions, and has presented the highest R value of ≈338 N/mm.

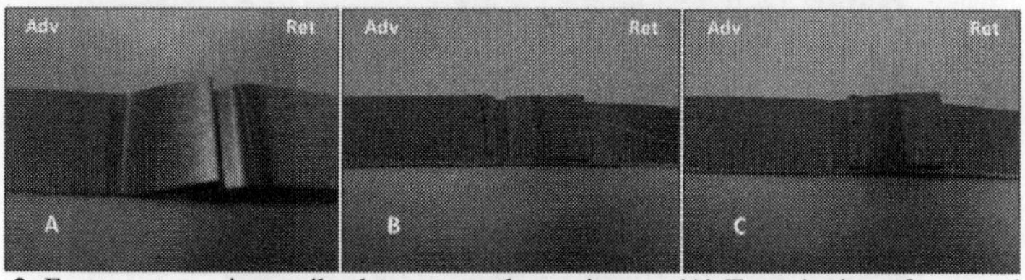

Figure 3: Fracture types in tensile shear strength specimens. (A) Type 1, shear fracture through the interface. (B) Type 2, tensile fracture through the hook feature. (C) Type 3, tensile fracture through HAZ.

Macro/Microstructural Characterization

Some representative cross sectional views of FSW lap joints produced under different welding conditions are shown in Figure 4. In FSW lap joints performed with a probe length that equals the top sheet thickness (Figure 4-A and Figure 4-B) a continuous un-welded interface remains where the original faying surfaces were located. This interface is formed as a result of a non-efficient elimination of the oxides present in the faying surfaces before welding. The use of a threaded probe promotes a slightly higher deformation in the un-welded interface due to a higher stirring action of the tool (Figure 4-B).
FSW lap joints carried out with probe penetration in the bottom sheet and "cold weld" conditions (Figure 4-C and Figure 4-D) feature a similar un-welded interface as FSW lap joints represented

in Figure 4-A and Figure 4-B. However, a more prominent deformation and bonding effect has been observed, primarily in the interface located in the advancing side.

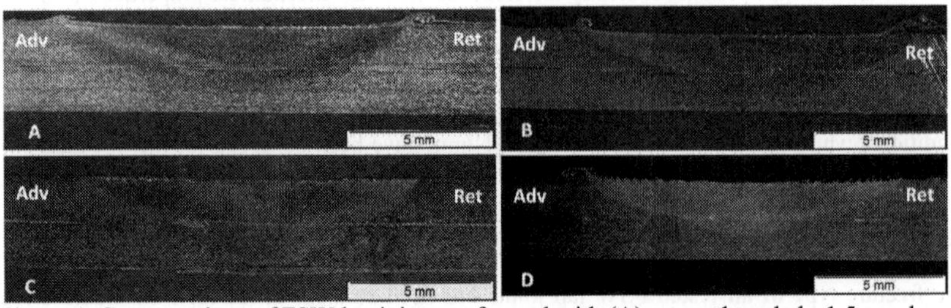

Figure 4: Cross-sections of FSW lap joints performed with (A) a non-threaded - 1,5 mm long probe tool at 1500 rpm – 150 mm/min; (B) a threaded - 1,5 mm long probe tool at 1500 rpm – 150 mm/min; (C) a non-threaded - 2 mm long probe tool at 750 rpm – 300 mm/min; and (D) a threaded - 2 mm long probe tool at 750 rpm – 300 mm/min.

Figure 5 shows a cross-section of a FSW lap joint produced at "hot weld" conditions by a threaded 2,5 mm long probe tool. It can be seen that an interface pull-up effect is produced in both advancing and retreating sides of the FSW lap joint (see details in Figure 5-A and Figure 5-C). These pull-up effects result in a hook feature formation in the advancing side and an extended un-welded interface in the retreating side. The hook feature and extended un-welded interface produce an effective sheet thickness reduction on the top sheet which is reflected into the reduced R values of FSW lap joints produced at these welding conditions.

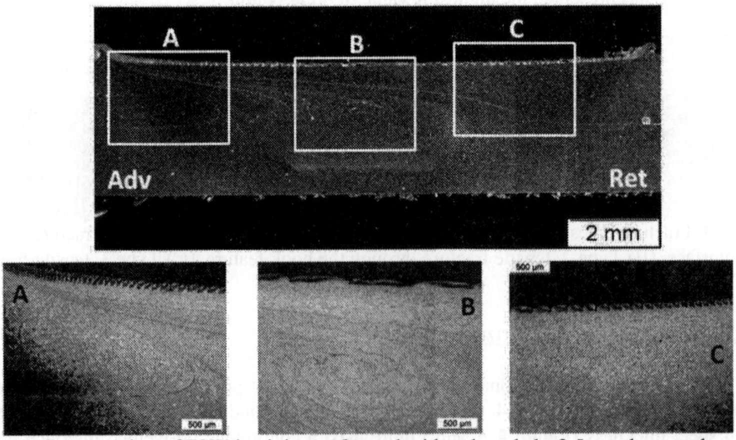

Figure 5: Cross-section of FSW lap joint performed with a threaded - 2,5 mm long probe tool at 1500 rpm - 150 mm/min.

A completely different joint microstructure has been observed in the study of the FSW lap joint cross section produced by using a non-threaded 2,5 mm long probe tool operated at "hot weld" conditions (Figure 6). Even though the only change in the welding conditions compared to the sample represented in Figure 5 is the design of the probe (threaded/non-threaded), a very different defect formation has been observed. A pull-down effect takes place in the advancing side resulting in a hook feature formation towards the bottom sheet (Figure 6-A). This hook feature produces a reduction in the effective sheet thickness of the bottom sheet but the thickness of the top sheet remains the same in the advancing side. On the other hand, a very small pull-up effect has been observed in the retreating side having practically no influence on the effective sheet thickness of the top and bottom sheets. However an un-welded interface extension continuing the top and bottom sheet interface towards the stir zone has been observed in the retreating side (Figure 6-B). This un-welded interface crosses the stir zone and turns down towards the bottom sheet in the proximities of the hook feature observed in the advancing side (Figure 6-A).

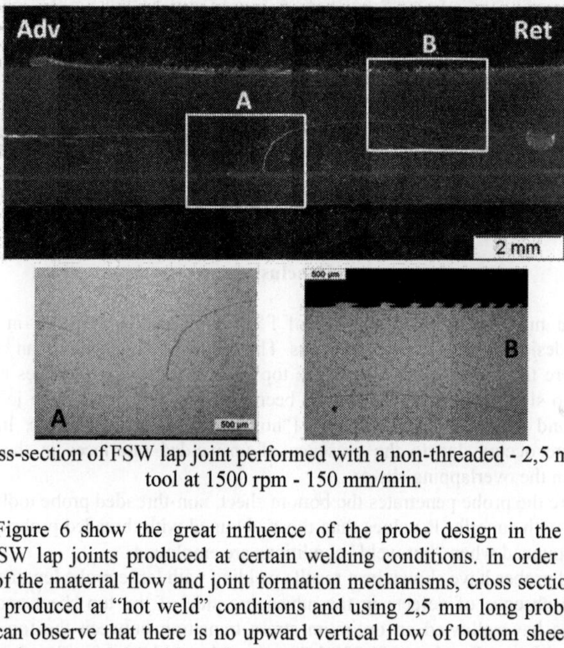

Figure 6: Cross-section of FSW lap joint performed with a non-threaded - 2,5 mm long probe tool at 1500 rpm - 150 mm/min.

Figure 5 and Figure 6 show the great influence of the probe design in the microstructural properties of FSW lap joints produced at certain welding conditions. In order to gain a better understanding of the material flow and joint formation mechanisms, cross sections of dissimilar FSW lap joints produced at "hot weld" conditions and using 2,5 mm long probes are shown in Figure 7. One can observe that there is no upward vertical flow of bottom sheet material while some of the top sheet material is driven downwards into the bottom sheet when a non-threaded probe is used (Figure 7-A). On the contrary, vertical flow of bottom sheet material towards the top sheet is promoted by the thread in case that a threaded probe is used (Figure 7-B). This vertical flow is responsible for the development of hook features in the advancing side and extended un-welded interfaces in the retreating side, resulting in a reduction of the effective sheet thickness of the top sheet.

As a summary, it can be stated that FSW lap joints produced under conditions that do not promote vertical flow of the bottom sheet material (Figure 4) show no effective sheet thickness reduction of the top sheet. FSW lap joints produced in these conditions show a reasonably high R values (270-300 N/mm) and present an interfacial type 1 fracture as they contain un-welded interfaces in the original faying surfaces. On the other hand, FSW lap joints produced under conditions that promote upward vertical flow of bottom sheet material develop defects such as hook features in the advancing side reducing the effective sheet thickness of the top sheet (Figure 5). When these joints are loaded in the advancing side, they present a type 2 fracture at very low R values (≈114 N/mm). On the contrary, when conditions that promote downward vertical flow of top sheet material are used, there is no reduction of the effective sheet thickness of the top sheet and hook features as well as the un-welded interfaces are driven towards the bottom sheet (Figure 6). When these FSW lap joints are loaded in the advancing side of the top sheet, the absence of defects in the areas where the stresses are transmitted from the top to the bottom sheet produces a type 3 fracture in the HAZ at the highest recorded value of ≈338 N/mm.

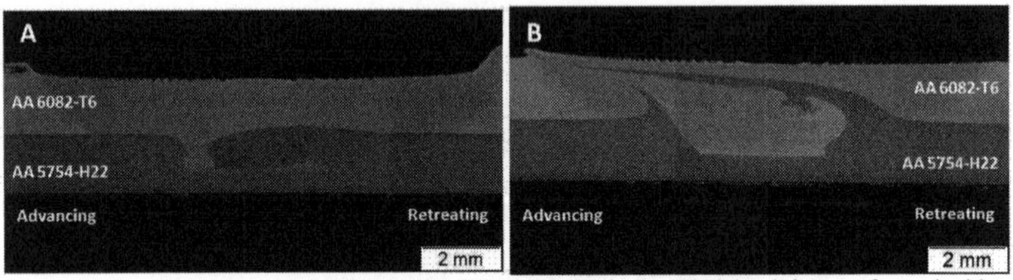

Figure 7: Cross-sections of AA6082-T6/AA5754-H22 dissimilar FSW lap joints performed with (A) a non-threaded - 2,5 mm long probe tool at 1500 rpm – 150 mm/min; and (B) a threaded - 2,5 mm long probe tool at 1500 rpm – 150 mm/min.

Conclusions

Mechanical and microstructural properties of FSW lap joints have been investigated using different probe designs and welding parameters. The following conclusions can be presented:

1. In cases where the probe length equals the top sheet thickness, i.e. it does not penetrate the bottom sheet, no significant differences have been found in R values of the joints for different probe designs and other welding conditions ("hot welds" or "cold welds"). In these cases an interfacial failure occurs due to the presence of un-welded interfaces in the original faying surfaces between the overlapping sheets.
2. In cases where the probe penetrates the bottom sheet, non-threaded probe tools have produced FSW lap joints with superior R values than those obtained with threaded probes. This effect has been more pronounced when "hot weld" conditions are employed.
3. Non-threaded probes have shown best results welding with large probe lengths and "hot weld" conditions. The absence of defects in the advancing side of the top sheet and a reduced un-welded interface have allowed an optimum stress transition between the top and the bottom sheets. Thus the highest R values of ≈338 N/mm have been obtained with a 2,5 mm long probe operated at 1500 rpm and 150 mm/min, presenting a fracture in the HAZ in advancing loading conditions.
4. Using these same conditions to perform FSW lap joints with a threaded probe has produced the lowest R values of ≈114 N/mm. The formation of a hook feature in the advancing side reduced the effective sheet thickness of the top sheet. Thus under advancing loading conditions a fracture through the hook feature occurs at a very low R value.
5. Different probe designs promote different flow of plasticized material during FSW of lap joints resulting in different defect formation. It is possible to optimize the strength of the FSW lap joints by using the appropriate probe design and welding conditions depending on the loading requirements of the joint.

References

1. W. M. Thomas et al., European Patent Specification EP 0 615 480 B1.

2. Threadgill et al., "Friction Stir Welding of Aluminium Alloys", Int. Mat. Rev., Vol59 N°2 49-93 (2009).

3. L. Cederquist, A. P. Reynolds, "Factors affecting the properties of friction stir welded aluminum lap joints", Weld J Res Suppl 80, 281, (2001).

4. S. B. Jung et al, "Lap joint properties of FSWed dissimilar formed AA 5052 and AA 6061 Al alloys with different thickness", J Mat Sci 43, 3296, (2008).

5. S. Yazdanian, Z. W. Chen, G. Littlefair, "Effects of friction stir lap welding parameters on weld features on advancing side and fracture strength of AA6060-T5 welds", J. Mater. Sci., Published online (2011).

6. M. Ericsson, L. Jin, R. Sandström, Int. Jour. Fatigue 29-57, (2007).

7. M. K. Kulekci, A. Sik, E. Kaluç, "Effects of tool rotation and pin diameter on fatigue properties of friction stir welded lap joints", Int. J . Adv. Manuf. Technol. 36: 877-882 (2008).

8. R. Kovacevic et al, "Investigation of the friction stir lap welding of aluminium alloys AA5182 and AA6022", J Mat Eng Perf, 16 (4), 477, (2007).

9. G. M. D. Cantin et el., "Friction Skew-stir welding of lap joints in 5083-O aluminium", Sci. Tech. Weld. Join., Vol10 N°3 268-280 (2005).

10. M. J. Brooker et al., "Applying Friction Stir Welding to the Ariane 5 Main Motor Thrust Frame", Proceedings of the Second International Symposium on FSW, Gothenburg, (2000).

11. C. Fuller, M. Mahoney, W. Bingel, "Friction Stir Weld Tool and Method", U.S. Patent N° US2005/0121497A1.

MECHANCIAL PROPERTIES OF REPAIRED 7075-T73 FRICTION STIR WELD BUTT WELDS

Christian A. Widener[1], John Franklin[1], Bharat K. Jasthi[1], and Michael K .West[2]

[1]Advanced Materials Processing and Joining Laboratory
South Dakota School of Mines and Technology
501 E. St. Joseph St., Rapid City, SD 57701 USA

[2]Department of Materials and Metallurgical Engineering
South Dakota School of Mines and Technology
501 East St. Joseph Street, Rapid, South Dakota, 57701

Keywords: 7075 T-73, Friction stir processing; Friction stir welding; Welding defect; Welding Repair

Abstract

For production parts joined by friction stir welding, there is a need to identify the effects of rework in a shop environment, or later repair of damage in service, on mechanical properties. Effects on strength or microhardness have been previously characterized; however, often design loads may not be purely driven by static strength allowables, but instead by fatigue limits, which may be less sensitive to rework than ultimate or yield tensile strengths. To investigate those effects, defective butt welds were created in 0.25 in thick 7075-T73 Aluminum alloy plates. Welds were then repaired using similar parameters directly over the original weld. The mechanical properties of up to four passes after the initial weld were also investigated with a focus on fatigue life. An analysis of the results and a discussion of the implications for repair and rework of friction stir welded structures and assemblies are included.

Introduction

As the use of friction stir welding in manufacturing and in particular aerospace continues to increase, there is a growing need to understand effects and limitations of repairs in friction stir welded components. Therefore, the purpose of this investigation was to determine the quality of a repair to friction stir butt welded in 0.25-in (6.3mm) thick 7075-T73 aluminum alloy plates which contained volumetric defects within the welded region. The repair process was achieved by applying an additional weld pass over the original defect containing weld. Up to three additional passes after the repair were also applied to simulate the effects of additional rework welds on the plate. The aluminum alloy 7075-T73 is a non-weldable, high strength, heat treatable alloy containing mainly zinc and magnesium, which is commonly used in airframe structures and other high stress parts [1]. Friction stir welding, however, is an effective joining process for this alloy since it creates a solid-state joint with strengths above 80% joint efficiency [2]. A survey of reported tensile strength data for friction stir welded Al 7075-T7 material is summarized in Table 1.

Table 1: Reported Tensile Results for As-Welded 7075-T73

Material Thickness	Ultimate Tensile Strength	Yield Strength	Weld Elongation (%)	Ref.
0.040-in. (1 mm)	65.4 ksi (451 MPa)	49 ksi (338 MPa)	10.0	[3]
0.100-in. (2.5 mm)	69 ksi (476 MPa)	- -	- -	[4]
0.125-in. (3.1 mm)	69.2 ksi (477 MPa)	49.2 ksi (316 MPa)	13.8	[5]
0.240-in. (6 mm)	69.6 ksi (480 Mpa)	- -	- -	[2]
0.250-in. (6.35 mm)	68.6 ksi (473 MPa)	- -	6.1	[6]
0.320-in. (8 mm)	66.4 ksi (458 MPa)	- -	- -	[7]
Average	68.0 ksi (469 MPa)	49.5 ksi (341 MPa)	9.97	

The authors also presented some fatigue data for as-welded 7075-T7, showing a reduction of fatigue life for samples in the as-welded condition when the weld flash was not removed. Additionally, Mahoney et al. [8], when investigating properties of friction stir welded 7075-T6, presented an excellent depiction of the variation of strain in the weld zone, which is typical of FSW. The variation in strain occurs in the heat affected zone (HAZ), where the hardness minimum occurs in friction stir welds in precipitation strengthened aluminum alloys such as Al 7075. This loss of hardness correlates well with a small drop in tensile and yield strength, which can also correspond to a reduction of fatigue strength, especially at higher load levels were strain localization can occur in the HAZ. At lower stress levels, however, where the strains are predominantly elastic, fatigue strengths can be nearly equivalent to parent material because, unlike the tensile strength, the elastic modulus is not appreciably affected by FSW. This phenomenon was demonstrated in another precipitation strengthened aluminum alloy, 2024 [9]. Additionally, the effect of multipass welds on the tensile properties of Al 7050, a similar 7xxx series aluminum alloy, was performed by Brown et al. [10]; however, the effect of multiple friction stir weld passes on fatigue was not investigated. This paper investigates the quality of repairs in defective friction stir butt welded .25in thick 7075-T73 Aluminum alloy plates.

Experimental

Welds were produced on an MTS ISTIR 10 5-axis friction stir welding system with a hydraulic spindle motor. The welds were fixtured on a steel anvil. The pin tool was made of H13 tool steel and is shown in Figure 1. Two 0.25-in. (12.7 mm) thick x 4-in. (102 mm) wide x 24-in. (610 mm) 7075-T73 Aluminum plates were set up in butt weld configuration. Weld parameters were 400 rpm, 13 ipm, with 5500lbs of force for the defective pass and 4400lbs of force for all subsequent passes. The experiment involved creating a defective friction stir butt weld with the first pass, and then repairing the defective weld with between one and four additional friction stir welds. Additional welds were performed directly over the previous welds. The welds were cooled in ambient air before additional passes were completed.

Figure 1: FSW pin tool

Metallographic Samples

Samples for metallographic analysis were prepared using a Leco PR-25 sample press to encase the samples in Bakelite. The samples were oriented such that samples were sectioned transverse to the welding direction and the view is in the direction of welding. A liquid cooled 120 grit Leco BG-32 belt sander then sanded the sample surface flat. Next, a Leco spectrum system polisher sanded and polished the samples. Each sample was sanded with 400, 600, and 1200 grit wet sand paper, and then with 9, 6, 3, and 1 micron diamond solution. Each step down in grit size included a 90 degree rotation of the sample about the central axis. Keller's etchant etched the samples to reveal the weld zones, and a microscope was used to photograph the samples.

Hardness Testing

Passes 1 through 5 received micro-hardness testing. A manual Vickers microhardness tester was used to evaluate the samples. Micro-hardness was done across the weld and heat affected zones, 0.125-in. (3.2 mm) from the bottom of the weld, parallel to the bottom of the weld, as seen in Figure 2. A hardness measurement was taken every 0.025-in. (0.64 mm) along this line, for a total of 40 measurements.

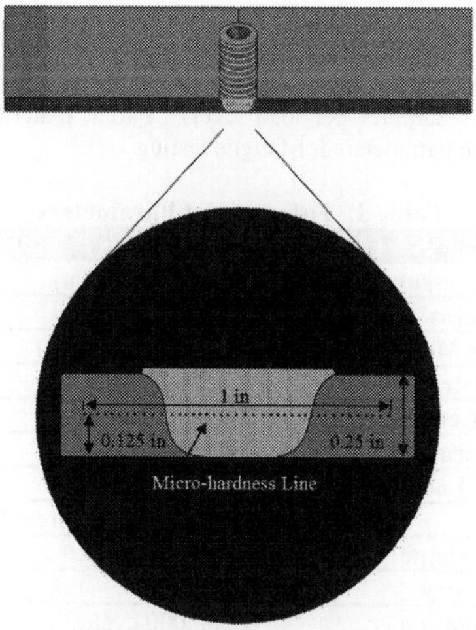

Figure 2: Microhardness measurement diagram

Tensile Testing

Tensile testing was performed per ASTM E8. The tensile samples were cut by waterjet for the parent material. A 2-axis CNC mill created all welded tensile samples. An MTS 810 hydraulic load frame pulled the tensile samples and recorded tensile properties. The parent material had six tensile samples. Passes one through five had three tensile samples each. Samples were lightly sanded to remove burrs, and were measured to document actual specimen dimensions. The exact centers of the gage lengths were then marked. To ensure data accuracy, an MTS extensometer measured strain in the samples. The standard used is shown in Table 2, which defines the parameters for tensile testing.

Table 2: Tensile Test Parameters

Tensile Test Setup Values	
Grip Separation	4.375-in. (111 mm)
Initial Rate	0.015in/min (0.4 mm/min)
Measured Elongation	1.18-in. (30 mm)
Nominal Gage Length	1-in. (25.4 mm)
Removal Point	0.0035in/in (0.09 mm/mm)
Secondary Speed	0.25-in/min (6.4 mm/mm)
Strain Endpoint	0.0035in/in (0.09 mm/mm)
Thickness	0.25in (6.35 mm)
Width	0.5in (12.7 mm)

Fatigue Testing

Fatigue samples were tested per ASTM E466. The parent samples were cut using a waterjet, and a 2-axis CNC mill created all welded fatigue samples. Since it has been shown that the fatigue life of friction stir welds can be compromised by not removing surface roughness in the weld track, the weld flash was removed for this testing [2,9]. The test plan called for sanded samples using 1200 grit and each coupon was measured to verify actual coupon dimensions. An MTS 810 hydraulic load frame was again used for the fatigue testing. The machine tested 9-12 samples in fatigue for each weld pass (3 samples per load level). Parent material was tested as well. The standard in Table 3 defines the parameters for fatigue testing.

Table 3: Fatigue Test Parameters

Fatigue Test Setup Values	
Grip Pressure	1.5 ksi (10.3 MPa)
Thickness	.25-in (6.35 mm)
Elastic Modulus	10,300 ksi (71 GPa)
Yield Strength	52 ksi (358 MPa)
Gage Length	2-in. (51 mm)
Temperature	73°F (23°C)
Cyclic Load Variation	R = 0.1
Stress	50% to 70% of Parent Y.S.
Wave Shape	Sine
Frequency	15 Hz
Termination	1,000,000 cycles

Results and Discussions

Four major sets of data were taken to characterize the friction stir weld repair of 7075-T73 butt welds they are: (1) tensile properties, (2) fatigue properties, (3) metallographic samples, and (4) microhardness.

Tensile Properties

Comparison of ultimate tensile strength and yield strength are shown in Figure 3. Each point represents a different sample, and the y axis indicates the number of weld passes the sample received. Tensile properties show a slight decrease of both ultimate tensile strength and yield strength with additional repair passes, which is consistent with previous work in a similar alloy [10]. The softening effect is related to the overaging of the precipitates in the HAZ; however, this change in the precipitate structure does not affect the overall elastic modulus of the material which depends on the alloying elements and is relatively insensitive to overaging.

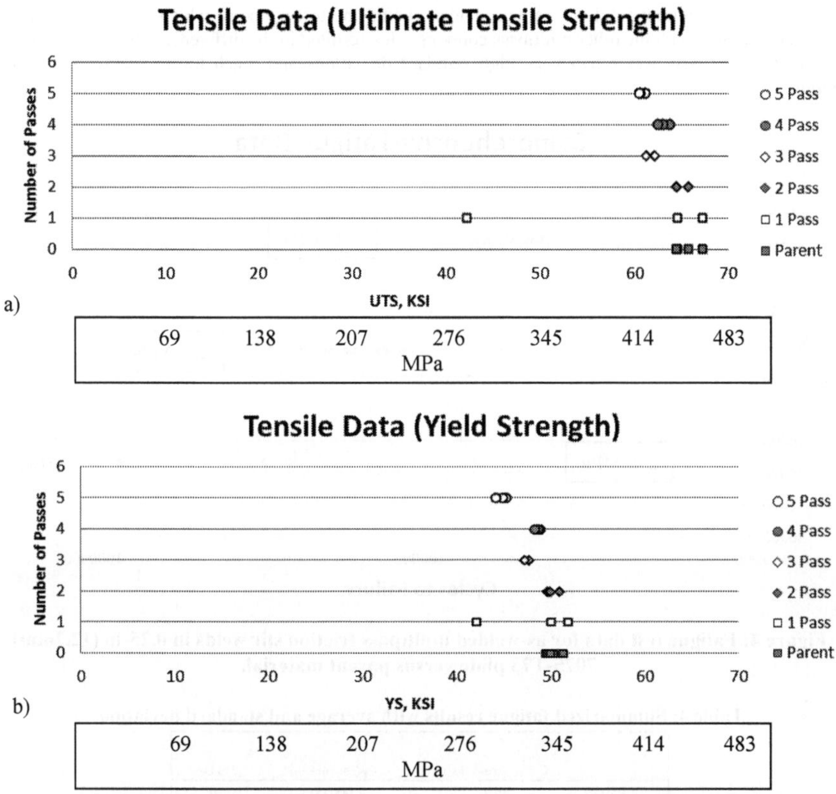

Figure 3: Tensile strength versus parent material and number of weld passes a) Ultimate Tensile Strength, and b) Yield Strength

Fatigue Properties

A plot of the fatigue test results for this investigation is shown in Figure 4. Each data point represents a unique sample, with logarithmic cycles to failure on the y-axis versus cyclic maximum stress on the x-axis. The shape and color of the points corresponds to the number of weld passes the sample received. A summarized table of the results is shown in Table 4. The friction stir welded samples investigated here had a decrease in overall average fatigue life and an increase in standard deviation of fatigue results, however, in most cases individual samples were identified that performed as well as or better than parent material and were in general at least comparable to parent material. Additionally, up to 5 passes over the same material appear to have little to no effect on the actual fatigue performance of the friction stir welded material. The strongest effect on fatigue life was found to be the presence of the small wormholes which were present in the first pass of the welded samples, and were observed to have the most detrimental effect on fatigue life at the highest stress loads. When detrimental defects were present in the fatigue coupons of the 1 pass welds, they were visible in the fracture surfaces of the fatigue coupons as the initiation site for fatigue failure. A few of the 1 pass weld coupons did not contain defects within the test coupons, which gave them fatigue lives comparable with the other multipass coupons. This resulted in the differences shown in Table 4, where the first pass welds had very high standard deviations and much lower average cycles to failure.

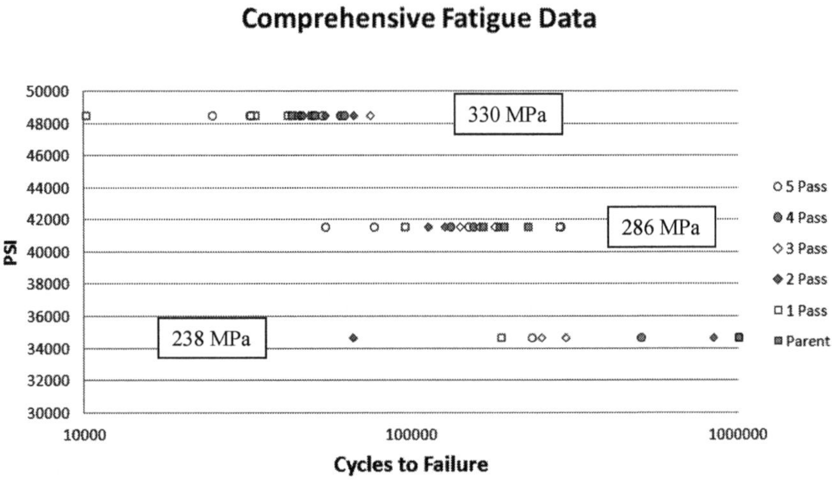

Figure 4: Fatigue test data for as-welded multipass friction stir welds in 0.25-in (12.7mm) 7075-T73 plate versus parent material.

Table 4: Summarized fatigue results with average and standard deviation.

	Cyclic Stress (ksi)	Cycles (mean)	Std Dev
Parent	34.7	1,000,000	0
	41.6	194,414	30,967.21
	48.5	47,783	4,185.94
1 Pass	34.7	132,566	529,444.92
	41.6	94,430.75	133,140.49
	48.5	29,379.00	13,533.97
2 Pass	34.7	725,308.00	446,110.45
	41.6	177,065.00	78,271.32
	48.5	53,511.00	9,745.38
3 Pass	34.7	514,203.00	421,321.26
	41.6	160,379.00	19,614.12
	48.5	56,388.00	16,170.76
4 Pass	34.7	467,929.00	288,320.64
	41.6	155,976.00	25,190.62
	48.5	56,344.00	9,125.99
5 Pass	34.7	744,254.00	442,965.07
	41.6	93,525.00	49,521.65
	48.5	36,673.00	14,916.90

Metallographic Samples

For metallography, all of the plates were sectioned at the beginning (M1), middle (M2) and end of the plate (M3), as shown in Figure 5. Metallography was used to identify the weld defects in the first pass, which were identified in the fatigue testing. Some of the samples from the 1 pass welds performed equivalent to the other welded samples, and failure did not originate from defects in the weld nugget. It was later confirmed by metallography that some of the coupons from the end of the plate did not actually contain weld defects, which explained their improved performance.

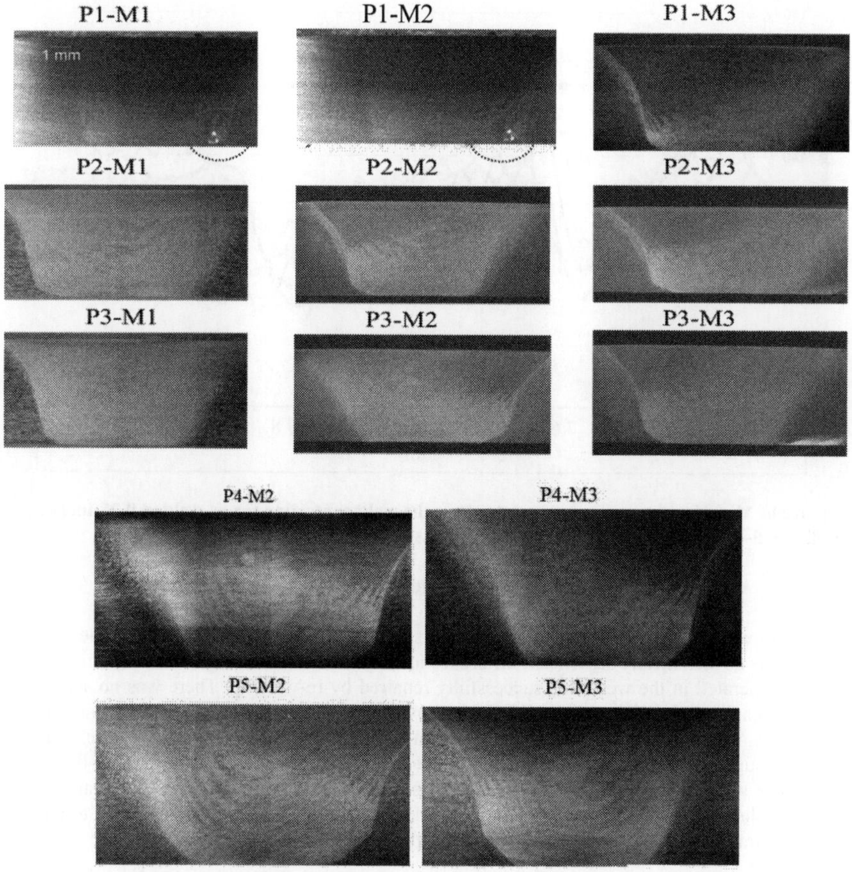

Figure 5: Metallographic samples showing defective welds in pass 1 (circled in Red), and subsequent defect free welds for passes 2-5.

Microhardness

Microhardness measurements revealed a typical softening impact on the HAZ, that worsened to some degree with subsequent weld passes, as shown in Figure 6. The highest hardness values were observed on the advancing side, which has been observed and often reported to reach the highest peak temperatures. Minimum hardness values were also observed on the retreating side, which corresponds to the slowest cooling rates, and is generally the location of maximum softening and failure in friction stir welds, as it was also in this investigation.

Advancing Retreating

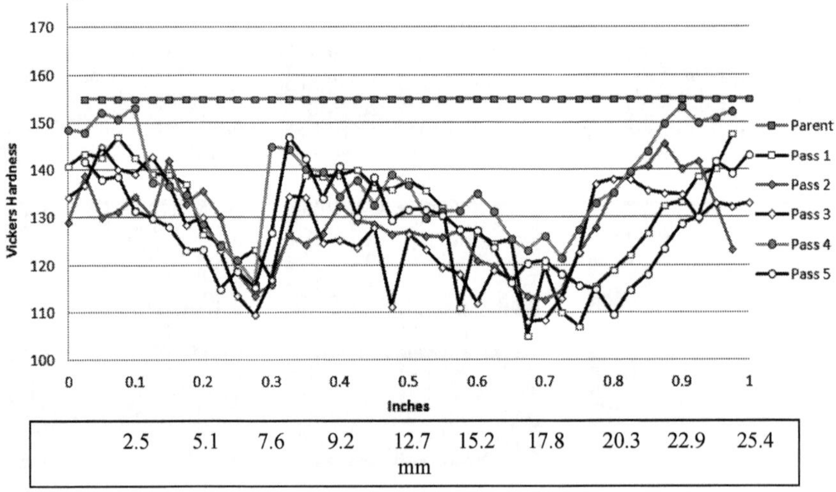

Figure 6: Microhardness measurements along the weld centerline for as-welded 0.25-in. (12.7 mm) 7075-T73 plate versus number of weld passes

Conclusions

Data indicated up to 10% drop in Ultimate Tensile Strength and Yield Strength by the 5th pass. There was no statistically significant difference in fatigue life for the limited test conditions. The defect generated in the weld was successfully repaired by re-welding. There was no significant drop in tensile strength or fatigue life in this investigation, as compared to the parent material which was found to be very close to its minimum specification for tensile strength. The results of this investigation suggest that for applications that are fatigue strength limited rather than yield or tensile strength limited, additional weld passes for repair or rework may not result in any appreciable reduction in design life, if care is taken to produce defect free welds and to remove the weld track and other surface stress concentrations after repair.

Acknowledgements

This work was completed as a portion of a project investigating repair of aluminum aerospace structures at the Center for Friction Stir Processing, a National Science Foundation Industry/University Cooperative Research Center, of which the South Dakota School of Mines and Technology is an active university member.

References

1) "Selection and Application of Aluminum Alloys." *Aluminum and Aluminum Alloys*. Ed. Joseph R. Davis. Materials Park, OH: ASM International, 1993. pp. 59-62.
2) Klæstrup Kristensen, J. et al. "Properties of Friction Stir Welded Joints in Aluminum Alloys 2024, 5083, 6082/6060 and 7075," 5th Intl. Friction Stir Welding Symposium, September, 2004.
3) Dalle Donne, C., Biallas, G., Ghindini, T., Raimbeaux, G., "Effect of Weld Imperfections and Residual Stresses on the Fatigue Crack Propagation in Friction Stir Welded Joints," Proceedings of the 2nd International Symposium on FSW, sponsored by TWI, Ltd., Gothenburg, Sweden, 26-28 June, 2000.
4) Tweedy, B., Allen, C., and Kumar, B., "Design, Analysis and Testing of Friction Stir Welded Thin Sheet 2024 and 7075 Aluminum Alloys," Aeromat Conference Presentation, Orlando, FL, 6-9 June, 2005.
5) C. Widener et. al, "Evaluation of Post-Weld Heat Treatments to Restore the Corrosion Resistance of Friction Stir Welded Aluminum Alloy 7075-T73 vs. 7075-T6," Materials Science Forum, vols. 539-543, 2007, pp. 3781-3788.
6) Lumsden, J., Pollock, G., and Mahoney, M., "The Effect of Thermal Treatments on the Corrosion Behavior of Friction Stir Welded 7050 and 7075 Aluminum Alloys," Materials Science Forum, vols. 426-432, 2003, pp. 2867-2872.
7) Reynolds, A.P., Lockwood, W.D., and Seidel, T.U., "Processing Property Correlation in Friction Stir Welds," Materials Science Forum, Vols. 331-337, 2000, pp. 1719-1724.
8) Mahoney, M.W., Rhodes, C.G., Flintoff, J.G., Spurling, R.A., and Bingel, W.H., "Properties of Friction-Stir-Welded 7075-T651 Aluminum," Metallurgical and Materials Transactions A, vol. 29A, July, 1998, pp. 1955-1964.
9) C. Widener, B. Tweedy, and D. Burford, "An Investigation of the Effects of Tool Design and Welding Parameters on Fatigue Life in Friction Stir Welded 2024-T3," 7th International Friction Stir Welding Symposium, Awaji Island, Japan, 20-22 May, 2008.
10) R. Brown, W. Tang, and A.P. Reynolds, "Multi-pass friction stir welding in alloy 7050-T7451: Effects on weld response variables and on weld properties," Materials Science and Engineering: A, Vols. 513–514, 15 July, 2009, Pages 115–121.

Friction Stir Welding and Processing VII
Edited by: Rajiv Mishra, Murray W. Mahoney, Yutaka Sato, Yuri Hovanski, and Ravi Verma
TMS (The Minerals, Metals & Materials Society), 2013

PROCESS DEVELOPEMENT OF INTEGRAL FASTENERS USING FRICTION STIR SPOT WELDING WITH "C-FRAME" END EFFECTOR ON AN AIRCRAFT CABIN DOOR MADE FROM AA6061-T6 and AA2024-T3

Alan Bruce Handyside[1], Vishwanath Iyer[1], Ron Preston[2], Enkhsaikhan Boldsaikhan[1], Mike McCoy[1]

[1]Wichita State University, Wichita, KS, USA.
[2]Hawker Beechcraft Corporation, Wichita, USA

Keywords: Friction Stir Spot Welding (FSSW), Integral Fastener, Aluminum alloy 6061-T6, Aluminum alloy 2024-T3

Abstract

Application of Friction Stir Spot welding (FSSW) to the creation of integral fasteners on an Aircraft Cabin Door using an industrial robot with a "C-Frame" end effector is explored. Plunge depth control, sheet lifting, pin temperature and anvil sticking issues are addressed. Tools with various shoulder concavities are investigated. Through optimization of forge load, spindle speed and process duration (dwell) time, integral fastener shear loads are found to be 1.8 times higher than resistance spot welding. Optimized anvil and spindle cooling time is found to be up to 9 seconds on the "C-Frame" end effector. Preliminary Welding Procedure Specification (pWPS) per AWSD17.3 is developed. Frame and skin assembly articles are fastened together using three unique joining sequences. A "go – no go" inspection gage is used to verify pWPS standards. Hard tooling to shorten process time, robot cells and multiple end effector studies are proposed for further study.

Introduction

The objective of this study was to apply friction stir spot welding (FSSW) to the creation of integral fasteners on a civil aircraft cabin door using a robot equipped with a "C" frame end effector. Some of the restrictions to overcome involved programming that are restricted to force control, over-plunging of the probe and sheet lifting seen in earlier internal lab studies. Solutions taken in this study involved plunge depth control by adding shims to hard stops and the exploration of sheet interaction with FSSW properties using tool design modifications.

Methodology

The airframe material design incorporated fastening a 0.063 inch thick AA6061-T6 frame to a 0.032 inch thick AA2024-T3 skin. Four FSW tools were considered including featureless probes and probes with scrolls near the joint interface. The tools included 10 and 15 degree shoulder concavities. The tool producing the best metallography with high ultimate tensile strength (UTS) is shown in Figures 1 and 2. The UTS tests for a 100% shear loaded single spot integral fastener were performed using the coupons shown in Figure 3. Initially a 4-spot integral fastener fixture was used, as shown in Figures 4 and 5. This was redesigned into a 10-spot integral fastener fixture to better understand the heat build-up during the fastener creation process, shown in Figures 6 and 7. This also better replicated the continuous process on the aircraft door panel.

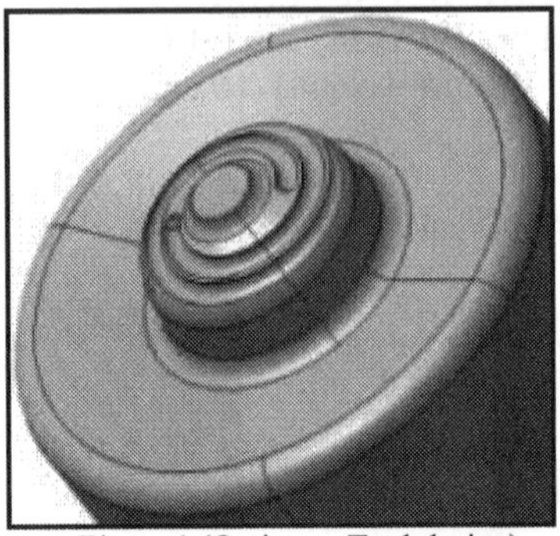

Figure 1 (Optimum Tool design)

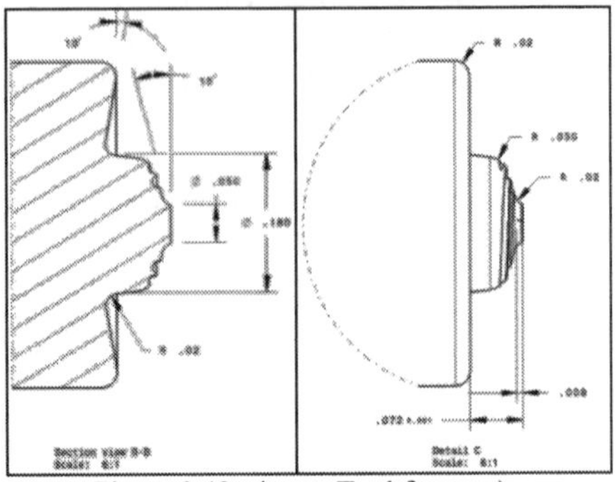

Figure 2 (Optimum Tool features)

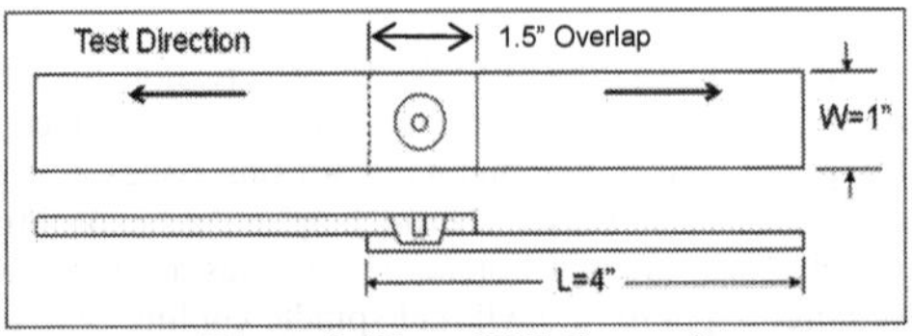

Figure 3 (Shear UTS Coupon)

Figure 4 (4 fastener setup)

Figure 5 (4 fastener coupon)

Figure 6 (10 fastener setup)

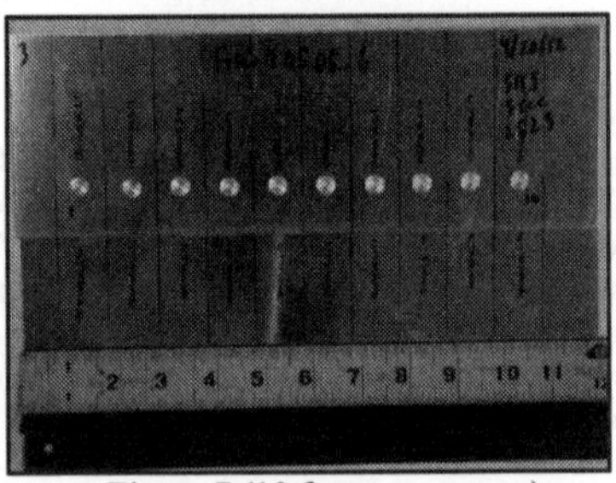

Figure 7 (10 fastener coupon)

Two Designs of Experiment (DOE) were performed using a 4 spot fastener setup. DOE-1 compared the four probes using load versus fastener parameter plots in Statgraphics as shown in Figures 8, 9, 10, and 11. DOE-2 compared the two probes with the highest average joint strength from DOE-1 as shown in Figures 12 and 13.

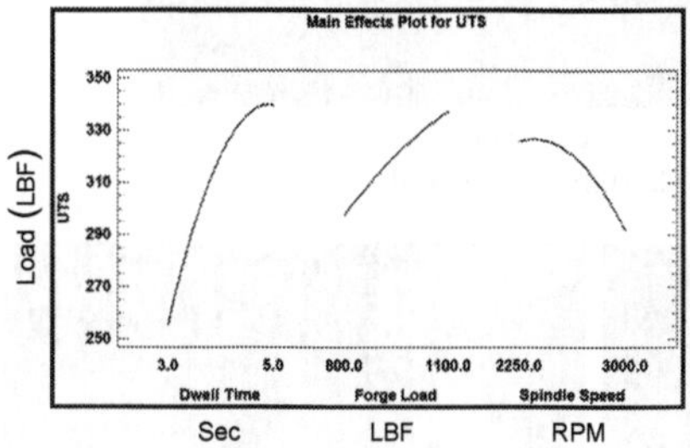

Figure 8 (DOE-1, Tool -1)

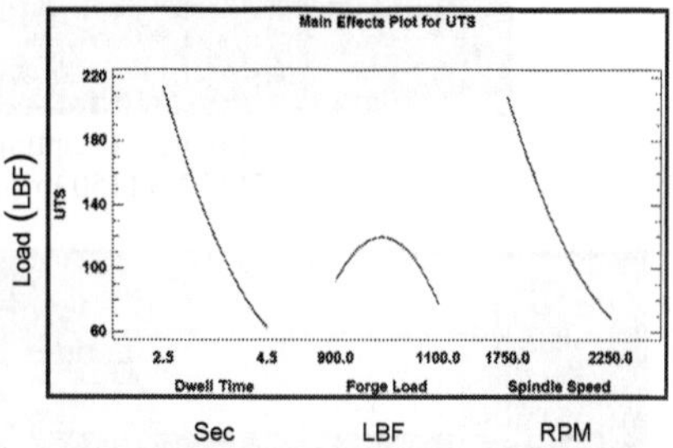

Figure 9 (DOE-1, Tool -2)

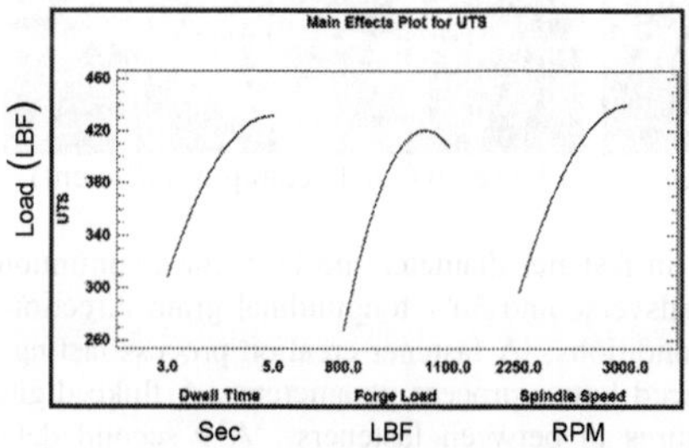

Figure 10 (DOE-1, Tool -3)

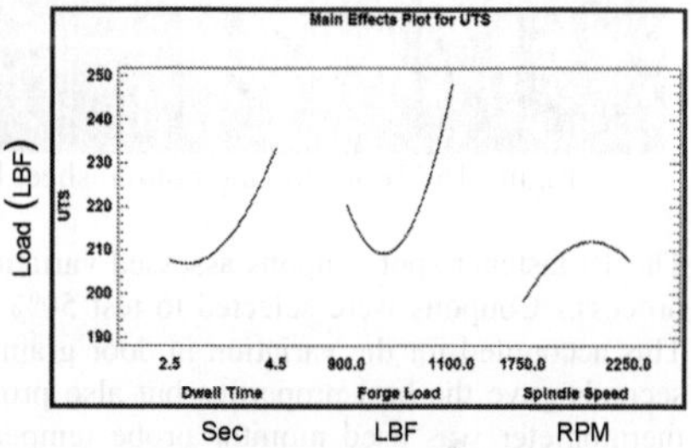

Figure 11 (DOE-1, Tool -4)

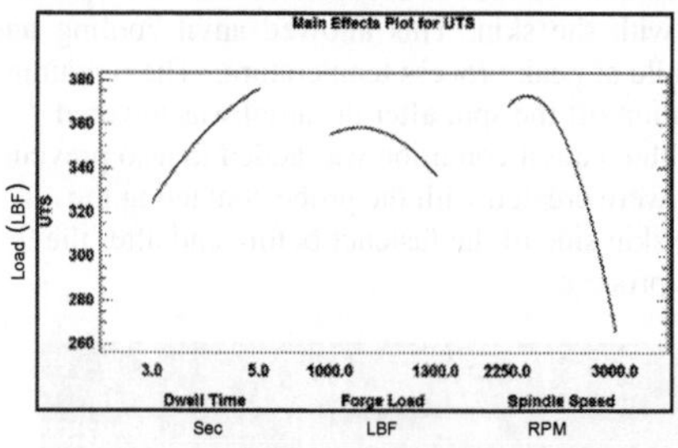

Figure 12 (DOE-2, Tool -1)

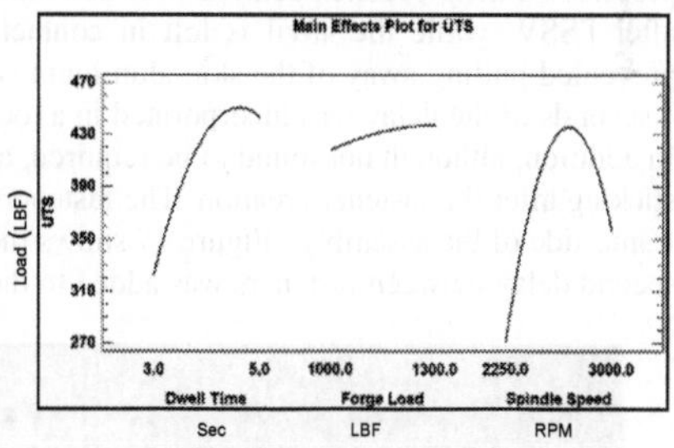

Figure 13 (DOE-2, Tool -3)

The optimized parameter for Tool-3 with a UTS load of 1.92 KN (432 lbf) produced the metallography shown in Figure 14. A small fixture ledge was added to the coupon fixture and used with clamping in the 10 fastener series to prevent unsupported droops and simulate the frame support in the door assembly. The clamps prevented observed sheet separation during the coupon process which was not seen on the stiffer frames during trial runs on the engineering geometry. This is shown in Figures 15 and 16.

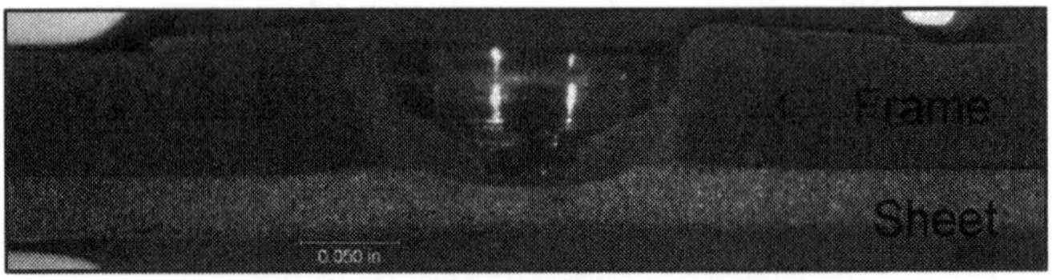

Figure 14 (Optimum Values – Tool-3)
5115 N (1150 lbf), 2625 rpm, 5 sec dwell

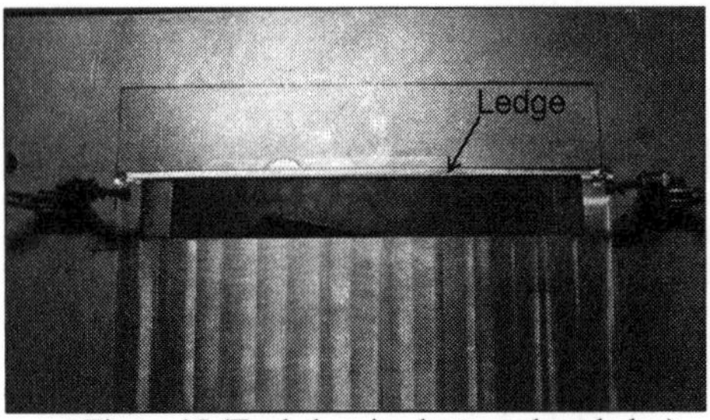

Figure 15 (Tool showing bottom sheet ledge)

Figure 16 (single clamp is sufficient)

The 10 fastener spot coupons assessed variation in fastener diameter and UTS for a continuous process. Coupons were selected to test 50% transverse and 50% longitudinal grain directions. This accounted for the variation in door grain conditions. A fastener creation process lasting 5 seconds gave the best properties but also produced hotter process parameters. A fluke digital thermometer was used monitor probe temperatures in between fasteners. A 9 second delay between fasteners was determined to give a consistent pin tool temperature during the creation process. Of the 9 seconds delay, it was determined that the first 5 seconds would be incorporated after FSSW while the anvil is left in contact with the skin. This allowed anvil cooling and prevented pulling away of the skin aluminum while at peak process temperature. The remaining 4 seconds of the delay was incorporated in a location off the spot after the anvil was lowered.
In addition, although not found to be required, a blued anvil condition was added to also prevent sticking after the fastener creation. The fasteners were created with the probe contacting the frame side of the assembly. Figure 17 shows the skin side of the fastener before and after the 9 second delay between fasteners was added to the process.

Skin side with pull away from immediate withdraw

Skin side after dwell and anvil bluing

Figure 17 (pull away at end of fastener creation)

Pressurized air was used to cool the pin and anvil during the fastener creation. A fastener inspection process was determined using a "go"-"no go" gage concept. The criteria used for the gage was a 0.365 in (9.27 mm) or larger diameter. The dwell time for fastener creation was also increased by 1 second on the first fastener in the series to heat up the probe and meet this diameter requirement. Figure 18 and 19 show the proposed production configuration for the gage and the actual gage.

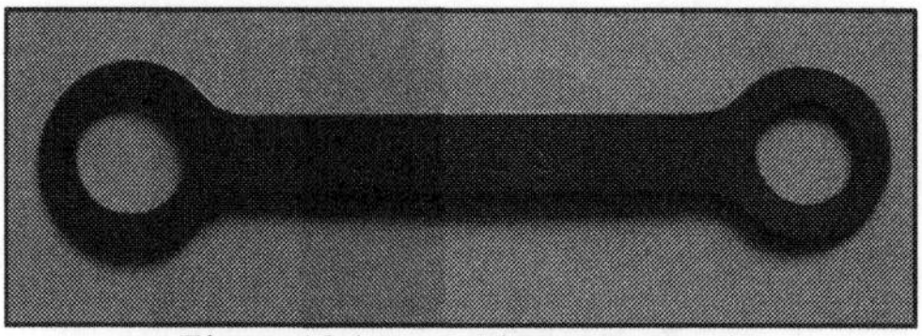

Figure 18 (proposed gage concept)

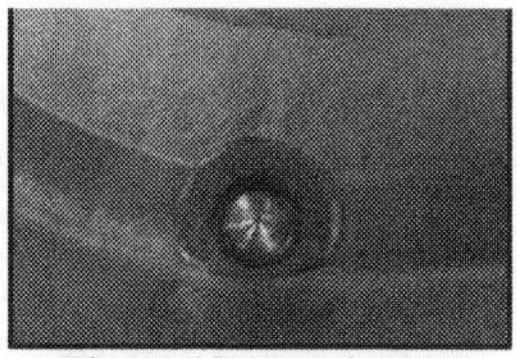

Figure 19 (actual gage)

A "Preliminary Welding Procedure Specification" Set-up sheet (PWPS) was developed per the American Welding Society Document AWSD17.3 [1] for the optimized Probe Tool-3 which is shown in Figure 20.

Preliminary Welding Procedure Specification (pWPS)	
Fabricator, Location: Advanced Joining & Processing Lab (NIAR), Wichita KS	
pWPQ Number: FSSW120513	Prepared by: Valeriano/ Vishwanath
Reference WPQ: AWS D17.3/D17.3M:2010	Date: 4/17/2012
Welding Operators: Vishwanath, Valeriano and Andrew	
Background	**Sketch of Joint Design a Test Specimen**
Part: HBC Door	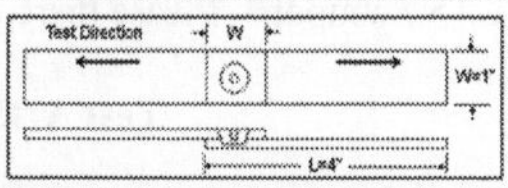
Weld Class: Class A	
Process Type: Friction Stir Spot Welding	
Friction Stir Spot Weld Method: Robotic C-frame	
Joint Type: Lap Joint	
Weld Type: Basic Poke Spot	
Aluminum Alloys	
Alloy 1: Top Sheet, Al 6061-T6, Bare, 0.063" thick	Alloy 2: Bottom Sheet, Al 2024-T3, Clad, 0.032" thick
Material Certification: AMS 4027	Material Certification: AMS 4041 / AMS_QQ_A 250/5
Manufacturer, Lot Number: N/A	Manufacturer, Lot Number: 74977/10
Grain Direction: Parallel with Test Direction	Grain Direction: Parallel with Test Direction
Preweld Cleaning: No	Preweld Cleaning: No
Machine Identification	**Machine, Weld Tool, Clamping Set-up Pictures:**
Machine: ZZX200S Series Kawasaki Heavy Duty Robot	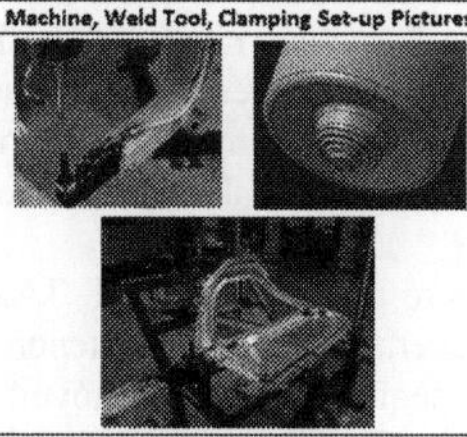
Weld Gun Model Number: P60B13200ICX23	
Weld Gun Manufacturer: Sanyo Denki	
C-Frame Anvil Material: Steel	
Welding Tool Identification	
Welding Tool Drawing Number: 11-0072-0425-2	
Welding Tool Type: Fix Probe Concave Shoulder	
Welding Tool Material: Maraging 300	
Welding Tool Coating: No	
Clamping Arrangement	
Clamping Fixture: Toggle Clamps	
Weld Procedures	**Welding Variables**
Test Piece Dimension: 12" width and 4" length	Axial Force (lbs [kN]): 1300 [5.782]
Number of Welds per Test Piece: 10	Plunge Depth: N/A
Weld Cycle Time: 16 sec	Spindle Speed (r/min): 2625
Tool & Anvil Cooling: Continuous Comp Air Cooled	Direction of Tool Rotation: Clockwise
Spacing between Welds: 1"	Dwell Time (s): 8
[redacted]	

Figure 20 (PWPS per AWSD17.3 Set-up sheet for Tool-3

The process specification calls for 3 Lots of 100 fastener shear coupons to be tested. This requires 10 of the 10 spot coupons to be run for each lot. 5 coupons were run transversely and 5 coupons were run longitudinally as previously mentioned. A force load plot at ultimate shear for the 100 samples in groups of 10 on a coupon is shown in Figure 21. UTS was shown to be 1.4 to 1.8 times the 250 lbf specification taken for resistance spot welding, which is the current process for the civil aircraft door chosen for this study.

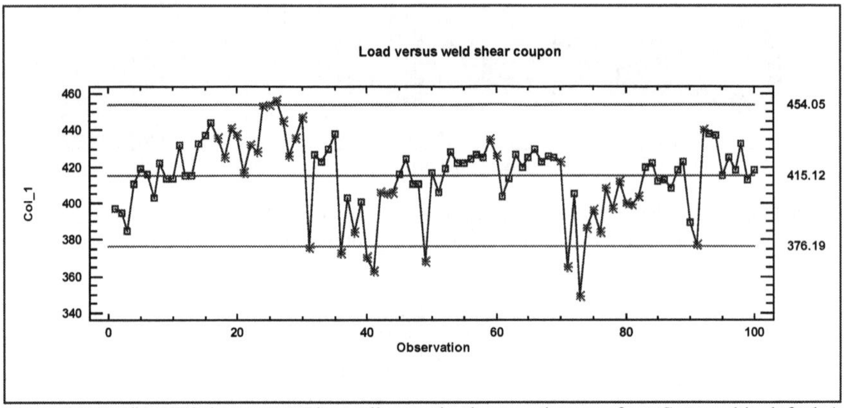

Figure 21 (Lot#1, UTS / coupon – Please disregard color or point type from Statgraphic defaults)

Test Article Results

The current requirements for resistance spot welding on the test article called for a 0.75 inch spacing. The lowest load from Lot #1 was calculated to maintain the cumulative door load with a 1.05 in spacing therefore the spacing parameter was set at 1 inch. The door was work hardened through forming operations therefore the dwell time in the fastener creation was increased to meet the gage diameter criteria. This resulted in a fastener creation time of 8 seconds per fastener with a 1 or 2 second increase for the first fastener in the series. The wait time in between fastener creations still remained at 5 seconds before disengaging the anvil and 4 seconds without any work piece contact to complete tool cooling. Two sequential approaches were used to try and mitigate the observed skin and frame contour mismatches.

Three test articles were identified as TA1, TA2 and TA3. TA1 was a sacrificial assembly which was also used for several swept spot fasteners using the MTS machine during a previous study phase. The fastener sequence for TA1 moved from the outside edges in and from the contoured area into the flat area. Tool holes were not installed in TA1 so the skin was best fit by visual means. For this study, some material was cut from the non-fastened areas to test fastener size parameters as a small coupon. TA2 was processed creating fasteners using the parameters developed from these small coupons in TA1. Initially, fasteners were created at a 6 inch spacing to tack the contour of skin to the frame. Fastener sequence moved from the inside of the assembly to the outside on the assembly. Fasteners were then created in between the tacked locations in the same sequence as the tacking was performed. TA3 was also tacked using 6 inch fastener spacing. The inside to outside ordering remained the same, but in response to the rippling seen in TA2, the sequence was changed to push the material toward the edges of the curved area last. It was hoped that this would wash the skin contour off the edge of the frame and produce a better fit.

TA1 sacrificial door fasteners and coupons are seen in Figures 22 through 24. Tab holes were not seen in the skin so after the best fit of TA1 3 holes were added to common tab locations found in TA2 and TA3 frame pieces. Two tooling locators were then made based on TA1 for locating these holes on TA2 and TA3.

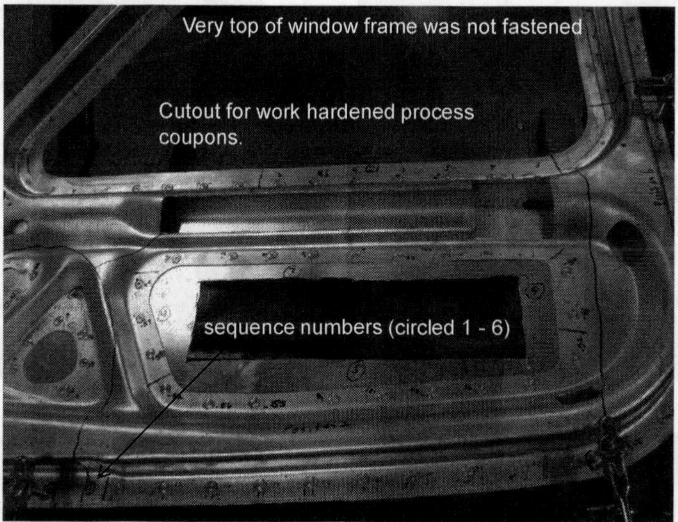

Figure 22 (TA1 Sacrificial Assembly)

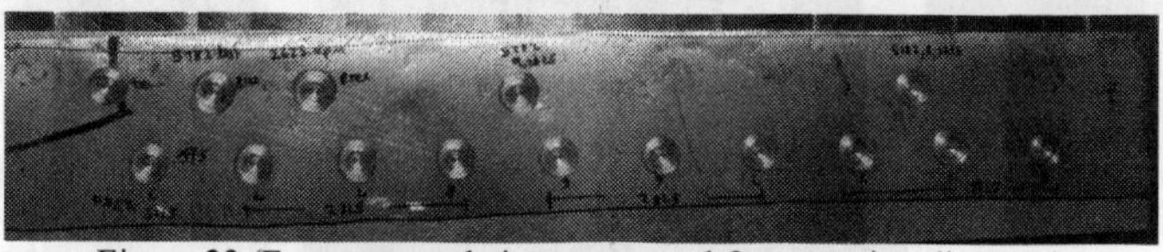
Figure 23 (Fasteners made in coupon tool for assessing diameter)
9 second fastener process time, 5782 N forge (Z), 2625 RPM with 0.418 in diameter

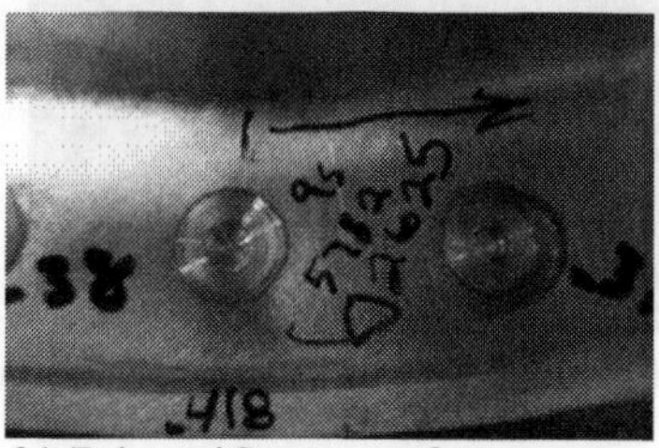
Figure 24 (Poke and Swept spot fastener comparison)

TA2 had only 2 tabs common to TA1 so was located using only those 2 tab holes. The 3rd tab hole in the frame common to a skin tab did not have holes that lined up for pinning. TA2 assembly is shown in Figures 25, 26 and 27. The numbers indicate the step wise assembly order.

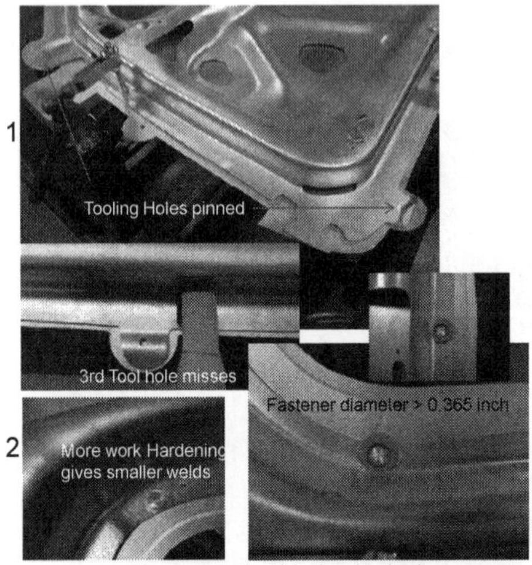

Figure 25 (Tool hole pinning)

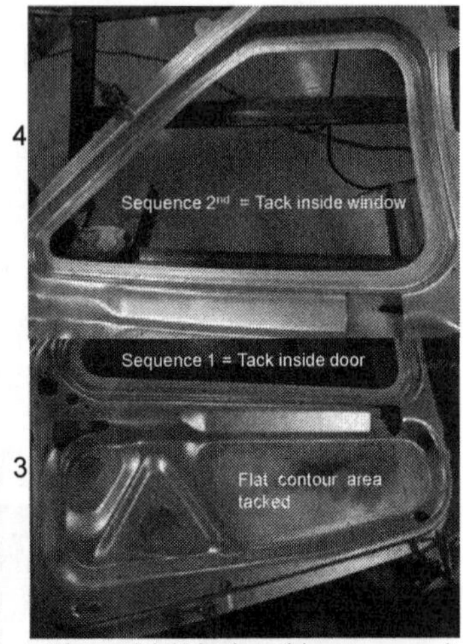

Figure 26 (6 inch tacking sequence)

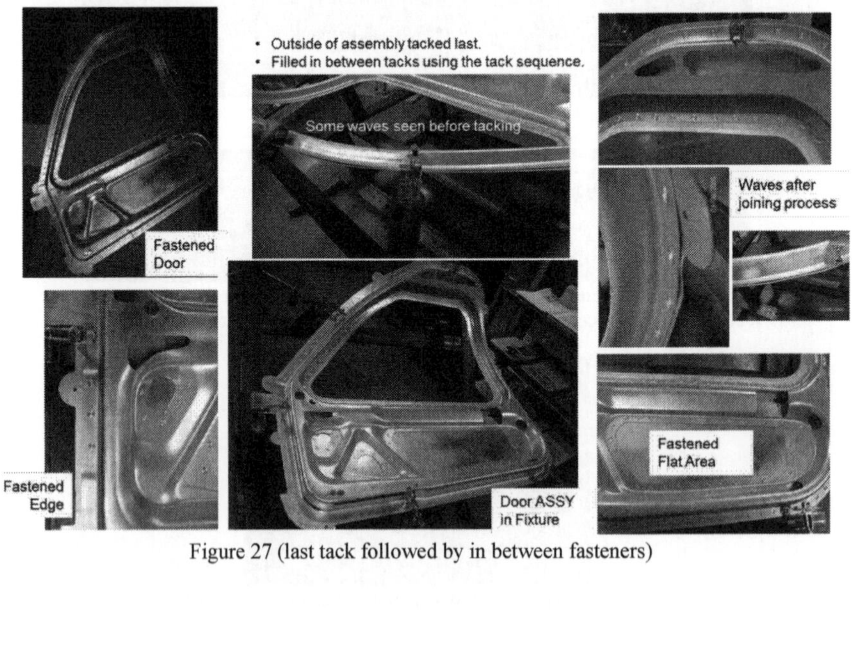

Figure 27 (last tack followed by in between fasteners)

For TA3, the desired result of washing the skin contour off the edge of the frame to produce a better fit was unsuccessful. An attempt to use a spacer in the last clamp area to see if the spacer would pull the skin back after fastening was also not successful. The frame included 2 of the 3 tab holes common to TA1 but the distance between the tabs did not match and all other tab locations were cut away from the frame. The fit up and fastener installation are seen in Figures 28 and 29.

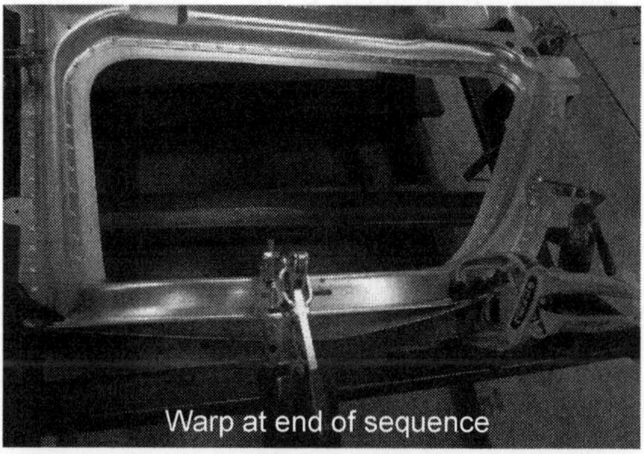

Figure 28 (TA3 warp)

Figure 29 (TA3 warping)

Conclusions

- Poke spot fastener ultimate loads are 1.4 to 1.8 times higher than resistance welding. This reduces the number of spots required.
- Parameters for optimized strength values required longer process times resulting in a hotter process temperature.
- The plunge and retract process of the "C" frame creates dimpling or flat spots on the outer skin surface based on the anvil geometry.
- The plunge spot "C" frame end effector required a process time up to 15 seconds per fastener in order to avoid the material pull off by the anvil while maintaining a stable pin tool temperature.
- Servo stops for plunge depth and fastener diameter using a "go"–"no go" gage can be used by production to inspect for allowable fasteners.
- Hard tooling could reduce process time (possibly by 2/3rds) through elimination of the time required to avoid material pull off.
- Coupon optimization may have been more accurate had the shear coupons been made of the work hardened material. This may have also reduced the creation time to a colder process with the chance of less distortion.
- Since locating tabs were missing and there was no hard tooling to check to, it was unclear how much distortion was caused by the fastener process.

Recommendations

- Design of a hard tooling surface ring should be investigated for following reasons
    - Maintains smooth contour of outer surface skin.
    - Increases speed of integral fastener process by avoiding material pull away.
    - Tooling could be designed to allow for rotation of engineering for easier robot movements and more efficient attitudes.
    - Multiple robot use could be investigated with one robot controlling the surface fixture and the other robot processing the fastener.

References

1. American Welding Society, Specification for Friction Stir Welding of Aluminum Alloys for Aerospace Applications, AWS D17.3/D17.3M:2010, 1st edition, 550 N.W. LeJeune Road, Miami, FL 33126, July 1, 2009.

Friction Stir Welding and Proceesing VII
Edited by: Rajiv Mishra, Murray W. Mahoney, Yutaka Sato, Yuri Hovanski, and Ravi Verma
TMS (The Minerals, Metals & Materials Society), 2013

EFFECT OF POST-WELD AGING ON THE CORROSION RESISTANCE AND MECHANICAL PROPERTIES OF FRICTION STIR WELDED ALUMINUM ALLOY 7475-T73

Bharat K. Jasthi[1], Erik Klinckman[1], Todd Curtis[1], Christian Widener[1], Michael West[1], Robert B. Ruokolainen[2], Ashish Dasgupta[2]

[1] Arbegast Materials Processing and Joining Laboratory
South Dakota School of Mines and Technology
501 E. St. Joseph St., Rapid City, SD 57701 USA

[2] Center for Advanced Technologies, Focus: HOPE
1400 Oakman Blvd, Detroit MI 48238

Keywords: Al 7475, Friction Stir Welding, Post-Weld Aging, Corrosion Testing

Abstract

The main objective of this work is to investigate the effect of post-weld artificial aging (PWAA) on the corrosion resistance and mechanical properties of friction stir welded Al 7475-T73 alloy. Friction stir welding (FSW) was performed on 0.22-in thick plates with varying heat inputs. The PWAA treatment performed was a stabilization treatment at 225° F for 24 hours, followed by aging at 325° F for 4 hours. Electrochemical and alternate immersion corrosion testing was performed to evaluate the effect of PWAA on the corrosion performance. Electrical conductivity measurements along with the tensile properties were evaluated as a function of weld heat input. Both corrosion resistance and mechanical properties of the PWAA specimens were comparable with the parent Al 7475-T73 material.

Introduction

Al 7475 is a high strength aluminum alloy with superior fracture toughness and good resistance to fatigue crack propagation. The fracture toughness values of this alloy are approximately 40 percent greater than alloy Al 7075 in the same tempers. This material can be used for fracture critical components of high performance aircraft, and in applications where high fracture toughness is a major design consideration. It is also found in the Navy weapon systems and it is commonly used in the manufacturing of shell casings and other structures. Many of these Navy support equipment and aging weapon systems are often subjected to wear and corrosion. Replacement of these old components which are low in quantity but high in value can be difficult and very expensive. Therefore, research regarding repair, refurbish and return to service of these components is extremely important. The main objective of this work is to investigate the effect of post-weld artificial aging on the mechanical and corrosion properties of friction stir welded Al 7475-T73 alloy. The broader objective of this study is to examine the suitability of FSW as a repair technique for a Navy cylindrical component that is subjected to localized corrosion. Therefore, this research focuses on evaluating the mechanical properties of the friction stir welded Al 7475 joints and optimization of post weld artificial aging cycles for restoring corrosion properties.

Prior researchers have shown that the resistance of 7XXX series alloys, particularly Al 7075, to exfoliation, intergranular corrosion, and stress corrosion cracking is greatly reduced by friction stir welding when left in the as-welded condition [1-5]. One potential method of restoring the corrosion resistance to 7XXX series alloys is through post-weld artificial aging after joining by FSW. The results of a variety of post weld heat treatments have been reported in literature; however, they deal almost exclusively with peak aged (T6) and underaged tempers (W,O,T3, and T4) [1,6-12]. General approaches thus far have been to: 1) leave the material in the as-welded condition, 2) apply a low temperature stabilizing heat treatment (e.g. 24 hours at 100-121°C), 3) apply a solution heat treatment to the material after welding and then age to the desired temper, 4) apply additional post-weld aging to material originally in a T6 or earlier temper to arrive at the final desired temper, or 5) to apply a localized post-weld heat treatment.

The option of applying a solution heat treatment to the material after welding is not a viable option for large built-up structures. The expense of solution heat-treating a large structure is prohibitive, but more importantly it would be difficult or impossible to maintain the dimensional stability of the built-up structure. In addition, the propensity for abnormal grain growth, reported by many researchers in the nugget and TMAZ, discourages the use of solution heat treatment following FSW as a primary mode of improving its resistance to corrosion. Another approach is to weld the 7XXX series material in T73 condition and then apply PWAA to restore the corrosion resistance and also improve the mechanical properties. Widener et al. [13] demonstrated that joining 7075 material originally in the T73 followed by a PWAA is preferable to welding in the T6 temper followed by aging to T73. This is primarily because of higher tensile and yield strengths and better exfoliation corrosion resistance.

PWAA at 225°F for 100 hours is the most commonly used on friction stir welds to restore the mechanical properties and corrosion properties of Al 7075. Widener et al. [13] demonstrated that a stabilization treatment at 225° F for 24 hours, followed by aging at 325° F for 4 hours gives equivalent or better corrosion and mechanical properties to the joints when compared with 225°F for 100 hours. PWAA at higher temperature (325°F) will reduce the total time for aging. On the other hand, it requires more care in keeping the material from over-aging. In this study a similar heat treatment was given to Al 7475 friction stir welds that were made with different heat inputs and the effect of the post-weld heat treatment on the mechanical and corrosion properties was evaluated.

Experimental

Bead on plate welds were made on 0.22-in (5.5-mm) thick alloy Al 7475-T73 plates. Welds were produced on an *I-STIR*-10 FSW system in a forge control mode. Tools with convex shoulder designs are better than flat or concave shoulder designs for FSW of curved or complex curvature shaped parts. Use of convex shoulder tools with additional scroll features on the shoulder contributes to a better quality weld surface with minimal flash generation. Since the end application for this project is to perform repair in cylindrical structures, convex shoulder step spiral pin tools were used. The pin tools were made of H13 tool steel and have hardness in the range of ~HRC 46-50. The convex radius of the pin tool shoulder is ~7-in. Two different pin designs (flat and rounded) as shown in Figure 1 were investigated to identify the degree of material flow as a function of process parameters. The differences in pin designs could account for different amounts of material mixing due to more or less aggressive pin features as this can impact weld quality and the travel speeds. A series of developmental welds were made in a forge control mode and by varying the essential processing parameters (forge force, rotational speed, and travel speed). The range of process variables investigated is tabulated in Table I. The final parameters chosen were 300 RPM, 15 IPM, and 0.5° tool tilt. Production welds were produced with three different heat inputs (varying forge force: 7000, 7500 and 8000 lb_f) with the objective of investigating the effect of varying heat input on the mechanical and corrosion properties.

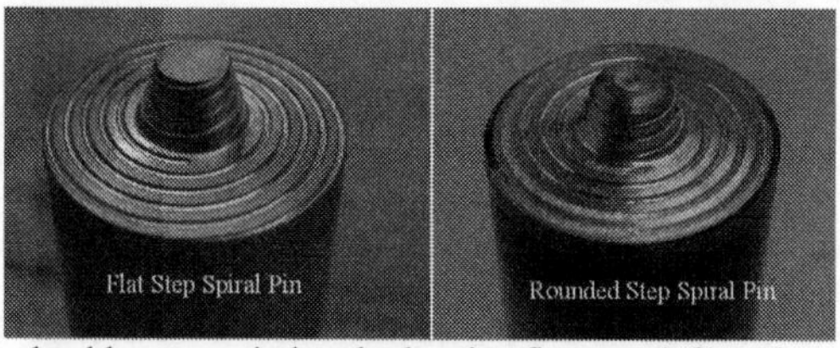

Figure 1: Convex shoulder step spiral tools showing flat step (left) and rounded step (right) designs.

Table I: Friction stir process parameter development.

Process Variable	**Range**
Forge Force (lb_f)	4000 – 9000
Rotational Speed (RPM)	300 – 600
Travel Speed (IPM)	10 – 35
Tool Tilt (degrees)	0.5 – 2.0

The welded plates were sectioned perpendicular to the welding direction and prepared for metallographic examination using standard metallographic polishing procedures. The polished specimens were etched with Keller's reagent (1.0 mL HF, 1.5 mL HCL, 2.5 mL HNO_3, and 95 mL H_2O) to observe the microstructure of the weld zone using the optical microscope. PWAA was performed on the welded panels for restoring the mechanical and corrosion properties of the friction stir welds. PWAA treatment was also performed on a few as-received Al 7475-T73 specimens to make sure the heat treatment did not affect the parent material negatively (over aging). Specimens were given a stabilization treatment at 225°F for 24 hours and aging was performed at 325°F for a period of 1-4 hours. Electrical conductivity measurements were performed on the specimens as a function of aging time. The final PWAA treatment was a stabilization treatment at 225° F for 24 hours, followed by aging at 325° F for 4 hours. Transverse tensile properties were evaluated in accordance with ASTM E8 standard.

Alternate immersion corrosion testing was performed for a period of 28 days on both parent and friction stir welds in accordance with ASTM G44 standard *(Standard Practice for Exposure of Metals and Alloys by Alternate Immersion in Neutral 3.5% Sodium Chloride Solution)*. Instead of neutral 3.5% NaCl solution, acidified synthetic sea water (as specified in *ASTM G85 Annex 3-Practice for Modified Salt Spray (Fog) Testing)* was used for corrosion testing. After 28 days the specimens were removed from the alternate immersion apparatus and cleaned with a wire brush to remove large portions of corrosion product that formed on the surface. The specimens were then cleaned in a heated solution (~90°C) of phosphoric acid and chromium oxide for 5-10 minutes followed by ultrasonic cleaning for 5 minutes. General corrosion rates were calculated from the weight loss measurements.

Electrochemical experiments were conducted using a potentiostat (Gamry Reference-600) in conjunction with a flat cell (K0235). The reference electrode used was a Ag- AgCl/KCl-saturated electrode, which has a potential of 197 mV relative to the normal hydrogen electrode (NHE). The test solution used for this electrochemical testing consisted of 58.5 g of NaCl and 9 mL of 30 % hydrogen peroxide per 1 L of aqueous solution (The solution is 1 M with respect to concentration of sodium chloride). The corrosion potentials (E_{corr}) (or open circuit potentials) were measured in accordance with ASTM G 69-03 standard (*Standard Test Method for Measurement of Corrosion Potentials of Aluminum Alloys).* The corrosion potentials were measured for a period of 60 minutes and the reported values in this report are the average values for the last 30 minutes. The E_{corr} values reported are an average of at least two tests; if the results were within 5mV after two tests no more were performed, if the average found after two tests was greater than 5mV more tests were performed until the average was within 10mV.

Results and Discussion

The microstructures of the friction stir welded specimens were characterized using optical microscopy. Figure 2 shows the macrographs of weld nuggets that were processed with the two different pin designs and at same processing conditions (7000 lb_f forge force, 300 RPM and 15 IPM). The figure clearly shows that rounded design produced better consolidation of the weld nugget than the flat step spiral pin design. Therefore, all the development welds and production welds were made using the rounded pin design. Welds that were made at higher travel speeds (25 and 30 IPM) produced worm hole defects, while the welds made at 20 IPM produced good quality welds with no visible defects. However, microscopic examination of the welds produced at 20 IPM showed an indication of "lack of mixing" beneath the pin tool. Therefore, all production welds for corrosion and mechanical properties evaluation were produced at 15 IPM, as this speed produced defect free and fully consolidated welds. Three production welds were produced at 15 IPM, 300 RPM, 0.5° tool tilt and varying forge force (7000, 7500, and 8000 lb_f) as the welding parameters.

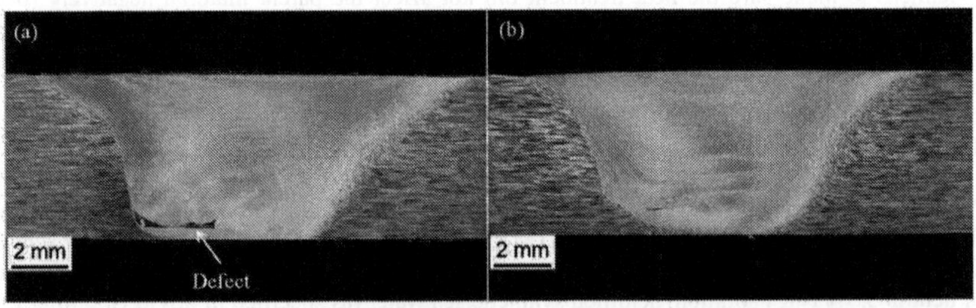

Figure 2: Macrographs of the weld nuggets processed with two pin designs: (a) flat step spiral pin; (b) rounded step spiral pin.

The relationship of electrical conductivity to aging in aluminum alloys is commonly used to identify whether or not a certain material has been properly aged. Electrical conductivity of precipitation-hardened aluminum alloys is strongly dependent upon the amount of dissolved solute in the matrix. As precipitation occurs, solute atoms are removed from the supersaturated solution, increasing its electrical conductivity as aging proceeds. Conductivity can also be influenced by the presence of precipitate free zones (PFZ) which commonly form along the grain boundaries where precipitates commonly nucleate. Guinier-Preston (GP) zones can also affect electrical conductivity by scattering electrons when the GP zones are extremely small and well distributed.

Even though FSW is a solid state joining process, the plastic deformation and friction heat generated from FSW process can re-solutionize the precipitates in the weld nugget, and partially dissolve or coarsen the precipitates in the thermo-mechanically affected zone (TMAZ) and heat affected zone (HAZ) regions in precipitation-hardened aluminum alloys [14]. Electrical conductivity measurements were taken from the Al 7475 parent and PWAA panels. PWAA treatment was also performed on a few as-received Al 7475-T73 specimens to make sure the heat treatment was not affecting the parent material negatively (over aging). These specimens were given the stabilization treatment at 225°F for 24 hours and aging was performed at 325°F for a period of 1-4 hours. Electrical conductivity measurements were performed on the specimens as a function of aging time as shown in Figure 3. The as received Al 7475-T73 had a conductivity of 40.5 % IACS (percent of the conductivity of the international annealed copper standard). The PWAA treatment performed on the parent material did not show any significant increase in electrical conductivities with increased aging time. Only a slight increase in the electrical conductivity in the parent material with aging time was observed. This suggests that a significant amount of precipitate coarsening did not happen in the parent material as a result of the PWAA treatment.

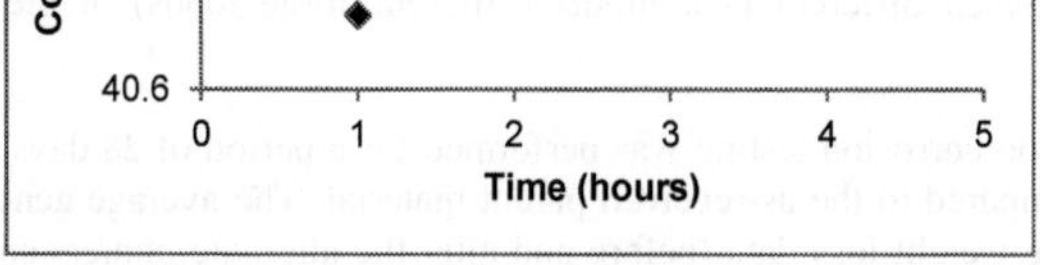

Figure 3: Conductivity of Al 7475 parent material after the stabilization treatment (225°F for 24 hours) followed by aging at 325 °F as a function of aging time.

A comparison of electrical conductivity across the weld nugget for as-welded and PWAA (stabilization treatment at 225° F for 24 hours, followed by aging at 325° F for 4 hours) specimens are shown in Figure 4. The as-welded specimens had a conductivity of 30 % IACS in the nugget region, suggesting a more supersaturated solid solution caused by dissolution of precipitates in the matrix during the FSW process. No significant variation was observed in the electrical conductivity of the different heat inputs (different weld forge forces) in the as-welded condition. Figure 4 also shows that the conductivity of the panels with PWAA increased in the weld nugget region and was comparable with the parent material. The average conductivity in the PWAA weld nuggets was in the range of 36.8 to 38.1 % IACS. It was also observed that the average conductivity in the PWAA weld nugget increased slightly as the heat input increased. This suggests that the response to the PWAA was better for the welds made with higher heat inputs.

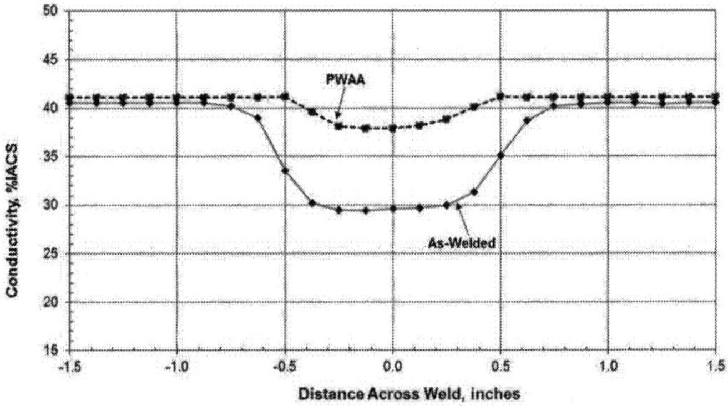

Figure 4: Conductivity across the weld nugget in the as-welded and PWAA condition (weld made at 8500 lb_f forge force).

Tensile testing results are shown in Table II. The reported tensile results are the average of three specimens for both parent and welded specimens. The predominant failure of the transverse tensile specimens is on the retreating side of the weld. As can be seen from the Table II, the transverse tensile strengths of the PWAA friction stir welded panels are approximately 90% of the parent material. However, a significant decrease in percent elongation of the transverse tensile specimens was observed. The comparison of the ultimate tensile strengths between the as-received Al 7475 and PWAA friction stir welds are shown in Figure 5. No significant difference was observed between different heat inputs (different forge loads) in the transverse tensile properties.

Alternate immersion corrosion testing was performed for a period of 28 days on PWAA friction stir welds and compared to the as-received parent material. The average general corrosion rates measured from the weight loss data (before and after the alternate immersion corrosion testing) are shown in Table III. The average corrosion rates measured for the parent Al 7475-T73 material is ~26 mpy, whereas for the welded specimens the corrosion rates changed from ~27 to 29 mpy. The results show the corrosion properties of the welded specimens after the PWAA treatment were comparable with the as-received parent Al7475-T73 material. However, no good correlation was identified between the various heat inputs and the corrosion rates.

Table II: Tensile properties of Al 7475 panels

Sample	YS (ksi)	UTS (ksi)	Elongation (%)
Al 7475-T73 (as-received)	62.5 ± 1	72.0 ± 0.7	19 ± 1
Al 7475-T73 + (225°F- 24 hr + 325°F- 1 hr)	61.9 ± 3	76.1 ± 0	18.0 ± 0
Al 7475-T73 + (225°F- 24 hr + 325°F- 2 hr)	61.2 ± 2	75.4 ± 0.5	19 ± 1
Al 7475-T73 + (225°F- 24 hr + 325°F- 3 hr)	58.2 ± 3	74.8 ± 0.2	18 ± 0
Al 7475-T73 + (225°F- 24 hr + 325°F- 4 hr)	58.7 ± 4	74.7 ± 0.2	18 ± 0
FSW-7500 lb_f + (225°F- 24 hr + 325°F- 4 hr)	54.9 ± 0.3	66.6 ± 0.2	9 ± 1
FSW-8000 lb_f + (225°F- 24 hr + 325°F- 4 hr)	54.5 ± 0.7	66.3 ± 0	10 ± 0
FSW-8500 lb_f + (225°F- 24 hr + 325°F- 4 hr)	49.3 ± 1	65.7 ± 0.5	8 ± 0

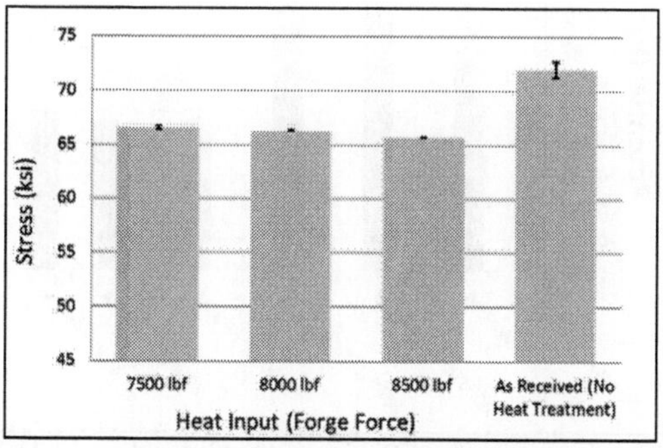

Figure 5: Comparison of ultimate tensile strengths between the as-received Al 7475-T73 material and PWAA friction stir welds

Table III: Average corrosion rates of Al 7475 Panels

Sample	Average Corrosion Rate (mpy)
Parent Al 7475- T73	25.8 ± 3.3
FSW (7500 lb_f) + PWAA	28.6 ± 0.2
FSW (8000 lb_f) + PWAA	29.4 ± 4.9
FSW (8500 lb_f) + PWAA	27.3 ± 1.2

Electrochemical corrosion potentials were measured for both parent and friction stir welded Al 7475 panels. PWAA specimens were also evaluated and compared to parent material. Figure 6 shows the corrosion potentials (with respect to the Ag-AgCl/KCl-saturated reference electrode) of the parent and welded specimens. The results show that the parent Al 7475-T73 specimens had a corrosion potential of about -705 mV, whereas the as-welded specimens had a potential of about -765 mV. No significant differences were observed in the potentials for the welds made with different heat inputs. The results indicate that the as-welded material showed a greater anodic (active) behavior compared to parent Al 7475-T73 material. Frictional heat generated and plastic deformation during the FSW process can re-solutionized the Mg and Zn rich precipitates (primarily $MgZn_2$) in the weld nugget. This causes large amounts of Zn and Mg to be retained in solid solution in the as-welded nugget region. Therefore, the decrease of corrosion resistance in the as-welded condition is most likely due to the greater degree of supersaturation that exists in the weld nugget than in the parent material. The PWAA specimens showed corrosion potentials in the range of -700 mV, which are comparable to the Al 7475-T73 parent material. This is most likely due to the re-precipitation of Mg and Zn rich precipitates during the post-weld aging process. Re-precipitation of these precipitates leaves the solid solution matrix with less Mg and Zn content, thus creating a lower anodic potential compared to the as-welded condition.

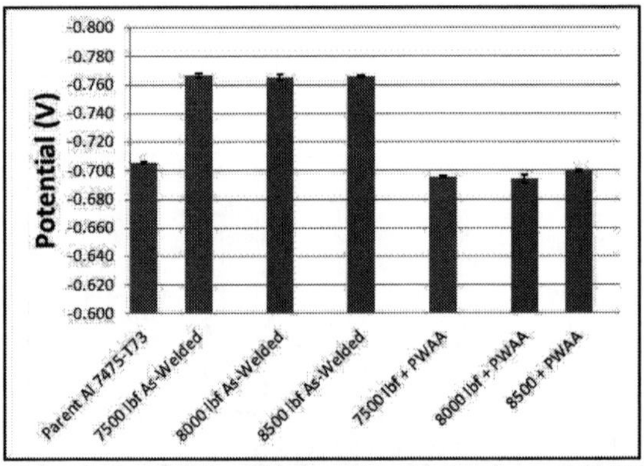

Figure 6: Corrosion potentials of friction stir welded and parent Al 7475 panels

Corrosion potentials were also measured using the Tafel extrapolation method. The potential scan was started after the sample had been exposed for one hour. The specimens were scanned in the range ± 250 mV with respect to the measured open circuit potential, at the rate of 0.167 mV/sec. Tafel scan comparisons for parent, as-welded and PWAA specimens are shown in Figure 7. This figure shows that the as-welded specimens showed a more anodic corrosion potentials when compared to the parent Al 7475-T73 material. Furthermore, the corrosion potentials for the PWAA specimens are comparable to parent Al 7475-T73 material. The overall corrosion potentials measured using the Tafel method are slightly lower by ~ 15-30 mV when compared to the ASTM G69 method for all specimens. However, both methods showed a similar corrosion behavior. This suggests that corrosion properties were effectively restored with the post-weld artificial aging process used.

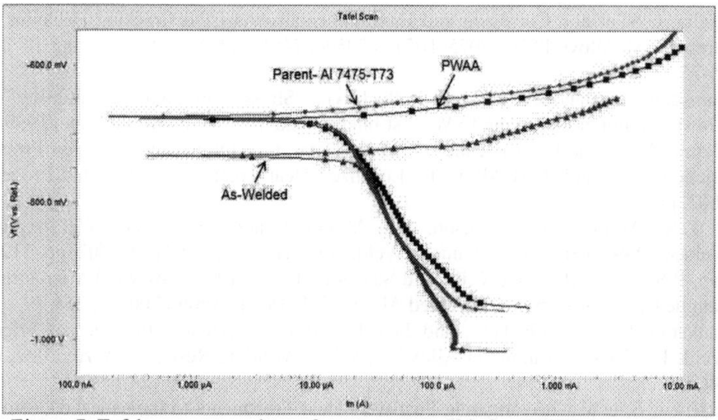

Figure 7: Tafel scan comparisons for parent, as-welded and PWAA specimens.

Conclusions

The effect of Post-weld artificial aging on the corrosion resistance and mechanical properties of friction stir welded 7475-T73 alloy was investigated. PWAA treatment of friction stir welded panels showed ~90% of the parent material strengths. Weld heat inputs had no effect on the resulting corrosion and mechanical properties of PWAA friction stir welded specimens. The corrosion potentials and corrosion rates of the PWAA specimens were comparable with the parent material. Finally, the PWAA treatment at 225 °F for 24 hours followed by 325 °F for 4 hours was effective and restored both corrosion and mechanical properties of friction stir welded Al 7475-T73 alloy.

Acknowledgements

Financial support for this project was provided by the Office of Naval Research under the contract number N00014-10-C-0258. The authors also gratefully acknowledge the assistance of Timothy Johnson and Matthew Carriker at the Arbegast Materials Processing Laboratory.

References

1. Leonard, A.J., "Corrosion Resistance of Friction Stir Welds in Aluminum Alloys 2014A-T651 and 7075-T651," 2nd International Friction Stir Welding Symposium, 26-28 June, 2000.
2. Anonymous. "Corrective Measures to Restore Corrosion Resistance Following FSW," Office of Naval Research, Report no. A580234, Rockwell Scientific, 2004.
3. Kumar, B. et al. "Applicability of FSW for Aircraft Applications," 46th AIAA SDM Conf., 2005.

4. Paglia, C.S. et al. "Corrosion and Environmentally Assisted Cracking Behavior of High Strength Al Alloys FSW: 7075-T651vs.7050-T7451," FSW and Processing II, 2003, pp. 65-75.
5. Lumsden, J., Mahoney, M., and Pollock, G., "Stress Corrosion Susceptibility in 7050-T7451Aluminum Following FSW," 1st International FSW Symposium, June, 1999.
6. Merati, A., Sarda, K., Raizenne, D., Dalle Done, C., "Improving Corrosion Properties of Friction Stir Welded Al Alloys by Localized Heat Treatment," FSW and Processing II, 2003, pp. 77-90.
7. Li, Z.X., Arbegast, W.J., Wilson, A.L., Moran, J., and Liu, J., "Post-Weld Aging of FS Welded 7249 Extrusions," Trends in Welding Research, 15-19 April, 2002, pp. 312-317.
8. Pao, P.S., Gill, S.J., Feng, C.R., and Sankaran, K.K., "Effects of Weld Microstructure on Fatigue Crack Growth in FS Welded Al 7050," TMS Aluminum 2001, p. 265-79.
9. Sankaran, K.K., Smith, H.L., and Jata, K., "Pitting Corrosion Behavior of Friction Stir Welded 7050-T74 Aluminum Alloy," Trends in Welding Research, April, 2002, pp. 284-286.
10. Paglia, C.S. et al. "Investigating Post-weld Heat Treatments to Increase the Corrosion and Environmental Cracking Behavior of 7075-T6 FSW," Trends in Welding Research, 2002, p. 279-283.
11. Lumsden, J., Pollock, G., and Mahoney, M., "Effect of Post Weld Heat Treatments on the Corrosion Properties of FSW AA7050," FSW and Processing II, TMS, March, 2003, pp. 99-106.
12. Dunlavy, M., and Jata, K.V., "High-Cycle Corrosion Fatigue of Friction Stir Welded 7050-T7451," FSW and Processing II, 2-6 March, 2003, pp. 91-98.
13. Widener et al., Evaluation of Post-Weld Heat Treatments to Restore the Corrosion Resistance of Friction Stir Welded Aluminum Alloy 7075-T73 vs. 7075-T6, Mat. Sci. Forum, 2007, Vols- 539-545, pp 3781-3788.
14. Li, Z.X., Arbegast, W.J., Wilson, A.L., Moran, J., and Liu, J., "Post-Weld Aging of Friction Stir Welded 7249 Extrusions," Trends in Welding Research Conference, 15-19 April, 2002, pp. 312-317.

FRICTION STIR WELDING AND PROCESSING VII

Friction Stir Welding: Light Materials – II

Friction Stir Welding and Proceesing VII
Edited by: Rajiv Mishra, Murray W. Mahoney, Yutaka Sato, Yuri Hovanski, and Ravi Verma
TMS (The Minerals, Metals & Materials Society), 2013

ON FRICTION STIR WELDING OF Mg-Zn-RE-Zr ALLOY USING THREADED TOOLS FOR AEROSPACE APPLICATION

S.P.Madavan[1], Manas Mohan Mahapatra[2] and Pradeep Kumar[3]

[1]Research Student, Mech. & Ind. Eng. Department, I.I.T Roorkee
[2]Assistant professor, Mech. & Ind. Eng. Department, I.I.T Roorkee
[3]Professor, Mech. & Ind. Eng. Department, I.I.T Roorkee
Roorkee, Uttrakhand,India-247667

Keywords: Friction stir welding, Tool geometry, Magnesium cast alloy

Abstract

Magnesium alloys are widely used for manufacturing aerospace engineering components due to their light weight, excellent high temperature properties and corrosion resistance. In the present investigation, Friction Stir Welding (FSW) has been investigated as a solid state joining method for Mg-Zn-RE-Zr cast alloy. The Mg-Zn-RE-Zr alloy (commonly known as RZ-5) of 6 mm thickness was fabricated by sand casting method in an aeronautical engineering component manufacturing environment. The cast Mg-Zn-RE-Zr alloy plates were then joined using FSW process by utilizing in-house developed tools with cylindrical threaded pin profiles and shoulder surface geometries. The effects of FSW process parameters such as rotational speed and welding traverse speed on the cast Mg-Zn-RE-Zr alloy welds were investigated. The tensile strength, percentage elongation, hardness and microstructure of the cast Mg-Zn-RE-Zr alloy welds were also investigated.

Introduction

The magnesium alloys are the light weight structural material with high specific strength and stiffness. In addition to this they have good damping capability, castability and machinability. These properties make magnesium alloys most attractive materials for application in aerospace engineering to reduce the effective weight [1]. The weight reduction is one of the most important and critical measure in ground and aerospace vehicles to improve fuel economy and to reduce environmentally damaging emissions [2]. Especially in the aircraft industry, in which the effective cost of weight savings is many times higher than in the automotive industry, magnesium alloys are preferred to aluminium alloy wherever possible[3].

Presently most of the magnesium alloy parts are produced by casting processes due to better fluidity and lesser cycle time, and the use of other manufacture technologies, such as plastic forming and arc welding, is still limited [4]. Most of the cast magnesium alloys are not joined by fusion welding due to the relatively poor quality of welds. The porosity present in the magnesium alloy casting itself leads to weld defect, and there is a tendency for formation of brittle inter-metallic phases and eutectics in the weld. High power density (beam) welding processes like Electron Beam Welding (EBW) and Laser Beam Welding (LBW) of cast magnesium alloys have provided better welds, but are very expensive. Hence, these joining methods have not yet been found very attractive for commercial use on magnesium components

[5]. The increasing demand of magnesium alloy imposes the development of effective and inexpensive welding techniques for industrial applications of magnesium alloys.

The Friction Stir Welding (FSW) which is a solid state welding process can be an alternative method for joining magnesium cast alloys. FSW is considered as the most significant development in metal joining in the recent past and it is a ''green'' technology, because of its energy efficiency, environment friendliness and versatility. FSW consumes considerably less energy than the conventional welding methods [6]. This method was invented at The Welding Institute (TWI) of UK in 1991, and it was initially applied to aluminum alloys and patented [6]. Recently, research is going on to use FSW technique for materials like magnesium, steel, titanium and metal matrix composites.

During FSW a non-consumable rotating tool with a specially designed shoulder and profile pin is inserted into the abutting edges of sheets or plates to be joined and traversed along the interface of plates as shown in Figure 1. The FSW tool serves two primary functions: (1) heating the workpiece due to the friction generated between the workpiece and the tool interface, and (2) stirring (deforming) and moving the material to produce the joint, which also produces considerable amount of heat. This localized heating softens the material around the pin, under the shoulder and combination of tool rotation and translation leads to movement of material from the front side of the tool to the back side of the tool [6]. As a result a joint between the plates is produced below the solidus temperature of the particular alloy system [7].

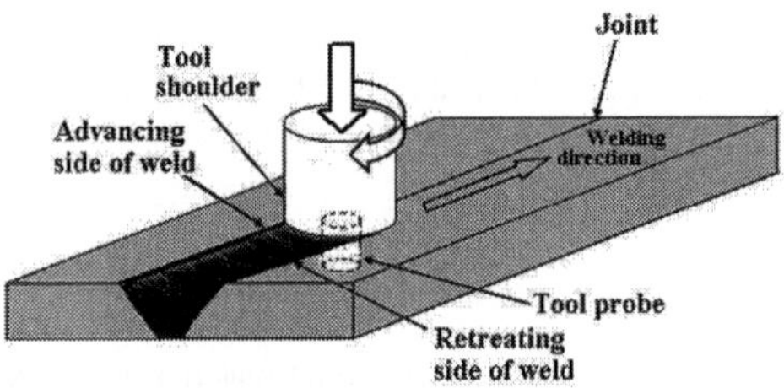

Figure 1. Schematic of FSW process

The friction stir welding of magnesium alloys has attracted investigators and many papers have been reported and published in recent years. However, the friction stir welding study on Zirconium and Rare-Earth (RE) containing ZE magnesium cast alloys which are widely used in aircraft industry are rarely found in the literature. In general, the findings related to friction stir welding of magnesium alloys emphasize that sound and pore-free welds are achievable on all magnesium wrought alloys and there is a fine, recrystallised microstructure in the weld zone. Different magnesium alloys have successfully been welded together. The tensile properties approach the properties of the base metal, except that the elongation seems to be reduced [8].

For aerospace applications high strength and creep resistance are the prime governing factor. The Mg-Zn-RE-Zr alloy, Elektron RZ5, is one among the well proven magnesium cast alloy, which is used for high integrity castings, operating at ambient temperatures or upto 150° C. Apart from the excellent castability, this alloy is weldable and exhibits pressure tightness [9]. Hence, this alloy is widely used for helicopter gearboxes, military equipment and aircraft components. The reported physical properties of Elektron RZ5 is listed in Table1, for this alloy optimum properties are achieved in T5 (Solutionised and artificially aged) condition. This cast alloy is widely used in aircraft industry but exhibits poor weld quality when joined by fusion welding techniques. The solid state welding of RZ5 is thus preferred to avoid the defects those occur due to fusion welding. Investigators have also studied the effects of texture and process variable on mechanical behaviour of friction stir processed magnesium alloys [10-11]. The development of FSW tool and process variable combinations to have the RZ-5 joint of adequate strength and quality is the aim of the present research.The objective of this work is to study the feasibility of producing the FSW joint of RZ-5 magnesium cast alloy with adequate mechanical strength.

Table 1. Physical properties of Elektron RZ5 [9]

Density	1.84 g/cc
Melting Range	510 - 640 °C
Coefficient of Thermal Expansion	27.1 X 10^6 K^{-1}
Thermal Conductivity	109 $W \cdot m^{-1} \cdot K^{-1}$
Specific Heat capacity	960 $J \cdot Kg^{-1} \cdot K^{-1}$
Modulus of Elasticity	44 X 10^3 MPa

Experimental setup

In this study, the Mg-Zn-RE-Zr cast alloy was sand-casted in-house. The castings were solutionised at 330 °C for 2 hours, air cooled to room temperature then aged at 180 °C for 16 hours, air cooled to room temperature. The nominal chemical composition and tensile properties at solutionised and aged condition (T 5) is listed in Table 2. The plates of size 150 mm X 75 mm X 8 mm were silted from the milled cast block, and then each plate was milled to get strips of 6 mm thick with parallel smooth surface. At the end of this milling operation the machined, smooth and clean surface on all the six faces of the plates were achieved.

Table 2. Chemical composition and tensile properties of base material.

Values in weight %		***Obtained***
Zinc	3.5 - 5.0	4.25
Rear Earth	0.8 - 1.70	1.20
Zirconium	0.4 - 1.0	0.54
Magnesium	Remainder	Remainder
Tensile properties		
0.2% YS (MPa)	135	153
UTS (MPa)	200	208
% Elongation	3	3

A longitudinal seam Friction Stir Welding was carried out on the material, with square-butt joint configuration. The faying surfaces of the plates were thoroughly cleaned just before starting the

welding operation to remove any oxidized and contaminated layers. The following processing conditions were used : spindle rotational speed of 550 rpm and traverse speed of 72 mm/min without tilting the tool and the process was carried out in atmospheric condition.
The schematic of a FSW tool with straight cylindrical threaded pin and concave shoulder used for FSW of the coupons is presented in Figure 2. This tool was machined out from the H-13 die steel and heat treated to get the required strength and hardness. The pin height is kept little lesser than the plate thickness to facilitate full penetration weld and allow the root weld formation [10].

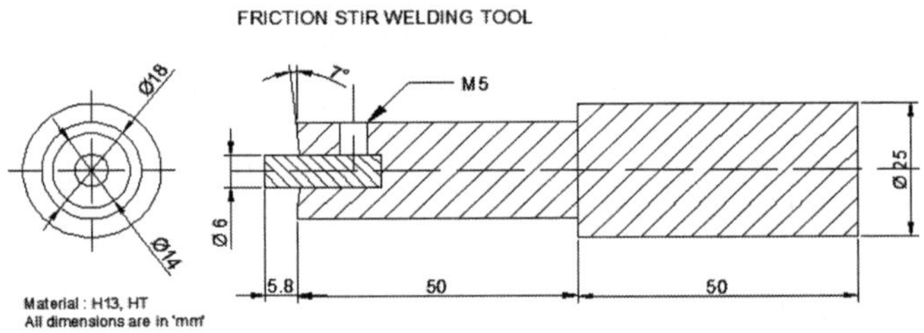

Figure 2. FSW tool drawing.

The welded sample was subjected to non-destructive radiograph test to ensure the soundness of the joint and found acceptable. The transverse tensile test specimens were machined across the weld joint in such a manner that the entire joint was in the gauge length of the specimen. The tensile test was carried-out using computer controlled Universal Testing Machine in accordance with ASTM E-8M. The fractured surface was studied using Scanning Electron Microscope (SEM). The metallographic samples were then mechanically polished using disc polishing machines with different grades of polishing papers to get scratch free surface. The polished weld surface was etched with 10% Nital. The micro and macro structure of the welded joint was studied. The Vickers hardness values with 5 Kg load were measured across the weld, at near the weld crown and at the middle of weld thickness.

Results and discussion

The welding trials confirmed the reproducibility of the Friction Stir Welding of magnesium cast alloy RZ-5. Acceptable quality weld was achieved over the length of the weld, but some spatter / flashing of the material observed which adhered to the surface. This flashing can be avoided by optimizing the process parameters like rotational speed, transverse speed and tool geometry.

The tensile test of the joint demonstrated ultimate strength of 185 MPa which is 88.9% of the base material strength, 0.2% proof stress of 135 MPa which is 88.2% of the base material proof stress and the percentage elongation of 1% which is 33% of the base material. From the tensile test results a moderate loss of ductility of the joint is observed. The fracture occurred at the interface of nugget and thermo mechanically affected zone. The friction stir welded joint strength levels are comparable with the base material but the ductility is comparatively less, this may be due to the formation of slightly harder weld nugget and the Thermo Mechanically Affected Zone (TMAZ). The Scanning Electron Microscopic analysis was carried out to understand the failure mechanism, revealed combination of brittle and ductile type of fracture, shown in Figure 3, the dimple fracture may be due to the micro-void coalescence rather than the cleavage of grains.

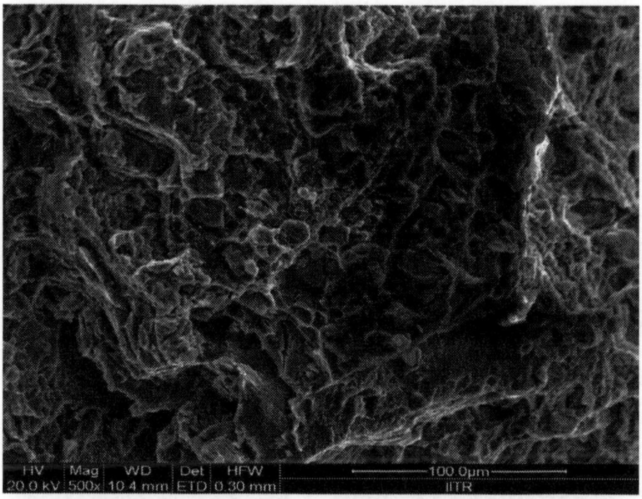

Figure 3. The fractured surface of FSW RZ-5 alloy.

The weld macroscopic photo taken at 7.5X magnification on the transverse section of the weld joint using stereomicroscope shown in Figure 4 revealed wide weld nugget. The weld zone has macroscopically visible pores at the advancing side (AD) on the joint, which may be due to the higher gas content of the base material. The presence of this has not affected the joint tensile strength much, which was discussed earlier. Similar results were also reported earlier for FSW of AM60 magnesium cast alloy [11].

Figure 4. Transverse section of FSW joint of two RZ-5 alloy.

The microstructure of the weld joint was examined using optical microscope and the micro photographs were taken and shown in Figure 5. The base material shown coarse grains having size 44 microns and eutectic segregations on the grain boundaries. The weld nugget zone shown the fine grains which are due to the dynamic recrystallisation happened during the welding operation due to combined effect of heat generated and the deformation due to the transverse force. This extremely high density of the precipitates inhibited the ability to distinguish between the separate phases present in it by optical microscope. The Figure 5 (c) shows the onion layer type of flow lines this shows the reflection of the material flow from the cooler walls of the HAZ. The induced circular motion leads to circles that decrease in radii and form the tube system [12].

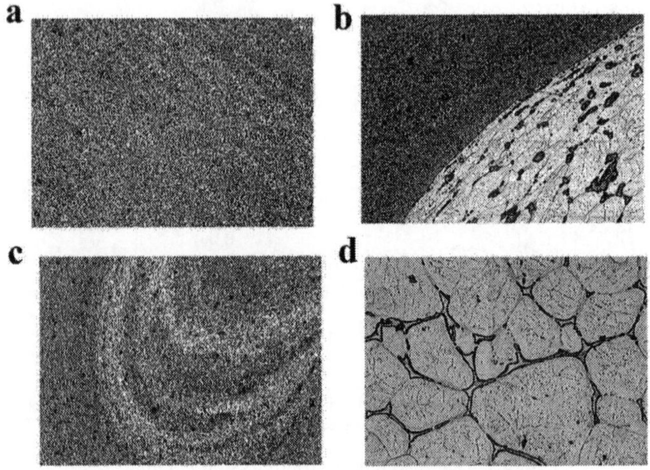

Figure 5 . The microstructure of FSW joint of RZ-5 Mg alloy at 100X: (a) Weld nugget zone; (b) Weld nugget and TMAZ;(c) Flow lines in Nugget;(d) Base Material.

The hardness measurements were made on the transverse side across the weld at two locations, one nearer to the weld crown and another at the middle of plate thickness with the 5kg load vikers hardness testing machine. This is was to find out the significance of change in hardness between base material, weld nugget and TMAZ. The changes in hardness were observed between the top of the weld nugget where the tool shoulder affects the crown of the weld and the middle region where the pin dictates the formation of stirred zone. The hardness profile plotted in Figure 6 indicates the scattering of hardness value. The hardness value was highest at the TMAZ interface due to the deformed material. The hardness value in the nugget zone is little higher than the base material. The hardness profile presented in Figure 6 indicates hardening of TMAZ and slight increase in hardness of weld nugget as compared to the base material. Overall a moderate increase in the hardness of TMAZ and weld nugget is observed in FSW of RZ-5 Mg alloy.

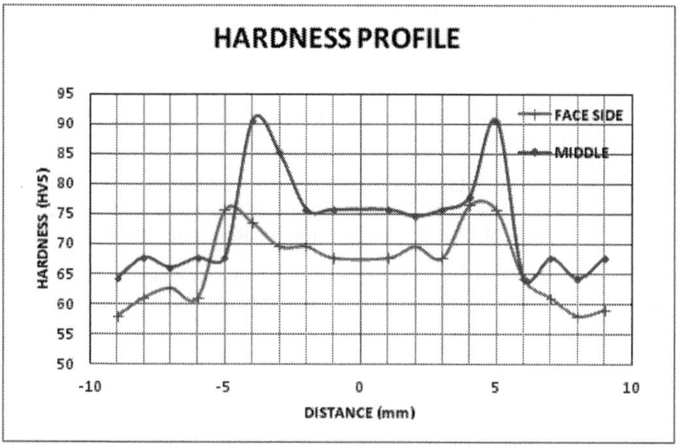

Figure 6. The hardness profile on transvers weld.

Conclusion

The Mg-Zn-RE-Zr, RZ-5 cast alloy was successfully friction stir welded and a brief investigation has been carried out. The results shown, that the transverse strength of the welded joint is comparable with the sand casted base material with decrease in ductility. The microstructure of the friction stirred zone was very fine with increase in the hardness. The nugget and TMAZ interface having higher hardness and the failure also occurred in the same location revealed mixed failure mode.

References

1. Su-Hyeon Kim, Bong-Sun You, Chang Dong Yim, and Young-Myoung Seo, "Texture and microstructure changes in asymmetrically hot rolled AZ31 magnesium alloy sheets," Materials Letters, 59 (2005), 3876–3880.

2. S.M. Chowdhury, D.L. Chen, S.D. Bhole and X. Cao, "Tensile properties of a friction stir welded magnesium alloy: Effect of pin tool thread orientation and weld pitch," Materials Science and Engineering A, 527 (2010), 6064–6075.

3. T.A. Freeney and R.S. Mishra, "Effect of Friction Stir Processing on Microstructure and Mechanical Properties of a Cast-Magnesium–Rare Earth Alloy," metallurgical and materials transactions A, 41A (2010), 73-84.

4. G.M. Xie, Z.Y. Ma and L. Geng, "Effect of microstructural evolution on mechanical properties of friction stir welded ZK60 alloy," Materials Science and Engineering A, 486 (2008), 49–55.

5. Jan Ivar Skar, Haavard Gjestland, Ljiljana Djapic Oosterkamp and Darryl L. Albright, “Friction Stir Welding of Magnesium Die Castings,” Magnesium Technology 2004 Ed. Alan A. Luo (TMS –The Minerals, Metals & Materials Society, 2004), 25-30.

6. R.S. Mishra and Z.Y. Ma, “Friction stir welding and processing,” Materials Science and Engineering, R 50 (2005), 1–78.

7. H. K. Mohanty, M. M. Mahapatra, P. Kumar, P. Biswas and N. R. Mandal, “Study on the effect of tool profiles on temperature distribution and material flow characteristics in friction stir welding”, Proc ImechE Part B: J Engineering Manufacture, 226(9) (2012), 1527-1535.

8. R. Johnson and P. Threadgill, "Friction stir welding of magnesium alloys", Magnesium Technology 2003 Ed. Howard I. Kaplan (TMS –The Minerals, Metals & Materials Society, 2003), 147-152.

9. Magnesium Elektron, Manchester, United Kingdom, Datasheet: 452: www. Magnesium-elektron.com.

10 G. Bhargava, W. Yuan, S.S. Webb and R. S. Mishra, "Influence of texture on mechanical behavior of friction-stir-processed magnesium alloy," (2010) Metallurgical and Materials Transactions A 41 (1), 13-17.

11 T. A. Freeney and R.S. Mishra, "Effect of Friction Stir Processing on Microstructure and Mechanical Properties of a Cast-Magnesium-Rare Earth Alloy," (2010) Metallurgical and Materials Transactions A, 41 (1), 73-84.

12. D. Venkateswarlu, N. R. Mandal, M. M. Mahapatra and S. P. Harsh, "Tool Designs Effects for FSW of AA7039", Welding Journal, 2012, Accepted for publication & in press.

13. Naiyi Li, Tsung-Yu Pan, Ronald P. Cooper, Dan Q. Houston, Zhili Feng and Michael L. Santella, “Friction stir welding of magnesium AM60 alloy”, Magnesium Technology 2004 Ed. Alan A. Luo (TMS –The Minerals, Metals & Materials Society, 2004),19-23.

14. K.N. Krishnan, “On the formation of onion rings in friction stir welds”, Materials Science and Engineering A, 327 (2002), 246–251.

Friction Stir Welding and Processing VII
Edited by: Rajiv Mishra, Murray W. Mahoney, Yutaka Sato, Yuri Hovanski, and Ravi Verma
TMS (The Minerals, Metals & Materials Society), 2013

MAGNESIUM BASED COMPOSITE VIA FRICTION STIR PROCESSING

S. Das[1], R. S. Mishra[1], K. J. Doherty[2], K. C. Cho[2] B. Davis[3] and R. DeLorme[3]

[1]Center for Friction Stir Processing, Department of Materials Science and Engineering, University of North Texas, Denton, TX 76203, USA
[2]Army Research Laboratory, Adelphi, MD 20783, USA
[3]Magnesium Elektron, Madison, IL 62060, USA

Keywords: Friction stir processing, metal matrix composites, reinforcements.

Abstract

Friction stir processing (FSP) was used to incorporate 6 vol.% of micron sized B_4C (6μm) reinforcements in WE43 alloy to form a surface composite. Multiple passes were utilized in order to achieve homogenization. Better distribution of the reinforcements, grain size refinement and an increase in hardness was observed after every FSP pass. The composite was 4-6 mm thick and showed higher values of hardness and modulus than that of the base material. Post FSP aging was performed at 210 °C for 48 h. The tensile properties were evaluated at room temperature for samples in as-FSP and after FSP+aging. The yield strength remained unaffected by the reinforcement addition.

Introduction

Composites have increasingly gained technological importance because they provide the opportunity to fabricate materials by a combination of a large variety of reinforcements and matrices. These result in achievement of hybrid properties in a single material which are otherwise not achievable in a conventional material. Although, the origin of metal matrix composites (MMCs) dates long back in history, it was not until the 1920s that dispersion strengthened alloys attracted scientific attention as the first composite materials [1]. Next came interest in continuously reinforced fiber composites in the 1960s, which displayed positive results but were discontinued owing to the high costs associated with the reinforcements and production [2]. This was followed by the era of discontinuously reinforced composites in which the costs of the reinforcements became much more competitive. It also permitted higher volume fractions of the reinforcements than that of a dispersion strengthened systems resulting in greater tailorability. As a result, discontinuously reinforced MMCs have recently attracted considerable attention in the automotive and aviation industries due to their remarkable properties and marked development in various processing techniques for MMCs [3, 4]. Today, composites have led to substantial advancements in various spheres of mechanical properties like strength, stiffness, wear resistance, creep resistance, fatigue resistance, etc. This has been possible due to marked improvements in various processing techniques. Various liquid phase or solid state processes exist for the fabrication of discontinuously reinforced composites. Though liquid phase processes have been quite successful in fabricating composites, there are still a few shortcomings associated with them [5, 6]. The liquid phase processes involve melting of the materials which can lead to formation of detrimental phases, segregation of second phase particulates, presence of a large number of defects, or difficulties associated with the wetting of the second phase particulates. These shortcomings may be overcome to a considerable extent by the use of solid state processes to manufacture composites [4, 7].

Friction stir processing (FSP), a solid state process developed by Mishra et.al [8] based on the principles of friction stir welding (FSW) is getting increased attention. It involves the traverse of a high strength rotating tool to locally heat the work piece and produce intense plastic deformation. The interplay between temperature and strain leads to a fine grained recrystallized grain structure. FSP is also stated as a 'green' process and consumes much less time, energy and money as compared to other thermomechanical processes [9]. Mishra et al. [10] have used FSP to fabricate a surface composite with SiC particulates in aluminum. This work was followed by other reports of composite fabrication in aluminum alloys with various reinforcements, like Al_3Ti, Al_2Cu, Al_2O_3 etc. via FSP [11,12,13]. Although aluminum forms an important part of lightweight materials, increasing demand for weight reduction has expanded the interest of the automotive and aircraft industries to magnesium alloys as well. Efforts have been made to fabricate various Mg based MMCs [14,15,16] with various ceramic reinforcements through processes like disintegrated melt deposition, powder metallurgy, and casting.

Fabrication of various composites can have different objectives. The goal for the addition of reinforcement to magnesium is to increase the stiffness so that light weighting of the structure can be done effectively. Significant focus has recently been given to the precipitation strengthened Mg-RE alloys, especially the Mg-Nd-Y system, owing to their enhanced mechanical properties along with improved castability, creep resistance and corrosion resistance. However, there are no reports of composites with WE43 as a matrix. B_4C on the other hand is the lightest ceramic reinforcement with the highest elastic modulus. Hence, an attempt was made in this study to merge the structurally efficient properties of WE43 alloys with high stiffness of B_4C through FSP.

Experimental Procedure

Processing

A commercial hot rolled F temper WE43 (30 cm x 5 cm x 1.6 cm) plate and commercial purity B_4C powder (6 μm) were used in this study. FSP was carried out with a stepped spiral conical tool with a featureless shoulder and a pin length of 6.5 mm, which was made of H13 tool steel. A set of holes with a depth of about 6 mm were drilled into the plate in the pattern shown in Fig.1 (a) and the B_4C powder was then filled into these holes. The FSP was carried out by employing one capping pass and four overlapping passes. The capping pass was performed with a tool with no pin at a tool rotation speed of 600 rpm, traverse speed of 101.6 mm/min and a tilt angle of 2.5°. The capping pass was performed to close the holes and thus reduce the loss of reinforcement. There was a complete overlap between all the overlapping passes. All the overlapping passes were performed with the same process parameters – a tool rotation rate of 1100 rpm, a traverse speed of 25.4 mm/min and with the tool tilt angle of 2.5°, but each subsequent pass was in opposite direction to ensure higher mixing of particulates. The plate was allowed to cool to room temperature after every pass.

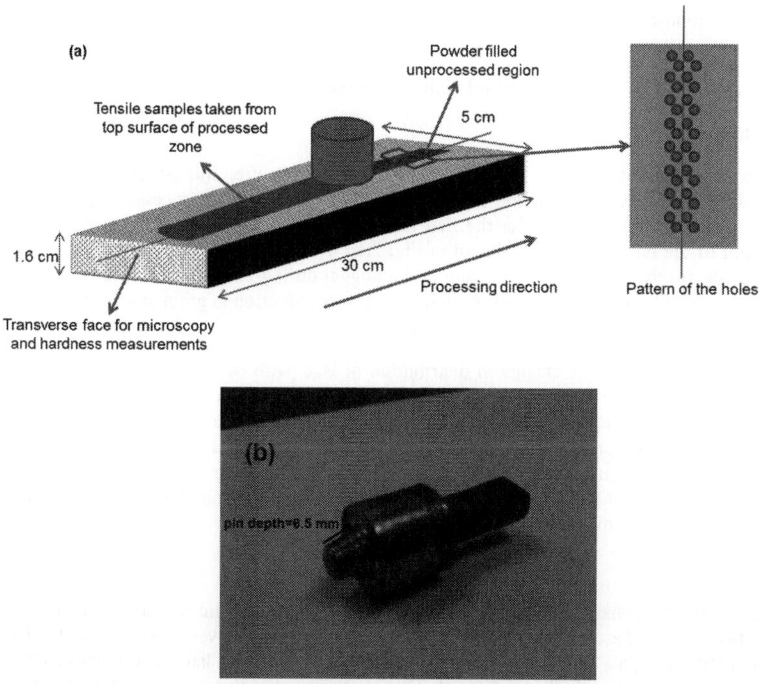

Fig. 1. (a) Schematic showing the FSP process, plate dimensions, pattern of the drilled holes, location of samples for characterization and (b) the tool used for processing.

Microscopy and Mini-tensile testing

The FSP plate was cut in the transverse direction, ground and polished for optical microscopy and scanning electron microscopy. Mini tensile samples with gage dimensions (2 mm x 1 mm x 1 mm) were cut out from the plate with tensile axis parallel to transverse direction for room temperature testing. Aging at 210°C for 48 hours was carried out after FSP. The tensile tests were performed both before and after aging to evaluate the change in strength and ductility with aging. The initial strain rate in these tensile tests was 10^{-3} s^{-1}. The tensile results were averaged over 2-3 samples.

Microhardness and Modulus Measurements

The microhardness and modulus measurements were done on samples from the transverse face of the plate. There samples were ground and polished. The hardness measurements were recorded at an interval of 5 mm across the transverse face of the weld as shown in Fig. 2(a). Composite stiffness values were obtained by instrumented Vickers hardness equipment, by running the equipment in 'load-unload' mode with a test force of 500 mN and a hold time of 10 seconds.

Results and Discussions

Microstructure

The macrostructure of the nugget after four passes is shown in Fig. 2(a) which depicts a basin shaped nugget. It is evident that the nugget is free of porosity and processing defects. The average grain size in the nugget after the first pass was 3.37±0.48 μm. This is much finer than 29±2.80 μm of the base material as shown in Fig. 2(b). Hence, significant grain refinement has taken place due to dynamic recrystallization in the nugget during FSP. Fig. 2(c) shows the grain size to be 2.71±0.69 μm after four passes. Hence, a further reduction in grain size is evident after four passes as compared to that after a single pass.

Figs. 2(c) and 2(d) depict the change in distribution of B_4C with overlapping passes. It is seen that the distribution of the reinforcements is banded after the first pass (Fig. 2(c) which becomes more uniform after four passes (Fig. 2(d)). This suggests that the tool geometry and the processing parameters were effective in distributing the reinforcements in the matrix resulting in a basically even microstructure. The SEM image in Fig. 2(e) showing a uniform microstructure also further confirms that FSP is an effective way of distributing second phase reinforcements in the matrix to modify the microstructure locally.

Microhardness

Fig. 3 shows the microhardness results after every pass. The dotted lines mark the ends of the nugget separating the base material from the nugget. The various curves represent the hardness evolution after every pass. The nugget shows a significant increase in hardness of about 19% in the nugget as compared to the base material. This is attributed to the presence of the hard B_4C ceramic reinforcements in the nugget. Apart from this, the grain size refinement brought about by dynamic recrystallization also contributes to the hardness improvement.

Table I lists the hardness values and scatter in the hardness values with increasing number of passes. An improvement in hardness is observed up to the third pass. The scatter in hardness readings within the nugget are considered to be secondary and could be related to minor microstructural variations. This scatter is also seen to reduce after every pass up to the third pass.

Table I. A summary of the range of hardness values after every pass.

	1 pass	2 pass	3 pass	4 pass
Microhardness range ($HV_{0.2}$)	86 - 99	88 - 98	93 - 95	88 – 93

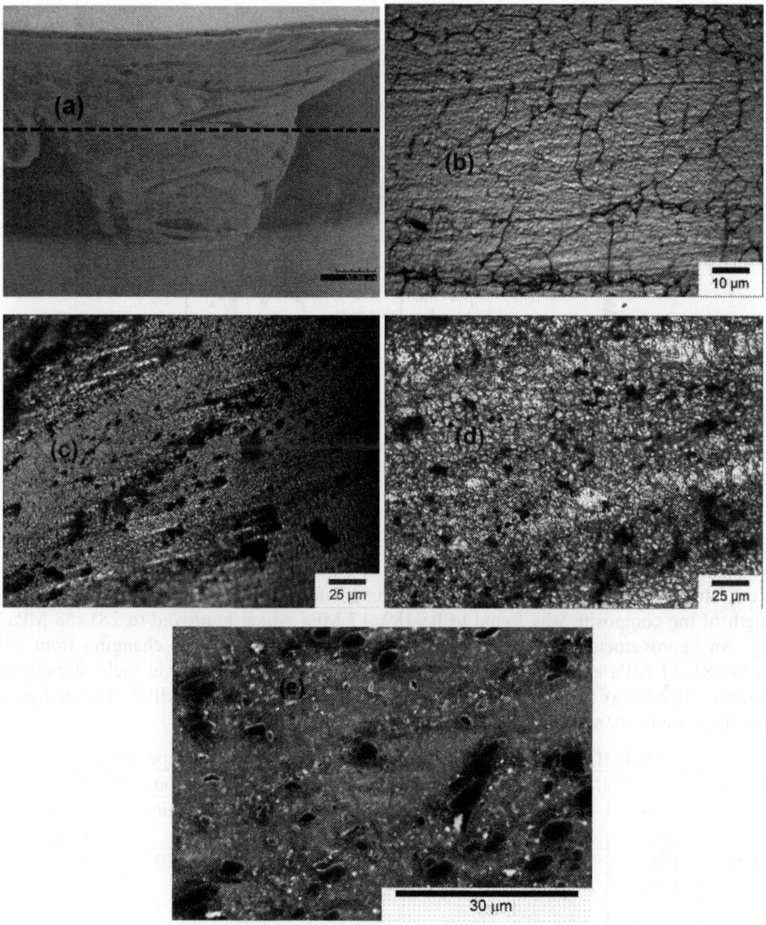

Fig. 2(a) Macrostructure of the nugget after FSP – the microhardness results were taken across the black dotted line, Optical micrographs of (b) the base material, (c) nugget after one pass, and (d) nugget after four passes. (e) A backscattered SEM micrograph showing the B_4C particles (darker).

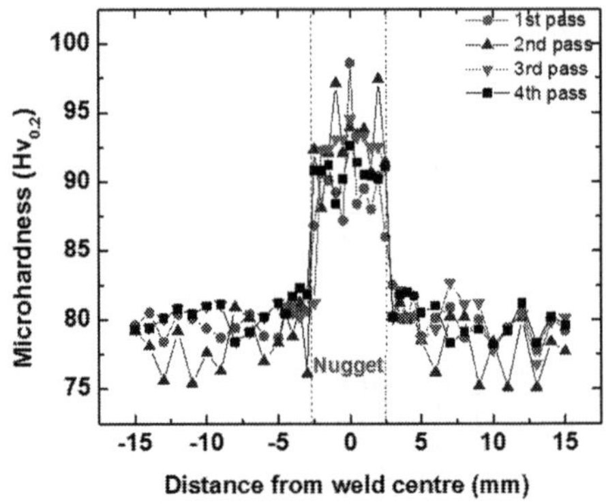

Fig 3. Variation in the microhardness across the processed region after every pass.

Tensile Properties

Table II shows the tensile properties for the composites tested at room temperature. The yield strength of the composite was found to be 189±17 MPa which improved to 281±58 MPa after aging. An improvement in ultimate tensile strength was also observed changing from 193±18 MPa to 281±51 MPa after aging. There was however no improvement in yield strength of the composite (189 MPa) as compared to that of the base material (185 MPa). The elongation to failure decreased very significantly in the composite specimens.

Table II. A summary of the tensile properties at room temperature.

Material	YS (MPa)		UTS (MPa)		Elongation to failure (%)	
	Before aging	After aging	Before aging	After aging	Before aging	After aging
WE43	185± 8.66	270±15 [17]	293±8	348±6 [18]	17.7±1.02	16±3
WE43+6 vol% B_4C	189±17	281±58	193±18	281±51	1.2±0.5	0

Modulus Measurements

Table III compares the elastic modulus of the alloy with that of the reinforced composite. The modulus of WE43 obtained in this study matched with the values reported in another study [18] and is listed in Table 3. A marked improvement of the modulus in the composite (about 41%) was observed as compared to the base material. The B_4C phase has an elastic modulus which is an order of magnitude higher than that of the matrix alloy and this accounts for the increased stiffness of the composite. The rule of mixture calculation suggest the WE43-6 vol.% B_4C composite modulus should be ~ 51 GPa (assuming, WE43 ~ 25 GPa, B_4C ~ 460 GPa). The current results show the improvement in modulus is consistent with the rule of mixture approach and indicate an effective load transfer. The force vs indentation depth curves for various readings are shown in Fig. 4, and these were used for modulus measurement of WE43 with and without reinforcement.

Table III. A summary of the modulus values for both the base material and the composite.

Material	WE43	WE43+6 vol.% B_4C
Elastic modulus (GPa)	26.54±0.23 25.0 – (F temper) [18] 25.8 – (T5 temper) [18]	43.47±2.87

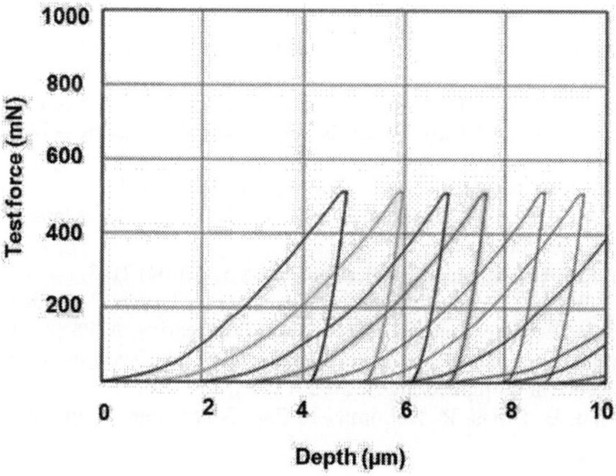

Fig. 4. A graph showing force vs indentation depth for various modulus readings of WE43+6 vol.% of B_4C composite.

Conclusions

Friction stir processing fabricated the composite successfully with a uniform dispersion of the reinforcements in the matrix which resulted in uniform and consistent properties. Following conclusions can be drawn based on foregoing discussion.

I. Grain refinement was observed as compared to base material. A decrease in grain size was also observed with overlapping passes.
II. An appreciable increase in hardness of about 19% as compared to the base material.
III. An improvement in modulus of about 41% was observed in the composites as compared to the base material by addition of 6 vol% of reinforcements.
IV. Though an improvement in ultimate tensile strength was observed for the composites, the yield strength remained unaffected.

V. Considerable reduction in ductility was observed for the composites.
VI. The composites show a similar aging response as that of the base material.

References

1. T.W. Clyne, P.J. Withers in ' An Introduction to Metal Matrix Composites ", edited by E.A. Davis, I.M. Ward, FRS (1993) 3.
2. I.A. Ibrahim, F.A. Mohamed, E.J. Lavernia, Journal of Materials Science 26 (1991) 1137.
3. T.W. Chou, A. Kelly, A. Okura, Composites 16 (1986) 187.
4. H.J. Rack, in " Processing and Properties of Powder Metallurgy Composites ", edited by P. Kumar, K. Vedula, A. Ritter (The Metallurgical Society, Warrendale, PA, 1988) 155.
5. P.K. Rohatgi, R. Asthana, S. Das, Int. Met. Rev. 31 (1986) 115.
6. R. Mehrabian, Materials Research Society Symposium Proceedings 120(1988) 3.
7. A.L. Geiger, M. Jackson, Advanced Materials and Processes 7(1989) 23.
8. R.S. Mishra, M.W. Mahoney, S.X. McFadden, N.A. Mara, A.K. Mukherjee, Scripta Materialia, 42(2000) 163.
9. R.S. Mishra, Z.Y. Ma, Materials Science and Engineering R 50 (2005) 1.
10. R.S. Mishra, Z.Y. Ma, I. Charit, Materials Science and Engineering A341 (2003) 307.
10. C.J. Hsu, C.Y. Chang, P.W. Kao, N.J. Ho, C.P. Chang, Acta Materialia, 51 (2006) 5241.
11. C.J. Hsu, P.W. Kao, N.J. Ho, Scripta Materialia, 53 (2005) 341.
12. A. Shafiei-Zarghani, S.F. Kashani-Bozorg, A. Zarei-Hanzaki, Materials Science and Engineering A500(2009) 84.
13. S.F. Hassan, K.F. Ho, M. Gupta, Materials Letters 58 (2004) 2143.
14. X. Zhang, H. Wang, L. Liao, X. Teng, N. Ma, Materials Letters 59 (2005) 2105.
15. H. Ferkel , B.L. Mordike, Materials Science and Engineering, A298 (2001) 193.
16. S.K. Panigrahi, W. Yuan, R.S. Mishra, R. DeLorme, B. Davis, R.A. Howell, K. Cho, Materials Science and Engineering A530(2011) 28.
17. T. Sano, J. Yu, B. Davis, R. DeLorme, K Cho, Magnesium Technology, TMS (2011) 345.

EFFECT OF INITIAL MICROSTRUCTURE ON THE MICROSTRUCTURAL EVOLUTION AND JOINT EFFICIENCY OF A WE43 ALLOY DURING FRICTION STIR WELDING

S. Palanivel[1], R.S. Mishra[1], B. Davis[2], R. DeLorme[2], K.J. Doherty[3], K.C. Cho[3]
[1]Center for Friction Stir Processing, Department of Materials Science and Engineering, University of North Texas, Denton, TX 76203, USA
[2]Magnesium Elektron North America Inc., Madison, IL 62060, USA
[3]U.S. Army Research Laboratory, Materials and Manufacturing Science Division, Aberdeen Proving Ground, MD 21005, USA

Keywords: Friction stir welding, Strain rate, Dynamic recrystallization, Joint efficiency, Stir Zone (SZ)

Abstract

The initial microstructure plays an important role in determining the spatial and temporal evolution of the microstructure during friction stir welding (FSW). The overall kinetics of microstructural evolution is location sensitive and the effect of the process strain, strain rate and thermal cycle creates complexities. In the present study, magnesium based WE43 alloy has been welded employing two different welding conditions. Joint efficiency has been subsequently evaluated. The results have been correlated with detailed microstructural information obtained from SEM and OIM-EBSD. The influence of microstructural evolution on strength has been analyzed. This framework provides an approach to maximize joint efficiency.

Introduction

The onset of many desirable features like fuel efficiency and high specific strength are driving Mg alloys towards potential structural applications in the future. Strengthening is often a key property for material selection in a majority of applications. As a result, the specific strength advantage of the Mg alloys over their counterparts has triggered intensive effort for their use in engineering applications. WE43 is currently touted to be one of the most promising Mg alloys due to its good elevated temperature properties, creep performance, flammability resistance and relatively higher strength [1,2]. However, welding of Mg alloys is a major challenge and limits the ability to fabricate large and complex components [3]. Therefore, joint efficiency can have significant impact on expanding its applicability towards various sectors in the near future. Extensive study on Al alloys and a lucid understanding of microstructural evolution during the friction stir process has been formulated and is available in the literature [4,5]. Starting from Al alloys, a variety of Mg alloys have also been studied.

Feng et al. [6] studied the effect of friction stir processing(FSP) on an AZ91 alloy and reported the dissolution and breakup of coarse eutectic $Mg_{17}Al_{12}$ phase. Park et al. [7] demonstrated the importance of texture and related it to the mechanical properties of an AZ61 alloy subjected to FSP. Ma et al. [8] suggested that FSP plays the role of solution treatment in AZ91 and saves time. Cao et al. [9] reported values of 69-78% for joint efficiencies in an AZ31-H24 alloy with failure occurring predominantly between stir zone(SZ) and thermo mechanically affected

zone(TMAZ). Several studies has been completed on the precipitation strengthened Mg alloys and indicate a strong dependency of strength on the evolution of strengthening precipitates [10,11,12,13] However, there is a strong demarcation with respect to the global value of strength, contributing factors and evolution path as a function of alloy chemistry during FSW. The thermal stability of the precipitates strongly determines the process conditions to be employed for achieving a desired microstructure. In comparison to the dissolution of precipitates in the AZ series, strengthening in Mg-Al-Ca and rare earth containing alloys like GWZ and WE43 is determined by the ability to control the intermetallic phases [6,11,12,13].
Freeney et al. [12] showed enhancement of the mechanical properties by breakage and dissolution of the second phases in a cast EV31 alloy. Xiao et al. [11] showed a large increase in the strength of a cast GW103 alloy after post FSP ageing. This was achieved by dissolving the Mg_5Gd intermetallics during processing. However, Zhang et al. [13] showed that the breakage and homogeneous distribution of Al_2Ca increased the strength and recommended faster welding speeds and lower heat input. Considering a variety of intermetallics (Mg_2Y, $Mg_{24}Y_5$, $Mg_{41}Nd_5$) that form in a WE43 alloy during casting and remain during subsequent shaping operations, it is interesting to study their response as a function of welding parameters [14,15,16]. Investigations indicate that different starting tempers can have significant impact during processing and result in different final values of strength. Freeney et al. [17] studied the effect of FSP on a cast WE43 alloy and obtained a nugget strength equivalent to 120% of the base material. Though a large extent of literature exists on the friction stir welding of Mg alloys, the industrially important WE43 alloy has not been evaluated much. To the best of our knowledge, no report exists on the microstructure-joint efficiency relationship in a wrought WE43 alloy.

The present study aims at correlating the effect of process parameters on the microstructure and mechanical property of a T5 temper WE43 alloy.

Experimental Details

For this study, 3.5 mm thick sheets of WE43 alloy in the rolled condition was provided by Magnesium Elektron. The nominal composition of the alloy was 4% Y-3% Nd-0.5% Zr–bal Mg (all in wt%). The alloy was subjected to ageing at 210 ^{0}Cfor 48h to maximize the base strength. The samples were longitudinally friction stir welded by employing a conical threaded pin made of tool steel with a concave shaped featureless shoulder. The pin height, pin diameter at the tip and root of the pin were 1.4 mm, 3.4 mm and 5.6 mm, respectively. A single pass FSW was performed at tool rotation rates of 600 and 800 rpm and a constant traverse speed of 102 mm/min.

The specimens for microstructural examination were cross sectioned perpendicular to the FSW direction. A standard sequential metallurgical procedure such as cutting, mounting and polishing with alcohol based diamond polishing compound was employed. Structural features were subsequently revealed by etching for duration of 10s in a mixture of acetic glycol (20 mL), (1 mL) HNO_3, (60 mL) ethylene glycol, and (20 mL) water, and washing with ethanol. Microstructural analysis and characterization was extensively carried out by employing a scanning electron microscope (FEI Nova NanoSEM 230). Samples for EBSD analysis were prepared using diamond based compounds to a final polishing step of 0.02 μm. These samples were subsequently electro-polished (40% H_3PO_4 and 60% ethanol at 20°C) using a voltage of 2–

3V for 10–15s and washed using ethanol. A FEI Nova NanoSEM 230 (20 kV, 3.1 mA, tilting 70°, 0.3 μm step size) was used to record the EBSD patterns for stir zone.

The Vickers microhardness measurements were done at a load of 1.961 N with a 10 s dwell time. These measurements were performed through the center of the transverse cross section in the SZ and extending into the base matrix on either sides for a distance of 15 mm. Quantitative analysis was done using the ImageJ software package. The grain size determination was done using the linear intercept method.

Results and Discussion

Initial microstructure: As rolled + T5

The as-rolled microstructure was subjected to ageing based on the optimal conditions reported in literature. This heat treatment has been reported to precipitate strengthening β" and β' phases [18,19,20]. Extensive analysis of the microstructure depicted three interesting features: shear bands, intermetallic clusters of RE and precipitation gradients. Fig (1a) shows patches of inhomogeneity in the rolled structure suggesting preferential shear localization during rolling. Jin et al. [21] studied the deformation behavior of a coarse grained AZ31 alloy and observed such inhomogeneous structures. The observation was reasoned based on dissimilar recrystallization magnitude of different grains owing to their orientation with respect to the compressive stress. Serajzadeh et al. [22] employed modeling and related the microstructural inhomogeneity to the inherent and inevitable strain gradients. The study also suggested that the extent of inhomogeneity would be directly proportional to the reduction per pass. In the present case, a greater reduction per pass at lower temperature substantiates the microstructural features. Another plausible reasoning could be the persistence of the solute that segregated to the interdendritic spaces during solidification. This mechanism of chemical partitioning can lead to a local supersaturation and subsequent precipitation of non-strengthening precipitates. Thus, the prior microstructure plays a crucial role in the subsequent evolution of microstructure and mechanical variables during isothermal ageing. The average grain size was 35±15 μm.

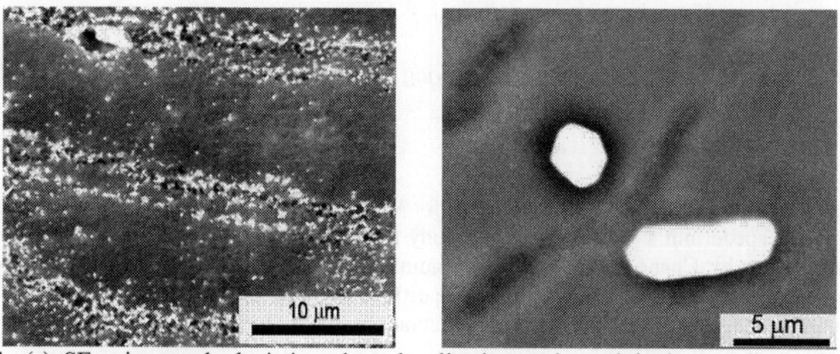

Fig. 1. (a) SE micrograph depicting shear localization and precipitation gradients (b) BSE micrograph showing cuboidal $Mg_2YZr_{0.3}$ and plate shaped $Mg_{41}Nd_5Zr$ intermetallics

Fig. 1(b) shows the presence of large cuboidal intermetallic of Mg_2Y and plate like $Mg_{41}Nd_5$.

These intermetallics have been reported to form during the solidification stage and persisted during subsequent rolling operations [14-15]. Such cuboidal precipitates have been reported by numerous researchers in Gd containing alloys [11]. High thermal stability of these precipitates has been reported and their melting point has been quoted to be around 620 ^{0}C.The presence of such coarse particles(~4-5 μm) would be inefficient in serving as an obstacle to dislocation motion. Therefore, it would be interesting to note the behavior of these particles subjected to FSW.

FSW

This study aims at reasoning the evolution of microstructure as a function of welding parameters and resultant thermal cycle. The major contributing factors to strengthening in a WE43 alloy are grain size and precipitate descriptors. Therefore, it is necessary to understand the competing aspects of strain, strain rate and temperature. The deformation strain, peak temperature and the grain size are correlated employing the Zener-Holloman parameter, Z (s^{-1}). Table I summarizes the mathematical equations employed to establish the Zener-Holloman parameter from the literature.

Table I. Equations for process parameters based calculations [23-26].

$Z = \dot{\varepsilon}\ \exp\left(\frac{Q}{RT}\right)$ (1)	$\varepsilon = ln\left(\frac{l}{APR}\right) + \left\|ln\left(\frac{APR}{l}\right)\right\|$ (2)	$l = 2r\cos^{-1}\left[\frac{r-x}{r}\right]$ (3)
$\dot{\varepsilon} = \varepsilon/t$ (4)	$l = \frac{APR}{v}$ (5)	$\ln d = A1 - A2\ln Z$ (6)

where $\dot{\varepsilon}$ is the strain rate, R the gas constant, T the temperature, Q the activation energy, l the maximum deformed length, APR the tool advance per revolution, r the tool pin radius, x the perpendicular distance to welding direction from retreating side to advancing side of the tool, v the welding speed (mm s^{-1}) and d is the grain size.

The peak temperature is determined by employing the Arbegast empirical relationship: [27]

$$\frac{T}{T_m} = K\left(\frac{\omega^2}{2.362v\,x\,10^4}\right)^{\alpha} \quad (7)$$

Where, T_m is the melting temperature (893 K for WE43), ω the tool rotation rate(rpm.) and v the tool traverse speed(mm s^{-1}). In the present study the values of K and α were averaged over the reported values by Chang et al. [28] and Commin et al. [29] for AZ 31 and Mg alloys. The activation energy was assumed to be the lattice diffusion of Mg and taken to be 135 $Jmol^{-1}$. Table II summarizes the process parameter based calculations employing equations from Table I.

Table II. Summary of process parameter based calculations.

Rotational speed (RPM)	Welding speed (mm/min)	Strain	Strain rate (s^{-1})	Peak temperature (K)	Zener-Holloman parameter, Z (s^{-1})
600	102	8.0	80	621	2×10^{13}
800	102	8.6	115	660	6×10^{12}

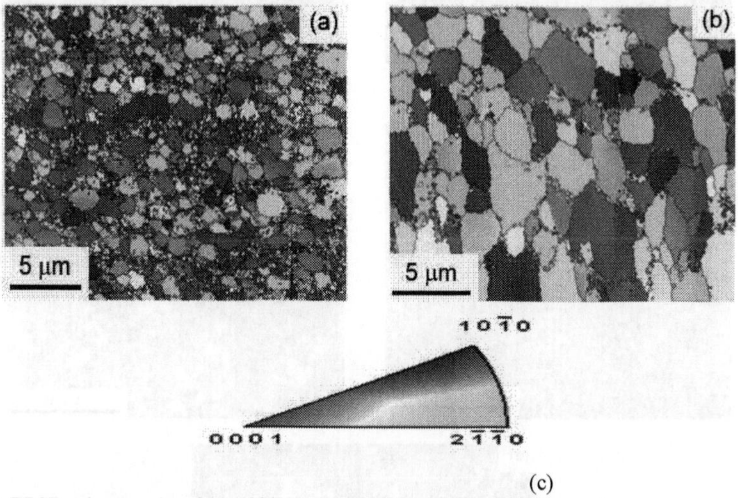

Fig. 2. EBSD micrographs of (a) 600/102, (b) 800/102 and (c) IPF color code

Examination of OIM micrographs reveal fine grained structures indicating extensive dynamic recrystallization. This is indicative of extremely high deformation rates calculated in Table II. Figures 2(a) and (b) show the grain size in specimens welded at 600/102 and 800/102 to be around 1.7 and 2.7, μm respectively. This observation can be rationalized based on the Z parameter and peak temperature. A welding parameter of 600/102 results in a lower peak temperature and a higher Z parameter. In comparison to the 800/102 sample, the increase in the Z magnitude for a 600/102 sample by an order results in a finer grain size. Freeney et al. [13] reported the grain size to be 6.1 μm for a cast EV31A processed at 400/102. Investigations for a T5 WE43 shows a grain size of 1.5-2.5 μm subjected to a higher heat input as compared to EV31A. A higher grain refining sensitivity in a WE43 alloy could be reasoned based on a greater extent of second phases and Zr content that would impact pinning and higher extent of solute elements. In comparison to the cast WE43 alloy processed at 400/102, T5 temper exhibited a finer grain size substantiating the importance of initial microstructure. Though, DRX is unanimously accepted, there is a conflict over the mechanistic mode triggering recrystallization.

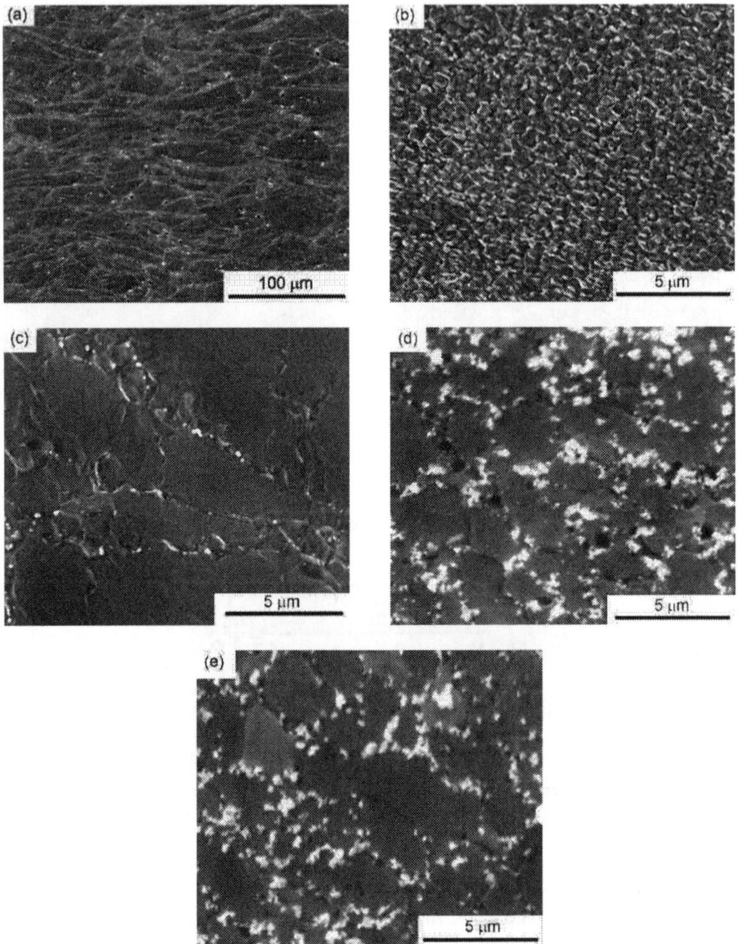

Fig. 3. SEM micrographs of different conditions ; (a) AR+T5 (210 ^{0}C/48h), (b) 600/4 SZ, (c) 600/102 TMAZ, (d) 800/102 SZ, (e) 800/102 TMAZ

Figures 3 (a-e) indicate the various zones and the characteristic features embedded in the microstructure obtained from SEM examination. The area fraction of the precipitates was evaluated using imageJ and decreased from 24.1% in the T5 state to 16.1 and 17.4% in 600/102 and 800/102, respectively. Friction stir welding has led to partial homogenization as compared to the T5 condition (Figs. 3b and d). For a sample processed at a higher rotational rate (Fig. 3d), it showed preferential precipitate coarsening at the grain boundaries as compared to 600/102. Since coarsening is dependent on the thermal profile and mechanical cycle, the evolution becomes a complex function of the competing factors. For an 800/102 sample, the peak temperature is above the dissolution temperature of strengthening precipitates and would soften the matrix to a greater extent. Non-equilibrium reprecipitation would then favor the formation of non-strengthening β precipitates at grain boundaries. Figs. 3(c) and (e) depict formation of necklace structures and fragmentation of prior grains leading to finer grains by recrystallization and break up. This is due to high bulk diffusivity of Mg at a particular temperature in comparison to Al alloys.

<u>Mechanical properties and effect of Post weld heat treatment (PWHT)</u>

Differences exhibited by 600/102 (Fig. 3b) and 800/102 (Fig. 3d) in terms of precipitate characteristics and grain size markedly affected the strength levels, by 6-7 HV in the HAZ, and by 5 HV in the nugget as shown in Fig. 4 (a). A W shaped curve was observed and can be reasoned based on the effect of shoulder. The shoulder diameter (d)/Sheet thickness (t) ratio was approximately three resulting in greater accumulation of heat. This is analogous to a heat pumping source possessing a greater temperature field over a smaller distance. Thus the HAZ would experience higher temperatures with a knockdown in properties. PWHT increased the SZ, TMAZ and HAZ strength and marginally lowered the strength of the base metal. Therefore, the joint efficiency values considerably increased after aging.

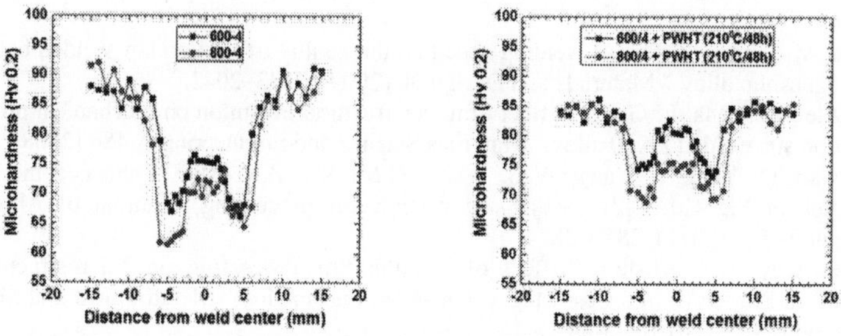

Fig. 4. Microhardness data for as processed and PWHT samples.

4. Conclusions

The effect of processing conditions on the joint efficiency of a T5 temper WE43 alloy was investigated. Based on the study, the following conclusions are made:

1) The FSW of WE43 resulted in grain refinement, breakage and dissolution of second phase particles leading to homogenization. The grain size reduced from 35 ± 15 μm in the T5 state to 1.7 ± 0.2 μm and 2.7 ± 0.26 μm in a 600/102 and 800/102 samples.
2) The shoulder diameter to sheet thickness ratio was 3 resulting in a concentrated temperature field. This lead to a wide HAZ extending from the periphery of the TMAZ to the base metal.
3) The HAZ served as the weakest link with a joint efficiency of 73.5% and 67% for a 600/102 and 800/102 samples. These values increased to 85 and 79.5% after PWHT respectively. A

better understanding of microstructure evolution coupled with aging optimization needs to be done.

5. References

1. B.L Mordike, "Development of highly creep resistant Magnesium alloys," Journal of Materials Processing and Technology, 117 (2001), 391-394.
2. I.J Polmear, "Recent development in light alloys," Materials transactions, 36 (1996), 12-31.
3. Frank Czerwinski, "Welding and Joining of Magnesium Alloys," Magnesium alloys-Design, processing and properties (intechopen.com).
4. A. Simar, Y. Bréchet, B. de Meester, A. Denquin, C. Gallais, T. Pardoen, "Integrated modeling of friction stir welding of 6xxx series Al alloys: Process, microstructure and properties," Progress in Materials Science 57 (2012), 95-183.
5. Threadgill PL, Leonard AJ, Shercliff HR, Withers PJ, "Friction stir welding of aluminium alloys," Int Mater Rev 54 (2009), 49–93.
6. A.H. Feng and Z.Y. Ma, "Enhanced mechanical properties of Mg–Al–Zn cast alloy via friction stir processing," Scripta Materialia 56 (2007), 397–400.
7. Seung Hwan C. Park, Yutaka S. Sato, Hiroyuki Kokawa, "Effect of micro-texture on fracture location in friction stir weld of Mg alloy AZ61 during tensile test," Scripta Materialia 49 (2003), 161–166.
8. Z.Y. Ma, A.L. Pilchak, M.C. Juhas, J.C. Williams, "Microstructural refinement and property enhancement of cast light alloys via friction stir processing," Scripta Materialia 58 (2008), 361–366.
9. X. Cao, M. Jahazi, "Effect of welding speed on the quality of friction stir welded butt joints of a magnesium alloy," Materials and Design 30 (2009), 2033–2042.
10. G.M. Xie , Z.Y. Ma, L. Geng, "Effect of microstructural evolution on mechanical properties of friction stir welded ZK60 alloy," Materials Science and Engineering A 486 (2008), 49–55.
11. B.L. Xiao, Q. Yang, J. Yang, W.G. Wang, G.M. Xie, Z.Y. Ma, "Enhanced mechanical properties of Mg–Gd–Y–Zr casting via friction stir processing," Journal of Alloys and Compounds 509 (2011), 2879–2884.
12. T.A. Freeney, R.S. Mishra, "Effect of Friction Stir Processing on Microstructure and Mechanical Properties of a Cast-Magnesium–Rare Earth Alloy," Metallurgical and Materials Transactions 41A (2010), 73-84.
13. Datong Zhang, Mayumi Suzuki, Kouichi Maruyama, "Microstructural evolution of a heat-resistant magnesium alloy due to friction stir welding," Scripta Materialia 52 (2005), 899–903.
14. Zaijun Su, Chuming Liu, Yingchun Wan, "Microstructures and mechanical properties of high performance Mg-4Y-2.4Nd-0.2Zn-0.4Zr alloy," Materials and Design (2012).
15. Kun Yu, Wenxian Li, Richu Wang, Bo Wang, Chao Li, "Effect of T5 and T6 Tempers on a Hot-Rolled WE43 Magnesium Alloy," Materials Transactions 49 (2008), 1818 -1821.
16. R. Arrabal, E. Matykina, F. Viejo, P. Skeldon, G.E. Thompson, "Corrosion resistance of WE43 and AZ91D magnesium alloys with phosphate PEO coatings," Corrosion Science 50 (2008), 1744-1752.
17. T.A. Freeney, R.S. Mishra, G.J. Grant, and R. Verma, "Friction Stir Welding and Processing IV, TMS, Warrendale," PA, 2007.

18. P. Mengucci, G.Barucca, G. Riontino, D. Lussana, M. Massazza, R. Ferragut, E. Hassan Aly, "Structure evolution of a WE43 Mg alloy submitted to different thermal treatments," Materials Science and Engineering A 479 (2008), 37-44.
19. Renlong Xin, Ling Li, Ke Zeng, Bo Song, Qing Liu, "Structural examination of aging precipitation in a Mg-Y-Nd alloy at different temperatures," Materials Characterization 62 (2011), 535-539.
20. J.F Nie, B.C Muddle, "Characterisation of strengthening precipitate phases in a Mg-Y-Nd alloy," Acta mater. 48 (2000), 1691-1703.
21. Qinglin Jin, Sung-Yong Shim and Su-Gun Lim, "Correlation of microstructural evolution and formation of basal texture in coarse grained Mg-Al alloy during hot rolling," Scripta Materialia 55 (2006), 843-846.
22. S. Serajzadeha, A. Karimi Taheria, M. Nejatib, J. Izadib, M. Fattahi, "An investigation on strain inhomogeneity in hot strip rolling process," Journal of Materials Processing Technology 128 (2002), 88–99.
23. Myshlyaev, M.M., McQueen, H.J.,Mwembela, A. Konopleva, E., "Twinning, dynamic recovery and recrystallization in hot worked Mg–Al–Zn alloy," Materials Science and Engineering A, 337 (2002), 121-133.
24. K.Dehghani, A. Chabok, "Dependence of Zener parameter on the nanograins formed during friction stir processing of interstitial free steels," Materials Science and Engineering A, 528 (2011), 4325-4330.
25. A.P Reynolds, "Flow visualization and simulation in FSW, Scripta Materialia," 58 (2008), 338-342.
26. M. Ghosh, K. Kumar, R.S. Mishra"Analysis of microstructural evolution during friction stir welding of ultrahigh-strength steel," Scripta Materialia, 63 (2010), 851-853.
27. W.J. Arbegast, "Hot Deformation of Aluminum Alloys III," TMS, Warrendale, PA (2003).
28. C.I Chang, C.J. Lee, J.C. Huang, "Relationship between grain size and Zener–Holloman parameter during friction stir processing in AZ31 Mg alloys," Scripta Materialia, 51 (2004), 509-514.
29. L. Commin, M. Dumont, J.E. Masse, L. Barrallier, "Friction stir welding of AZ31 magnesium alloy rolled sheets: Influence of processing parameters," Acta Materialia, 57 (2009), 326-334.

Friction Stir Welding and Proceesing VII
Edited by: Rajiv Mishra, Murray W. Mahoney, Yutaka Sato, Yuri Hovanski, and Ravi Verma
TMS (The Minerals, Metals & Materials Society), 2013

MICROSTRUCTURE IN DISSIMILAR FRICTION SPOT WELD OF Al TO Mg ALLOYS OBSERVED BY STOP-ACTION TECHNIQUE

Uceu Suhuddin[1], Vanessa Fischer[1], Jorge F. dos Santos[1]

[1]Helmholtz-Zentrum Geesthacht, Institute of Materials Research, Materials Mechanics, Solid-State Joining Processes, Max-Planck-Str. 1, 21502 Geesthacht, Germany

Keywords: Friction spot welding, dissimilar joint, Aluminum, Magnesium

Abstract

In the current study, the formation of intermetallic compounds in dissimilar friction spot welding (refill friction stir spot welding) of AA5754 aluminum alloy to AZ31 magnesium alloy has been investigated by applying a stop action technique. The welding cycle was forced to stop during the dwell time, and subsequently, the weld was quenched by pouring a mixed solution of ice and water to freeze the microstructure. A large volume of eutectic structure was observed in the weld center of the stop action sample due to mutual diffusion of Al and Mg atoms induced by plastic deformation. During the sleeve retraction, a redistribution of eutectic structure and Mg alloy, and extensive diffusion led to an inhomogeneous intermetallic phase distribution in the Al top sheet.

Introduction

Friction spot welding (FSpW) also known as refill friction stir spot welding (RFSSW) is a variant of the friction stir welding process for joining materials in a spot-like configuration, as shown in Figure 1. The tool consists of two rotating parts of pin and sleeve, and one stationary clamping ring. The clamping ring holds the materials against the backing bar. In the present study, the rotating sleeve plunges into the top material while the pin moves in the reverse direction. The tool introduces intense plastic deformation and generates frictional heating which plasticizes the materials. During sleeve plunging, the softened material is squeezed into the cavity left by the pin. Then, the softened material is extruded back into the joint as the sleeve and pin return to their initial position leaving the weld without a keyhole. Due to some advantages, such as good mechanical properties and good weld surface quality, i.e., no keyhole, FSpW is being considered for applications in the automotive and aircraft industries in order to reduce structural weight.

FSpW has been employed for joining similar Al alloys [1,2] and Mg alloys [3]. Sound welds of similar Al and Mg alloys have been successfully produced. This process was also considered to produce dissimilar welds of Al to Mg alloys [4,5]. Blosmo et al., [4] have studied the effect of interlayer materials (such as alumina, Sn and Cu) between the Al and Mg sheets and compared them with a sample without an interlayer. Their results showed the lap shear strength of the bare material to be slightly lower than the strongest weld produced using an alumina-coated Al sample. Several studies have shown that the dissimilar friction-based joining of Al and Mg alloys led to the formation of intermetallic compounds (IMCs), such as $Al_{12}Mg_{17}$ [5-8], Al_3Mg_2 [7,8], and Mg_2Si [8].

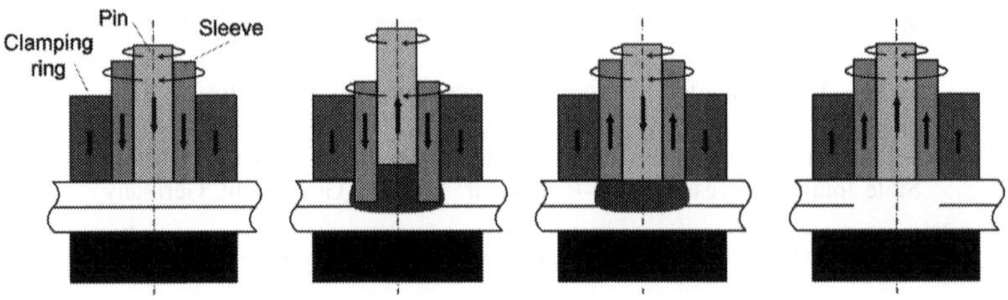

Figure 1. Schematic illustration of the friction spot welding process

In the present study, the microstructure of dissimilar friction spot welds of AA5754 and the Mg alloy AZ31 have been studied. To obtain an insight into the microstructure development, a stop action technique was adopted. In this technique, the process cycle is stopped during the dwell time. Subsequently, the weld was quenched to preserve the microstructure.

Experimental Procedure

2mm-thick AA5754 and Mg alloy AZ31 were used in this study. The chemical composition of both materials is shown in Table 1. The tool used for welding has dimensions of 14.5 mm, 9 mm, and 6 mm in diameter for the clamping ring, the sleeve, and the pin, respectively. The Al sheet was welded on top of the Mg sheet with the following welding parameters: 1900 rpm tool rotational speed, 1.6 mm plunge depth, 2s dwell time, and 12 kN clamping force. The plunging and retracting speeds of the sleeve were 0.8 mm/s. At the end of the process, the sleeve and the pin were maintained in the rotating state on the surface for 1 s to improve the weld appearance. To measure the temperature cycle during welding, the two K-type thermocouples were embedded in the stir zone at a depth of 2.5 mm from the bottom of the Mg sheet.

Table 1. The chemical composition in wt.% of Mg alloy AZ31 and AA5754.

Base Material	Al	Mg	Fe	Zn
AZ 31	2.52	Balance	-	2.03
AA 5754	Balance	3.66	0.59	0.43

The stop action experiments were carried out to obtain more insight into the grain structure development. The welding cycle was forced to stop during the dwell period. Subsequently, a mixture of ice and water was poured onto the sample to freeze the microstructure, hereafter referred to as a "quenched sample".

The welds were cut across the weld center, then ground and polished for microstructural analysis. Weld samples were observed using an optical microscope and a scanning electron microscope (SEM) equipped with an energy dispersive X-ray spectrometer (EDS).

Results and Discussion

As-welded Sample

A macrograph taken from a cross section of a welded sample is presented in Figure 2. No obvious thickness reduction is visible in the welded zone. Some areas with different contrast could be observed (highlighted by "a", "b", and "c") which are discussed in detail below. Regions "a" and "c" are located below the sleeve near the surface of the weld and in the weld center at the interface, respectively, whereas region "b" resides in a region between "a" and "c" as shown in Figure 2.

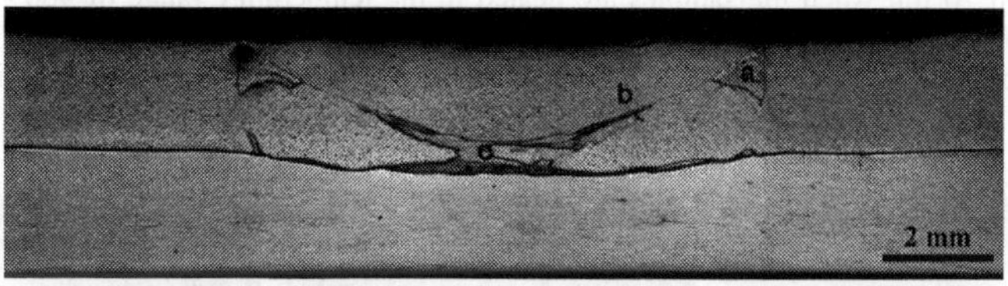

Figure 2. Macrograph taken from the cross section of the weld.

Region "a" has a chemical composition of 31.36 at.% Mg and 68.64 at.% Al. Region "b" has a non-homogenous microstructure (Figure 3a) which is associated with a different composition of Al and Mg alloys, as shown in Figure 3b. The layer in the middle has the highest Mg content compared to other neighboring layers. According to the Al-Mg phase diagram, region "a" should be comprised of α-Al and β-Al_3Mg_2, while the phases across the layer in region "b" should have a phase sequence of Al – (β Al_3Mg_2 + γ $Al_{12}Mg_{17}$) – (E + δ) – (γ $Al_{12}Mg_{17}$ + E) – (E + δ) – (β Al_3Mg_2 + γ $Al_{12}Mg_{17}$) – Al. Formation of such heterogeneous phases in the Al top sheet should be related with the transport of the Mg alloy (i.e., bottom sheet or Mg-enriched Al material) towards the Al top sheet induced by tool movement at high temperature, followed by mutual diffusion of Al – Mg atoms.

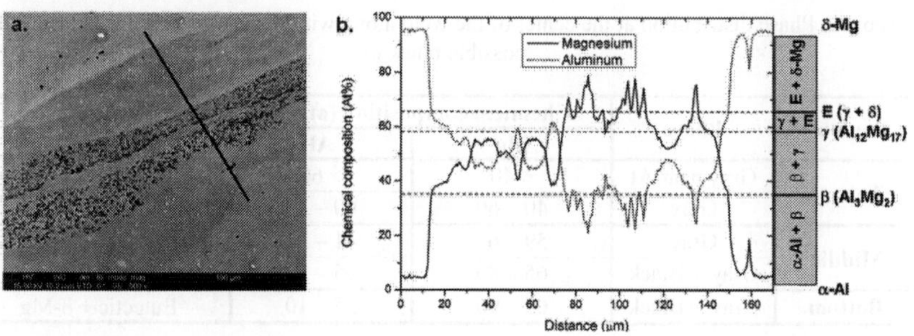

Figure 3. Micrograph taken from the region b as shown in Figure 2 (a) and chemical composition distribution taken from line as shown in (a) (b).

Region "c" has a different grain structure distribution compared to those observed in the previous regions (Figure 4). The region close to the Al base material (top) consists of a gray phase and the region near the Mg base material (bottom) has a mixed structure of gray and black phases. The middle region (i.e., between top and bottom regions) has a mixed structure of both those observed in the top and bottom regions. A chemical composition distribution of the phases and possible phases according to the Al-Mg phase diagram (and phase identification of some phases by integration of EDS and EBSD [5]) are summarized in Table 2. The total thickness of the phases is about 500 μm. In a previous study [9] on dissimilar Al-Mg alloys joined by the diffusion bonding process, the thickness of the intermetallic phases at the interface was shown to be less than 70 μm when the samples are heated in the temperature range of 460 - 480 °C for about 40-60 min. Apparently, the diffusion rate during FSpW is much faster than that during diffusion bonding. The effect of deformation on the diffusion rate during welding will be discussed later.

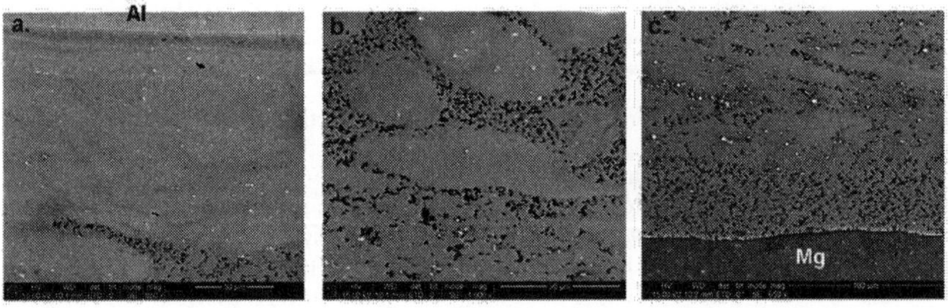

Figure 4. Microstructure taken from region c as shown in Fig.2 (a) in top, middle and bottom (a-c), and the chemical composition distribution taken from line as shown in (a-c) (d-f), respectively.

Table 2. Phase distribution at the center of the weld along with the chemical composition and possible phases.

Region	**Phase**	**Chemical composition (at.%)**		**Possible phases (Phase diagram)**
		Mg	**Al**	
Top	Gray near Al	< 40	> 60	β-Al_3Mg_2
	Gray	40 – 60	60 – 40	β-Al_3Mg_2 + γ-$Al_{12}Mg_{17}$
Middle	Gray	59 – 65	41 – 35	γ-$Al_{12}Mg_{17}$ + Eutectic
	Gray + Black	65 – 80	35 – 20	Eutectic + δ-Mg
Bottom	Gray + Black	65 – 90	35 – 10	Eutectic + δ-Mg

A schematic illustration of the thermocouple placements and the thermal cycle during FSpW are presented in Figure 5. During sleeve plunging, the temperature increases rapidly up to 450°C for a short period, and subsequently decreases to 370°C. Then, the temperature rises and decreases with different heating and cooling rates towards the second peak temperature of 440°C. The maximum temperature achieved during welding probably corresponds to the solidus temperature of the eutectic phase. According to the equilibrium Al-Mg phase diagram, the solidus temperature of the eutectic structure is about 437°C. Such a temperature profile during welding probably corresponds to the liquation and solidification of the eutectic structure during the process. The simultaneous liquation and solidification processes resulted in a non-equilibrium solidus temperature induced by high rates of heating and cooling.

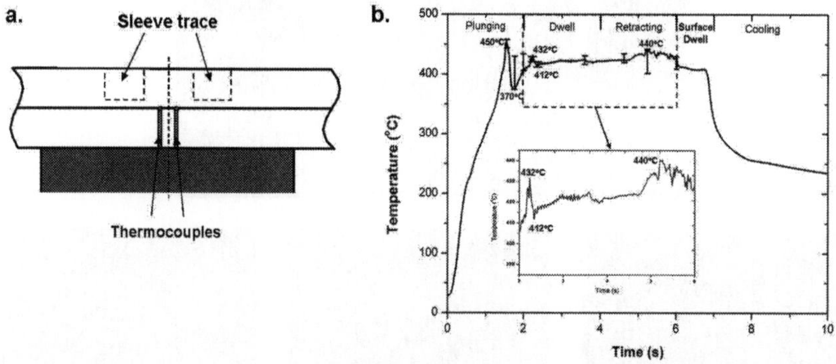

Figure 5. Thermal cycle measured during welding, (a) illustration of the thermocouple position during welding and (b) the temperature profile during welding.

<u>Quenched sample</u>

A macrograph taken from the cross section of the quenched sample is presented in Figure 6a. The Al alloy under the sleeve including the Mg alloy at the interface was squeezed into the cavity left by the pin. During sample removal from the welding tool, cracks developed beneath the pin across the weld center, probably along a brittle layer.

A large volume of eutectic structure and grey phase were formed in the center as shown in Figure 6b. Quantitative chemical analysis by EDS showed the composition of the grey phase in the region A as 63 at.% Mg and 37 at.% Al, while the eutectic structure has 73 at.% Mg and 27 at.% Al. According to the Al-Mg phase diagram, the phases in structure A and B correspond to β-$Al_{12}Mg_{17}$ and a eutectic structure consisting of γ-$Al_{12}Mg_{17}$ and δ-Mg, respectively. Apparently, during the process, the materials reached the solidus temperature of the eutectic structure. This is in agreement with the temperature measurement. The thermal cycle measurement reveals that the material exposed to the non-equilibrium solidus temperature, induced by high heating and cooling rates, was associated local melting. However in region c (crack area), no eutectic structure could be observed. This layer has a chemical composition of 42 at.% Mg and 58 at.% Al. Thus, during sleeve retracting, some of the plasticized Mg alloy was ejected into the Al sheet. According to Al-Mg phase diagram, this layer consists of β-Al_3Mg_2

and γ-$Al_{12}Mg_{17}$ phases. Meanwhile in region c, a 5-μm-thick layer was formed at the interface between the Al and Mg alloys. Line scanning across the interface shows a chemical composition gradient representing mutual diffusion of the Al and Mg atoms into Mg and Al base materials, respectively.

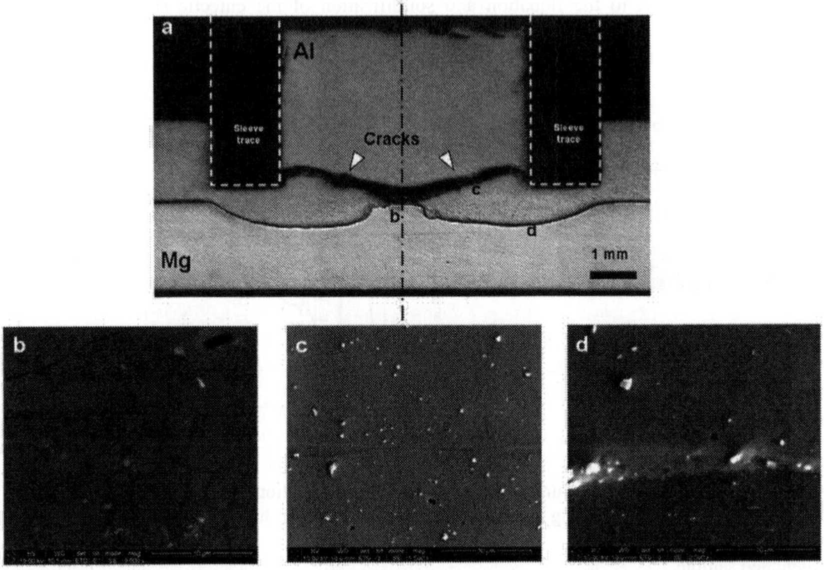

Figure 6. Quenched sample, (a) micrograph and (b-d) enlarged maps taken from regions as highlighted in (a).

In the dissimilar FSSW process, prior studies reported the presence of an intercalated layer of base material in the stir zone [10,11]. Figure 7a presents a SEM micrograph taken from the Al/Mg interface near the center of the quenched sample. Chemical composition along the white line is presented in Figure 7b. According to the Al-Mg phase diagram, the phase sequence across the line should be Al – eutectic structure – Mg-enriched Al – Mg layers. Likely, during the process a Mg layer has been transported into the Al base material due to deformation imposed by the axial and radial movement of the sleeve. Thus, the interfacial area between the Al/Mg alloys becomes greater accordingly enhancing the diffusion rate during the process. From Figure 7a, fine grains can be observed at the interface of the eutectic structure. Friction based joining processes produce welds having a fine grain structure due to simultaneously continuous and discontinuous recrystallization [12,13] and also a high density of low angle boundaries due to severe plastic deformation. Grain boundaries (low and high angle) are associated with a more open structure in which the migration of atoms is easier than diffusion through the lattice [14]. Therefore, the diffusion rate in the grain structure having fine grains should be higher than that having larger grains.

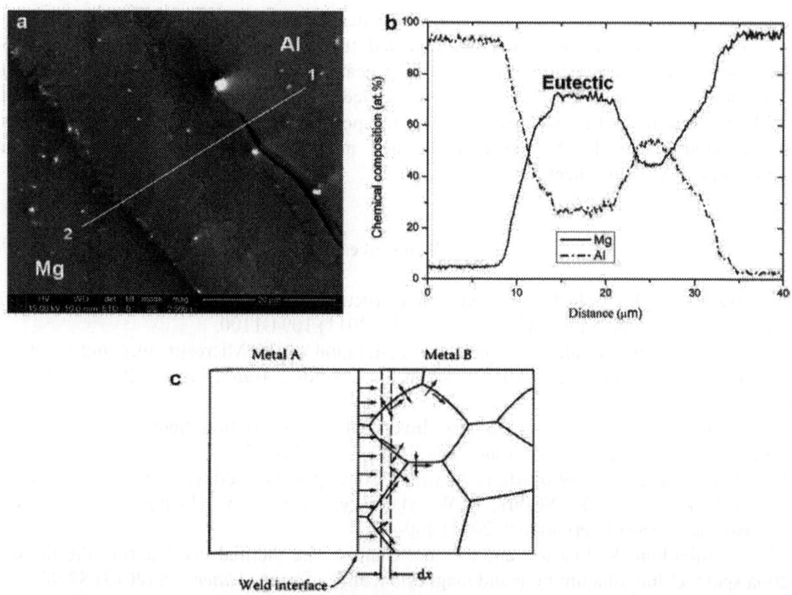

Figure 7. Enlarged map taken from Al-Mg interface (a), chemical composition distribution along line as highlighted in (a) (b), and effect of grain boundary diffusion combined with volume diffusion (c) (after R.E. Reed-Hill[15]).

Both formation of intercalated layers and a high density of grain boundaries enhance the diffusion rate during welding. In general, interfacial diffusion plays an important role, larger than grain boundary diffusion [14]. However, since the total grain boundaries should be much larger than the total interfacial area in the intercalated layer (Figure 7c), grain boundary diffusion becomes a more important factor affecting the diffusion rate.

During sleeve retraction, the soften material was extruded back into the weld by the pin. Some Mg alloy from base material and eutectic structure somehow was distributed into the Al top sheet, as shown in figure 2. During this stage, redistribution of the Mg alloy from base material or eutectic structure and extensive diffusion resulted in the typical structure shown in figure 2.

Conclusions

In the present study, the microstructure in dissimilar friction spot welds of Al to Mg alloys was observed by a stop-action technique. A large volume of eutectic structure was formed during the process in the weld center, due to mutual diffusion of Al and Mg atoms induced by plastic deformation. Formation of intercalated layers of base materials and a high density of grain boundaries including low and high angle boundaries enhances the diffusion rate at high temperature. Temperature measurement revealed that the material was exposed to a non-equilibrium solidus temperature induced by high heating and cooling rates due to solidification and liquation of the eutectic structure. During the sleeve retracting process, the eutectic structure and Mg material from the base material were transported into Al top sheet. Extensive diffusion and redistribution of the Mg material resulted in an inhomogeneous intermetallic phase distribution in the Al top sheet.

References

1. T. Rosendo et al., "Mechanical and microstructural investigation of friction spot welded AA6181-T4 aluminium alloy", *Mater. Design*, 32 (2011) 1094-1100.
2. S.T. Amancio-Filho et al., "Preliminary Investigation of the Microstructure and Mechanical Behaviour of 2024 Aluminium Alloy Friction Spot Welds", *Mater. Trans.* 52 (5) (2011), 985-991.
3. L.C. Campanelli et al., "Preliminary Investigation on Friction Spot Welding of AZ31 Magnesium Alloy", *Mater. Sci. Forum*, 706-709 (2012) 3016-3021.
4. D.J. Blosmo et al., "FSSW of Mg to Al with a Cold Sprayed Interlayer", *Friction Stir Welding and Processing VI*, ed. R. Mishra, M.W. Mahoney, Y. Sato, Y. Hovanski, and R. Verma, (Materials Park: ASM International, 2011) 409-416.
5. U.F.H. Suhuddin, V. Fischer, and J.F. dos Santos, The thermal cycle during the dissimilar friction spot welding of aluminum and magnesium alloy, *Scripta Mater.*, 68 (2013) 87-90.
6. Y.S. Sato et al., "Constitutional liquation during dissimilar friction stir welding of Al and Mg alloys", *Scripta Mater.*, 50 (2004) 1233-1236.
7. A. Kostka et al., "Microstructure of friction stir welding of aluminium alloy to magnesium alloy", *Scripta Mater.*, 60 (2009) 953-956.
8. Y.C. Chen and K. Nakata, "Friction stir lap joining aluminum and magnesium alloys", *Scripta Mater.*, 50 (2008) 433-436.
9. Y. Li et al., "XRD and SEM analysis near the diffusion bonding interface of Mg/Al dissimilar materials", *Vacuum*, 82 (1) (2007) 15-19.
10. P. Su et al., "Intermixing in dissimilar friction stir spot welds", *Metal. Mater. Trans. A*, 38A (2007) 584-595.
11. P. Su et al., "Formation and retention of local melted films in AZ91 friction stir spot welds", *J. of Mater. Sci.*, 42 (2007) 9954-9965.
12. J. Jeon et al., "Friction stir spot welding of single-crystal austenitic stainless steel", *Acta Mater.*, 59 (2011) 7439-7449.
13. U.F.H.R. Suhuddin et al., "Grain structure evolution during friction-stir welding of AZ31 magnesium alloy", *Acta Mater.*, 57 (2009) 5406-5418.
14. D.A. Porter and K.E. Easterling, Phase Transformations in Metals and Alloys, Van Nostrand Reinhold, Berkshire, 1981.
15. R.E. Reed-Hill, Physical Metallurgy Principles, 2nd ed., Van Nostrand, New York, 1973.
16. C. Schilling, J. dos Santos, "Method and Device for Linking at Least Two Adjoining Work Pieces by Friction Welding", (2004) US Patent 6,722,556 B2.

OPTIMIZATION OF WEAR RATE OF FRICTION STIR WELDED AL-B_4C COMPOSITE

K. Kalaiselvan[1], N. Murugan[2]

[1]Department of Mechanical Engineering, K.S. Rangasamy College of Technology, Tiruchengode Namakkal-*637215, Tamil Nadu, India. Mobile:919443952292*
Email:kalaiselvanmohit@gmail.com

[2]Department of Mechanical Engineering, Coimbatore Institute of Technology, Coimbatore Pin-641014, Tamil Nadu, India. Email:drmurugan@hotmail.com

Keywords: Al-B_4C MMC, Friction stir welding, Dry sliding wear

Abstract

Friction stir welding (FSW) is used to join metal matrix composites (MMCs) which are increasingly used as a structural material in aerospace, automotive, and marine systems. The welded zone of the composite offers a more homogeneous distribution of reinforcement particles with a fine recrystallized grain structure resulting in both enhanced mechanical and wear properties. In this paper, the dry sliding wear behavior of the stir zone of friction stir welded AA6061-B_4C composite is presented. The relationships between the wear rate of MMCs and friction stir welding process parameters were established using a mathematical model. The developed regression model was optimized to minimize the wear rate using the generalized reduced gradient (GRG) method.

Introduction

Aluminum metal matrix composites (AMCs) reinforced with a variety of hard ceramic particles exhibit improved wear resistant, good chemical and physical properties. The potential applications of AMCs are realized in aerospace, automotive, and marine industries [1]. Apart from traditionally used Al_2O_3 and SiC reinforcements, several new ceramic particles are being added to aluminum alloys to develop superior AMCs [1]. B_4C is a robust material having excellent chemical and thermal stability, high hardness (HV = 30 GPa), and low density (2.52 g/cm^3) and it is used for manufacturing bullet proof vests, tank armor, etc. In nuclear power plants, Al- B_4C composites (Boral and Metamic) are used as a neutron absorber [2]. In order to make larger structural components, joining to other components made of a similar composite or other metallic materials is often required. Joining of AMCs by fusion welding results in major weldability problems such as porosity formation or cracking, solidification shrinkage, and a deleterious reaction of the matrix with reinforcement [3,4]. Hence, a solid state joining process (FSW) is preferred to overcome the above problems. Studies on the wear behavior of friction stir welded B_4C reinforced aluminum AMCs are very limited. In the present study, the dry sliding wear behavior of a FSW zone of Al/B_4C composite is investigated. A mathematical model has been developed to predict the wear rate of a friction stir welded AA6061-B_4C composite by incorporating welding parameters such as rotational speed (N), welding speed (S), axial load (F), and % of reinforcement using design of experiments. The developed model is optimized to minimize the wear rate of the welded composite.

Experimental Procedure

The materials used in this study are AA6061-B_4C composites. The MMCs have different weight percentages (4, 6, 8, 10, and 12) of B_4C and were fabricated by a modified stir casting method. The details of fabrication are previously reported in the literature [5]. The fabricated AMCs were sliced into plates of size 100 mm x 50 mm x 6 mm to subsequently join by FSW. Four factors five level central composite design was used in this study. Based on the trial experiments, the range and level of process parameters shown in Table I were considered to produce defect free welds. Using a FSW machine, 31 butt joints were welded on the sliced composite plates according to the design matrix. The typical scanning electron microscope image of the composite before and after FSW is shown in Figure 1. The wear test pin specimens of size 6 mm X 6 mm X 30 mm were prepared from the weld zone of the friction stir welded plates. The test pin was kept perpendicular to the disc such that the surface of the stir zone contacts the disc surface. The tests were carried out at room temperature with fixed sliding wear parameters at a constant load of 20 N, sliding speed of 1 m/s for 50 min, and a distance of 3000 m for all tests. The wear rates are estimated and are presented in the Table II.

Development of a Regression Model

The response (wear rate) obtained in the design matrix is highly influenced by rotational speed, traverse speed, axial force, and percentage of B_4C reinforcement. The response function representing the wear rate (WR) of friction stir welded Al-B_4C joints can be expressed as follows:

$$WR = f(N, S, F, R) \tag{1}$$

The second order polynomial (regression) equation used to represent the response surface for four factors, could be expressed as:

$$WR = b_0 + b_1 N + b_2 S + b_3 F + b_4 R + b_{11} N^2 + b_{22} S^2 + b_{33} F^2 + b_{44} R^2 + b_{12} NS + b_{13} NF + b_{14} NR + b_{23} SF + b_{24} SR + b_{34} FR \tag{2}$$

All the coefficients (b_0, b_1, b_2,.... b_{34}) were evaluated and the initial model was developed using a statistical software package (SYSTAT 12). After determining the significant coefficients, the final model was developed by eliminating least and insignificant terms. The developed mathematical model for predicting the wear rate of Friction Stir Welded Al-B_4C composite is given below.

$$WR\ (mm^3/m) = 347.000 + 5.375\,N + 0.042S + 0.375F - 68.208R + 8.615N^2 + 7.240S^2 + 6.740F^2 - 5.510\,R^2 \tag{3}$$

The adequacy of the developed model is tested by analysis of variance (ANOVA) shown in Tables IV and V with the help of SYSTAT software. In this case, the value of R^2 is 0.983 and adjusted R^2 is 0.977. Higher values of R^2 indicate that the regression model is quite accurate.

Optimization of FSW Parameter

The developed regression equation was optimized using GRG method which is available in MS Excel solver module. The optimized welding parameters to minimize the wear rate are: tool rotational speed (N) = 968.80 rpm, tool traverse speed (S) = 1.29 mm/s, and axial force (F) = 9.94 kN, and % of reinforcement(R) = 12%; and the predicted wear rate (WR) =187.7003 x $10^{-5}mm^3/m$.

Results and Discussion

Based on the model developed, the various graphs are plotted to detail the significant effect of process parameters on the wear rate of the FSW composite joint. The trends obtained for each FSW parameter set on wear rate are represented in Fig. 2. The process parameters are independently influenced by the dry sliding wear behavior of the welded composites. As shown, the wear rate decreases with increasing wt. % of B_4C content in the MMCs. This may be attributed to the presence of the higher amount of particles which decreases the matrix grain boundary, resulting in brittleness as well as high hardness of the composites.

Conclusion

The dry sliding wear test has been successfully conducted on friction stir welded AA6061-B_4C composite with the following conclusions:

- A mathematical model was developed incorporating the friction stir welding parameters to predict the wear rate.
- The mathematical model was optimized using GRG method.

References

[1] Liu, H.J., Feng, .J.C., Fujii, H. and Nogi, K., " Wear characteristics of a WC–Co tool in friction stir welding of AC4A+30 vol% SiCp composite", *International Journal of Machine Tools & Manufacture*, 45 (2005), pp.1635–1639.

[2] Abenojar, J., Velasco, F. and Martinez, M.A., "Optimization of processing parameters for the Al+10% B4C system obtained by mechanical alloying", *Journal of Material Processing Technology*, 184 (2007), pp. 441-446.

[3] Ceschini, L., Boromei, I., Minak, G., Morri, A., and Tarterini, F., "Effect of friction stir welding on microstructure, tensile and fatigue properties of the AA7005/10 vol. %Al2O3p composite", *Composites Science and Technology*, 67 (2007), pp.605–615.

[4] Marzoli, L. M., Strombeck, A., Dos S.J.F Gambaro, C., and Volpone, L.M., "Friction stir welding of an AA6061/Al2O3/20p reinforced alloy," *Composites Science and Technology*, Vol. 66 (2006), pp. 363-371.

[5] Kalaiselvan, K., Murugan, N., Siva Parameswaran ,:"Production and characterization of AA6061–B_4C stir cast composite", *Materials and Design,* 32(2011), pp. 4004–4009.

[6] Shanmuga Sundaram, N., and Murugan, N., "Tensile behavior of dissimilar friction stir welded joints of aluminium alloys", *Materials and Design,* 31(2010), pp. 4184-4193.

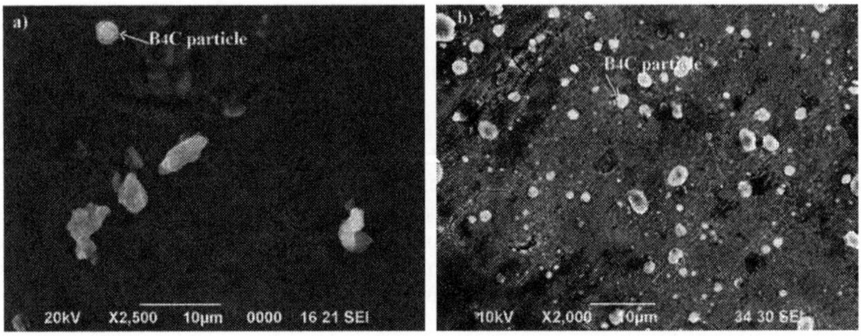

Fig. 1 SEM images of Al-4% B_4C composite: a) before FSW and b) after FSW.

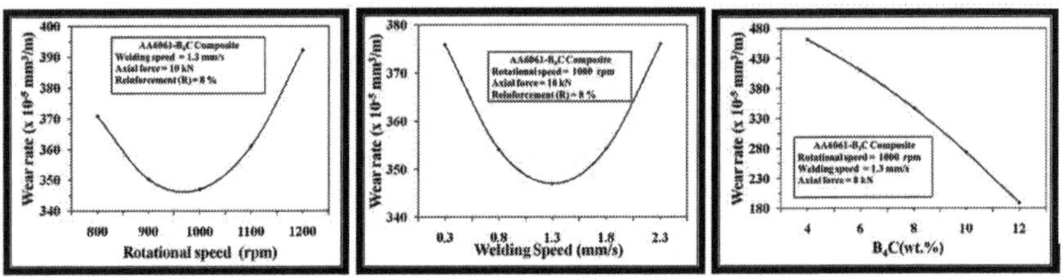

Fig. 2. Effect of Process parameters on wear rate.

Table I. FSW parameter range and levels.

Process parameter	Notation	Unit	Levels				
			-2	**-1**	**0**	**1**	**2**
Rotational speed	N	rpm	800	900	1000	1100	1200
Welding speed	S	mm/s	0.3	0.8	1.3	1.8	2.3
Axial force	F	kN	6	8	10	12	14
Reinforcement(B_4C)	R	wt.%	4	6	8	10	12

Table II. Design matrix and estimated wear rate.

Trial Run	Design matrix				Estimated wear rate ($x10^{-5}$ mm^3/m
	Process parameter				
	N	S	F	R	
1	-1	-1	-1	-1	434
2	+1	-1	-1	-1	418
3	-1	+1	-1	-1	423
4	+1	+1	-1	-1	444
5	-1	-1	+1	-1	417
6	+1	-1	+1	-1	437
7	-1	+1	+1	-1	429
8	+1	+1	+1	-1	440
9	-1	-1	-1	+1	294
10	+1	-1	-1	+1	290
11	-1	+1	-1	+1	295
12	+1	+1	-1	+1	301
13	-1	-1	+1	+1	282
14	+1	-1	+1	+1	299
15	-1	+1	+1	+1	278
16	+1	+1	+1	+1	300
17	-2	0	0	0	374
18	+2	0	0	0	400
19	0	-2	0	0	391
20	0	+2	0	0	372
21	0	0	-2	0	373
22	0	0	+2	0	386
23	0	0	0	-2	464
24	0	0	0	+2	197
25	0	0	0	0	344
26	0	0	0	0	356
27	0	0	0	0	340
28	0	0	0	0	359
29	0	0	0	0	354
30	0	0	0	0	334
31	0	0	0	0	342

Table III. Statistical result for the developed model.

Response	Multiple R	Squared Multiple R	Adjusted squared Multiple R	Standard Error of Estimate
Wear rate	0.991	0.983	0.977	9.666

Table IV. Analysis of variance (ANOVA).

Response	Source	Sum of square	Degree of freedom	Mean square	Calculated F-ratio	Tabulated F-ratio
Wear rate	Regression	118,077.794	8	14,759.724	157.964	2.40
	Residual	2,055.625	22			

FRICTION STIR WELDING AND PROCESSING VII

Friction Stir Welding and Processing: Modeling and Controls

Friction Stir Welding and Proceesing VII
Edited by: Rajiv Mishra, Murray W. Mahoney, Yutaka Sato, Yuri Hovanski, and Ravi Verma
TMS (The Minerals, Metals & Materials Society), 2013

THREE-DIMENSIONAL VISUALIZATION OF METALLIC FLOW AND CONTROL OF FSW JOINT PROPERTIES USING NEW FSP TECHNIQUE

Hidetoshi Fujii, Yoshiaki Morisada and Takuya Imaizumi

Joining and Welding Research Institute, Osaka University
11-1 Mihogaoka, Ibaraki, Osaka, 567-0047, Japan

Keywords: Friction stir welding, Material flow, X-ray radiography, Friction stir powder processing, FSPP, Recrystallization, Solid solution, Precipitation, Transformation

Abstract

Material flow is a key phenomenon to obtain sound joints by friction stir welding (FSW). This paper introduces the material flow three-dimensionally visualized during FSW using two pairs of x-ray transmission real-time imaging systems. The material flow was investigated by following the locus of a tiny tungsten tracer obtained at various tool rotation rates.

As the next step, a new method of freely designing the microstructure of a structure is introduced. The method called FSPP (Friction Stir Powder Processing), is the friction stir processing over a gap which is filled with a powder. In this method, depending on the requirements, the modification and alloying can be flexibly carried out through complicated metallurgical phenomena, such as stirring, dispersion, diffusion, solid-solution and precipitation.

Introduction

Friction stir welding [1] has a variety of excellent advantages that have already been used for various industrial applications, such as trains, ships, automobiles, and civil engineering structures involving Al alloys [2-10]. During this solid-state joining technique, a rotating tool is inserted into the interface at the butt line of metal plates and produces a highly plastically deformed zone. The metal plates are joined by the travel of the rotating tool along the interface. It is well known that the FSW joint shows excellent mechanical properties because of the recrystallized fine and equiaxed grains in the stir zone.

The material flow around the rotating tool during the FSW/FSP is the one of the most important phenomena for obtaining a sound joint. Accordingly, the material flow has been studied using various approaches because an accurate understanding of the material flow can lead to optimization of the process conditions.

The material flow was first investigated by Colligan using small steel shots placed at different positions in an aluminum sheet [11]. Many researchers have also investigated the material flow by the tracking of a tracer [12-15], observation of the flow pattern in a weld of dissimilar materials [16-19], analysis of the crystallographic texture in a weld [20], measuring the eutectic Si distribution [21] and using a transparent material [22]. The X-ray computer tomography technique has also been used to observe the dispersion of various tracers after the FSW [12, 23]. However, it is still not fully understood despite many such investigations and models. It is difficult to understand the accurate material flow using these approaches because the obtained experimental results show only one part of the process.

In-situ observation of the material flow by an X-ray radiography technique seems to be a very effective approach to reveal the material flow. Recently, the authors reported that the material flow during the FSW can be three-dimensionally visualized using a tiny spherical tungsten tracer [24]. The three-dimensional material flow was obtained using the locus of the tungsten tracer observed by two pairs of x-ray transmission real-time imaging systems.

Unlike the conventional welding methods, FSW is an asymmetric welding method. The side where the movement and rotational directions of the tool are the same is called the advancing side, and the side where the movement and rotational directions are opposite is called the retreating side [3, 4]. This unsymmetrical property may affect the flow pattern during the FSW.

3D visualization experimental setup

A schematic illustration of the experimental setup for the X-ray radiography during the FSW is shown in Fig. 1 [24]. A pure aluminum (A1050) plate was used as the work-piece. The thicknesses of the work-piece and the back plate were both 5 mm. The unthreaded tool, which has a columnar shape (φ15.0 mm) with a probe (φ6.0 mm, length: 1.9 mm), was used for the FSW. A2017 and Si_3N_4 were chosen as the materials of the back plate and the tool for the FSW,

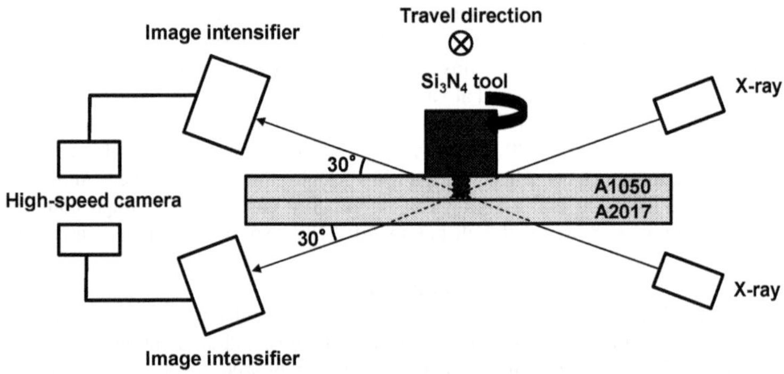

Figure 1. Schematic illustration of the experimental setup for the X-ray radiography during the FSW [24].

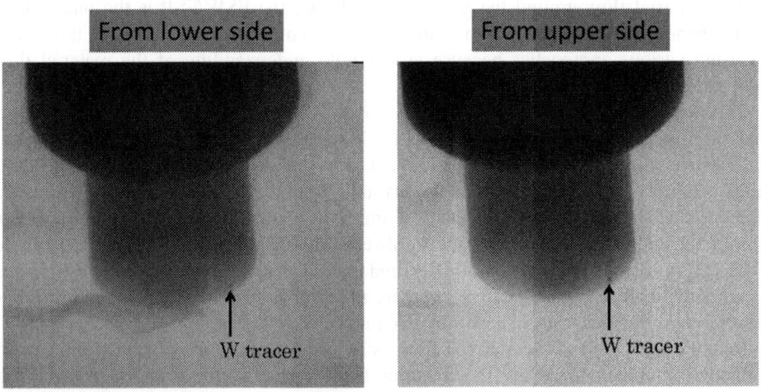

Figure 2. X-ray transmission images obtained by the x-ray transmission real-time imaging systems. (a): image from the upper side of the plate, (b): image from the lower side of the plate

respectively.

The X-rays can transmit through the work-piece, the back plate and the tool due to their low relative densities (Al: 2.7 g/cm^3, Si_3N_4: 3.4 g/cm^3) [24]. The two pairs of x-ray transmission real-time imaging systems were used in order to fix the perspective center in the middle of the stir zone. Both lines of the X-rays passed through the horizontal A1050 plate at an angle of 30°each. The X-ray transmission images were recorded by two high-speed video cameras at the frame rate of 500 fps. These video cameras were synchronized in order to obtain a three-dimensional graph using the locus of the tracer observed by the two imaging intensifiers.

Figure 2 shows the two X-ray transmission images. The tungsten tracer, which was under the tool shoulder, can be clearly observed in both images. It was set at 1 mm from the top surface of the A1050 plate before the FSW. The diameter of the tungsten tracer was about 300μm. Image (a) was taken from the upper side of the plate with the X-ray tube voltage of 100 kV and tube current of 8 mA. On the other hand, image (b) was taken from the lower side of the plate with the X-ray tube voltage of 180 kV and tube current of 1 mA.

Material Flow

Figure 3 shows the three-dimensional movement of the tungsten tracer at various tool rotation rates. WD, TD, and ND show the welding direction of the FSW, transverse direction of the A1050 plate, and normal direction of the A1050 plate, respectively. The FSW was carried out in the positional control mode. The travelling speed and the tilt angle of the tool were 400 mm/min and 3°, respectively. The rotating rate was changed from 200 rpm to 1000 rpm in order to investigate the effect of the heat input.

Surprisingly, the tungsten tracer rotated around the probe several times under all welding conditions. This result implied that the conjecture of the material flow based on the final position of the tracer contained inaccuracies. The velocity of the tungsten tracer and the peripheral velocity of the tool were not same as will be shown later. The flow zone of almost the same 2 mm width around the probe, regardless of the rotating and travel directions of the tool, could be confirmed. There was no significant difference between the advancing side (AS) and retreating side (RS).

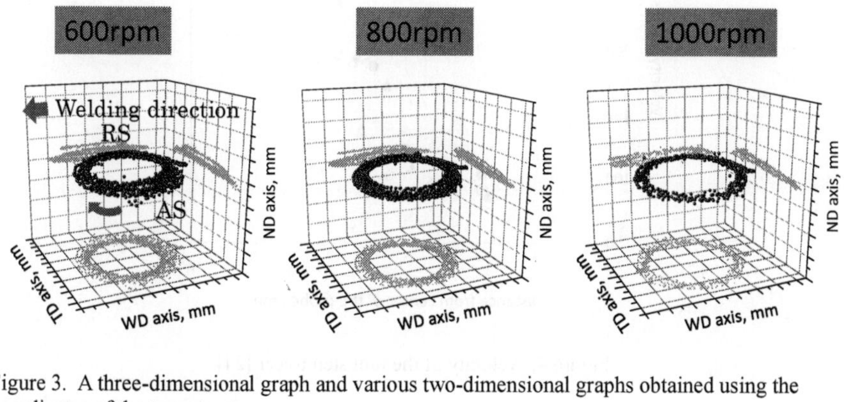

Figure 3. A three-dimensional graph and various two-dimensional graphs obtained using the coordinates of the tungsten tracer.

The velocity of the tungsten tracer at RS, AS, the backward side (BS), and the forward side (FS) were calculated by the change in the coordinate every 0.002 s on the WD-TD plane. Figure 4 shows the results for the 1000 rpm and 400 mm/min conditions. The tungsten tracer gradually moved to the outer side of the stir zone due to the centrifugal force. There was no significant difference in the velocity on all sides, and it increased with the increasing distance from the center of the probe. The velocity near the probe was lower than the peripheral velocity of the probe which was 314.0 mm/s. However, the velocity on the RS reached 339.0 mm/s which was higher than the peripheral velocity of the probe. It is considered that the velocity of the horizontal material flow was affected by both the probe and the shoulder [13, 24]. Therefore, the velocity on the outer side was higher than that of the inner side in the stir zone. The peripheral velocity at the periphery of the shoulder was 785.0 mm/s. Indeed, all the calculated velocities were lower than the peripheral velocity at the periphery of the shoulder.

The velocity on the RS was slightly higher than that on the AS at the same distance from the center of the probe. The difference of 6.7 mm/s (400 mm/min) can be easily explained by the relative velocity between the travel direction and the rotation of the tool. However, the difference between the velocity on the RS and AS seems to be ~20 mm/s which is larger than 6.7 mm/s. The relative high pressure by the tool traveling at the front of the probe should be one of the reasons for the lower velocity on the AS.

Figure 5 shows the two-dimensional tracer coordinates on the ND-WD plane. It is worth noting that the horizontal material flow around the probe was tilted against the travel direction. The flow goes upward when it goes to backside of the tool. This tilt direction is opposite to the direction of the tool tilt. The tilt of the horizontal flow can be explained by the resultant flow of the horizontal flow and the convectional flow which was generated at the rear of the probe by the shoulder [13, 24].

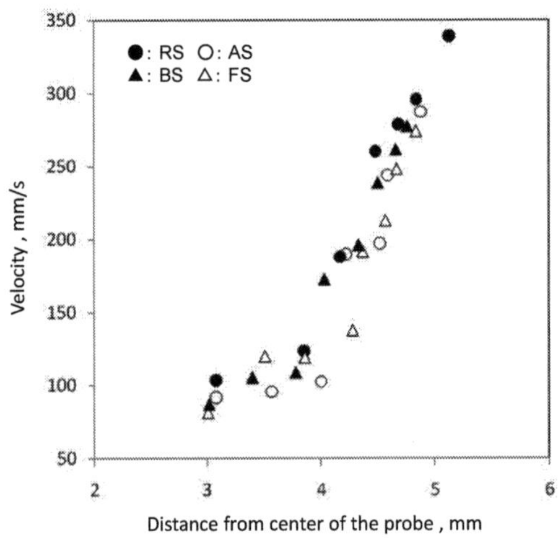

Figure 4. Velocity of the tungsten tracer [24].

Figure 6 shows the two-dimensional tracer coordinates on the TD-WD plane. The material flow zone had almost the same width around the probe from 600 - 1000 rpm, regardless of the rotating and travel directions of the tool. It is also worth noting that the horizontal material flow around the probe was also tilted on the ND-TD plane. The flow goes downward when it is going from the retreating side to the advancing side.

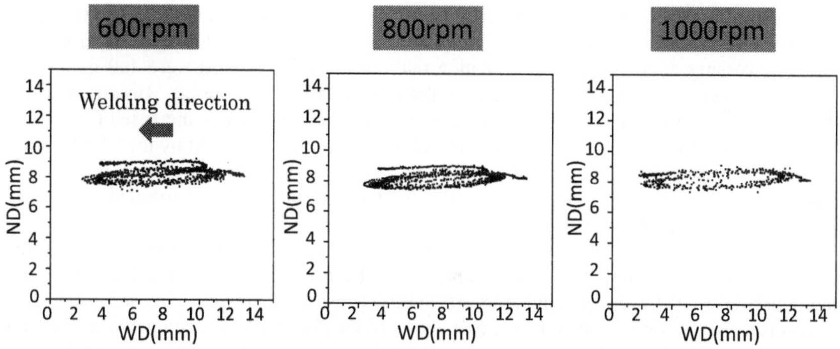

Figure 5. Two-dimensional tracer coordinates on the ND-WD plane (side view).

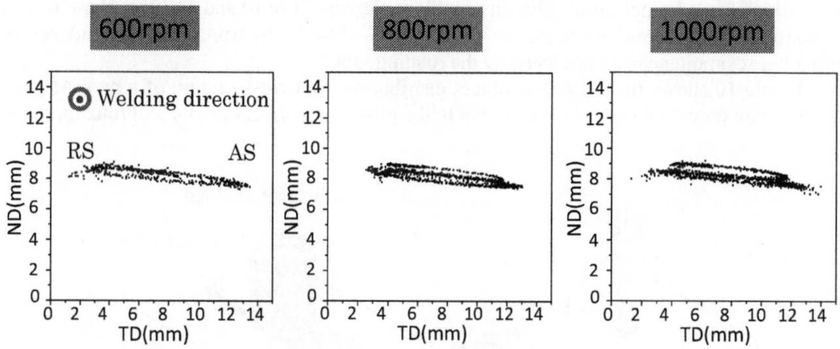

Figure 6. Two-dimensional tracer coordinates on the ND-TD plane (back view).

Friction stir powder processing (FSPP) for freely controlling the microstructure

Friction stir powder processing (FSPP) is a method of friction stir processing over a gap which is filled with a powder [25, 26], as shown in Fig.7. Using this method, modification and alloying can freely be carried out, depending on the applications, through complicated metallurgical phenomena. In this method, sufficient material flow, material diffusion and other metallurgical reactions are utilized, which is different from producing composite materials through the FSP [25, 26].

The addition of powder improves the mechanical properties of the processed region, but the material flow is very important in this process. Figure 8 shows the Vickers hardness distribution on a cross section of a pure Al plate (AA1050) with pure Cu powder perpendicular to the processing direction. The tool rotation rates and travelling speed were 400 rpm and 100 mm/min, respectively. The horizontal axis is the distance from the interface. Because the Al-Cu alloy is a typical precipitation strengthening system, the strength can be increased by dispersing fine precipitates in the matrix after the Cu is once dissolved in the pure Al by the FSPP. Whereas the Vickers hardness in the base material is about 40Hv, it increased to about 80Hv in the stir zone after the FSPP [25]. In particular, the hardness is uniformly distributed and doubled throughout the stir zone after the second pass.

Figure 8 is a TEM bright field image of the stir zone obtained from1500rpm-100mm/min-2pass during which the hardness significantly increased. Many fine intermetallic compounds caused by the precipitation are observed. Focusing on the size and shape of the precipitates, about 150nm bulk-shaped and about 10nm plate-shaped precipitates are observed. This result indicates that both a dynamic recrystallization and precipitation occur during the FSPP, and the precipitation is affected by the recrystallization behavior. Thus, in this method, the modification and alloying can be flexibly carried out through complicated metallurgical phenomena, such as stirring, dispersion, diffusion, solid-solution and precipitation.

Moreover, it is very interesting that a similar result is obtained even when using Fe powder which has a limited solid solubility in Al, as shown in Fig.10 and 11 [26]. These results are important as a unique phenomenon of the FSW (FSPP), in which heating and severe deformation is simultaneously produced by the rotating tool.

Figure 10 shows the Vickers hardness distribution on a cross section of a pure Al plate (AA1050) with pure Fe powder perpendicular to the processing direction. The tool rotation rates

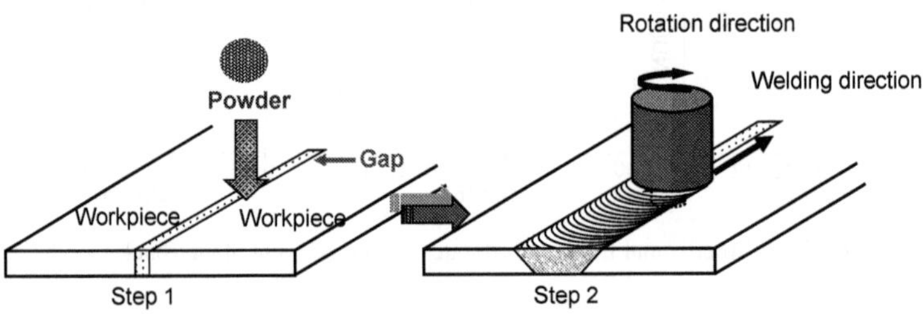

Figure 7. Friction stir powder processing (FSPP) [25].

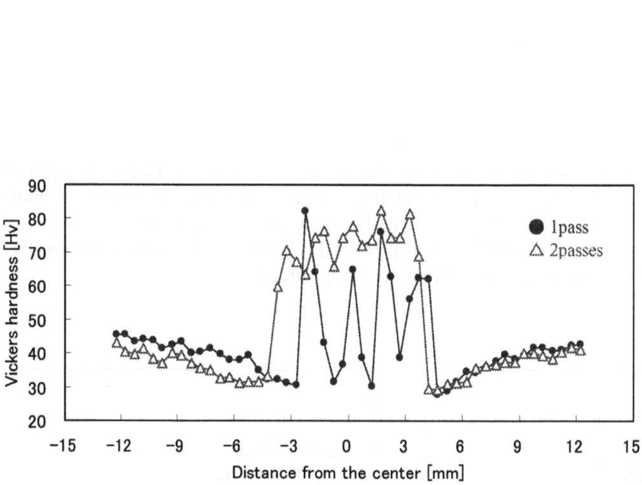

Figure 8. Hardness distribution on a cross section of a pure Al plate (AA1050) with pure Cu (1500rpm-100mm/min) [25].

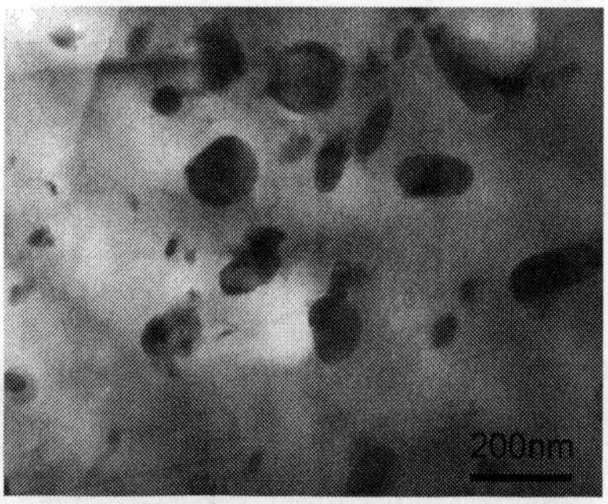

Figure 9. TEM bright field image of the stir zone after FSPP of a pure Al plate (AA1050) with pure Cu (1500rpm-100mm/min-2passes) [25].

and travelling speed were 400 rpm and 25 mm/min, respectively. The FSPP was performed twice. The horizontal axis is the distance from the interface. Regardless of the limited solubility, the hardness increased to about 70 Hv in the stir zone after the FSPP [26].

Figure 11 is a TEM bright field image of the stir zone obtained for the conditions of 1500rpm-25mm/min-2pass during which the hardness significantly increased. Many fine intermetallic compounds caused by the precipitation are observed, as in the Al-Cu system. Many fine precipitates are observed. This result indicates that the solubility was significantly increased due to the severe deformation by the rotating tool during the FSPP and many precipitates were formed during cooling. This is a very unique phenomenon for the FSW (FSPP).

These ideas can be extended to many industrial applications. Generally speaking, a strong material is difficult to shape. Accordingly, it is a good idea to construct a large structure first using deformable materials and then the structure can be partially strengthened after the construction by this new method depending on users' purposes. For steel, as another example, the processed part can be significantly hardened through the stir-dispersion-diffusion-transformation, when you add carbon powder [27]. The final composition can be changed by controlling the gap width or added powder compositions.

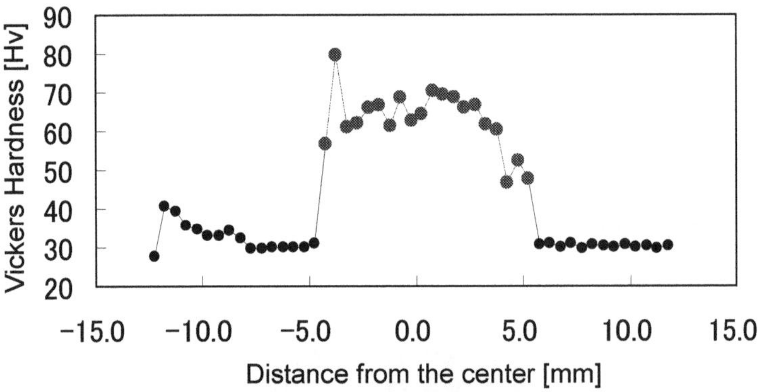

Figure 10. Hardness distribution on a cross section of a pure Al plate (AA1050) with pure Fe (1500rpm-25mm/min-2passes) [26].

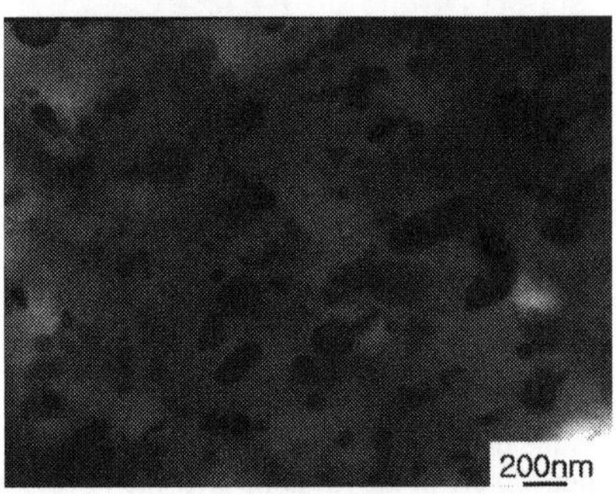

Figure 11. TEM bright field image of the stir zone after FSPP of a pure Al plate (AA1050) with pure Fe (1500rpm-25mm/min-2passes) [26].

Conclusions

The material flow during the FSW was successfully visualized by X-ray radiography. A three-dimensional graph was obtained using the locus of a tungsten tracer observed by two pairs of x-ray transmission real-time imaging systems. The profile of the material flow around the tool was clearly observed. The visualization produced the following conclusions.

1) There was a flow zone with a uniform width around the probe.
2) The horizontal round flow around the probe was tilted by the convectional flow.
3) The velocity of the material flow on the outer side was higher than that on the inner side in the stir zone.
4) The velocity on the RS was slightly higher than that on the AS at the same distance from the center of the probe, but the difference was very small.

FSPP (Friction Stir Powder Processing) is also introduced as a new technique for material designs. In this method, the modification and alloying can be flexibly carried out through complicated metallurgical phenomena, such as stirring, dispersion, diffusion, solid-solution and precipitation. Accordingly, a large structure is first constructed using deformable materials and the structure can then be partially strengthened after the construction, depending on users' requirements.

Acknowledgements

This study was supported by the Japan Science and Technology Agency (JST) under the Collaborative Research Based on Industrial Demand "Heterogeneous Structure Control: Towards Innovative Development of Metallic Structural Materials", Priority Assistance for the Formation of Worldwide Renowned Centers of Research - The Global COE Program (Project: Center of Excellence for Advanced Structural and Functional Materials Design), and Grant-in-Aid from the Ministry of Education, Culture, Sports, Science and Technology (MEXT), Japan.

References

1. W. M. Thomas, E. D. Nicholas, J. C. Needhan, M. G. Murch, P. Temple- Smith, and C. J. Dawes: *International Patent Application PCT/GB92/02203 and GB Patent Application 9125978.8*, UK Patent Office, London, December 6, 1991.
2. R.S. Mishra and Z.Y. Ma, *Mater Sci Eng R* 50 (2005), 1-78.
3. Japan Welding Society, H.Fujii and H.Kokawa, ed., *Friction Stir Welding*, (Sampo Shuppan, 2006).
4. R.S.Mishra, M.W.Mahoney ed., *Friction Stir Welding and Processing*, (ASM International, 2007)
5. R.Nandan, T.DebRoy and H.K.D.H.Bhadeshia, *Prog. Mater. Sci.*, 53 (2008), 980-1023.
6. H. Okamura, K. Aota, and M. Ezumi: *J. Jpn. Inst. Light Met.*, 50 (2000), 166-171.
7. G. Campbell and T. Stotler: *Welding J.*, 78 (1999), 45-47.
8. M. R. Johnsen: *Weld. J.*, 78 (1999), 35-39.
9. K. E. Knipstron and B. Pekkari: *Weld. J.*, 76 (1997), 55-57.
10. C. J. Dawes and W. M. Thomas: *Weld. J.*, 75 (1996), 41-45.
11. K. Colligan, *Welding Journal* 78 (1999), 229.
12. H.N.B. Schmidt, T.L. Dickerson, and J.H. Hattel, *Acta Mater* 54 (2006), 1199.
13. Y. Morisada, H. Fujii, T. Nagaoka, K. Nogi, and M. Fukusumi, *Composites A* 38 (2007), 2097.
14. A.P. Reynolds, *Scripta Mater* 58 (2008), 338.

15. O. Lorrain, V. Favier, H. Zahrouni, and D. Lawrjaniec, *J. Mater. Process. Technol.* 210 (2010), 603.
16. Y. Li, L.E. Murr, and J.C. McClure, *Mater Sci Eng A* 271 (1999), 213.
17. M. Guerra, C. Schmidt, J.C. McClure, L.E. Murr, A.C. Nunes, *Mater Character* 49 (2003), 95.
18. P. Su, A. Gerlich, T.H. North, and G.J. Bendzsak, *Metall. Mater. Trans. A* 38A (2007), 584.
19. B.C. Liechty and B.W. Webb, *J. Mater. Process. Technol.* 184 (2007), 240.
20. K. Kumar and S.V. Kailas, *Mater Sci Eng A* 485 (2008), 367.
21. H. Fujii, Y.G. Kim, T. Tsumura, T. Komazaki, and K. Nakata, *Mater. Trans.* 47 (2006), 224.
22. Y.Shimoda, M.Tsubaki, T.Yasui and H.Murakaw, *Q. J. Jpn. Weld. Soc.*, 3 (2011), 114s-118s.
23. R. Zettler, T. Donath, J.F. dos Santos, F. Beckman, and D. Lohwasser, *Adv. Eng. Mater.*, 8 (2006), 487.
24. Y. Morisada, H. Fujii, Y. Kawahito, K. Nakata, and M. Tanaka, *Scripta Mater.*, 65 (2011), 1085-1088.
25. K.Inada, H.Fujii, Y.S.Ji, Y.F.Sun and Y.Morisada, *Sci. Tech. Weld. Join.*, 15 (2010), 131-136.
26. H.Fujii, Y.F.Sun, K.Inada, Y.Yokoyama, H.Kimura and A.Inoue, *Mater. Trans.*, 52 (2011), 1634-1640.
27. K.Imagawa, Y.Morisada and H.Fujii, Proc. 9th Int. Symp. Friction Stir Welding, 15-17 May, 2012. USA, 5B-2.

Friction Stir Welding and Proceesing VII
Edited by: Rajiv Mishra, Murray W. Mahoney, Yutaka Sato, Yuri Hovanski, and Ravi Verma
TMS (The Minerals, Metals & Materials Society), 2013

System Parameter Identification for Friction Stir Processing

Dustin Marshall[1], Carl Sorensen[1]

[1]Brigham Young University
Department of Mechanical Engineering
435 CTB, Provo, UT, 84602, USA

Keywords:
Relay Feedback Test, Friction Stir Processing, System Identification, Control, Temperature

Abstract

Temperature control has been implemented in Friction Stir Processing and has demonstrated the ability to give improved process control. In order to have optimal control of the process, the parameters of the system must be accurately identified. The system parameters change with tool geometry and materials, workpiece materials, and workpiece holding system. This paper presents the use of the relay feedback test to determine the thermal parameters of the FSP system. The relay feedback test is easy to use and promotes system stability during its use. The results from the relay feedback test can be used to establish tuning constants for a feedback temperature control. The use of this method, as well as the quality of the resulting control is demonstrated in this paper.

Introduction

Friction stir welding is a solid state metal joining process developed by TWI in 1991. Control of the process can be very important for extending tool life and preventing defect formation. For this reason, temperature is one of the most important control variables.

Temperature is a process driven response. It is primarily affected by the spindle power entering the system. A controller must be used to maintain a steady temperature. Various controllers could be used, but a PID controller based on Ross's work was used because it is capable of providing a stable temperature, and is intuitive to use [1].

With a PID controller, the PID gains are the primary factors that affect how well it controls a system. The PID gains are often determined either through trial and error, or through tuning rules based on a system model. This paper demonstrates the use of the relay feedback test to determine the system model.

Background

Good process control is often essential to ensure defect free welds, and to promote tool life. Temperature seems to be especially important for this. Temperature control has been used by the Swedish Nuclear Fuel and Waste Management Company (SKB) to help reduce weld defects, and to help prevent tool failure. They found in welding large copper canisters, that if the tool temperature dropped below 790 degrees Celsius, they would produce welds with defects, and if they allowed the temperature to rise above 910 degrees Celsius, then failure of the pin commonly occurred [2].

Mayfield and Sorensen proposed a temperature control algorithm, which uses a controller to adjust the weld power to control the tool temperature [3] . This algorithm was later implemented by Ross [1].

Using a temperature control algorithm like Ross's can work well, assuming that the system is known and understood. With a PID controller, it is essential to pick controller gains which optimize the controller for that system. These gains can be found either through trial and error, or through system models. System identification is a field that works to identify the system model through experimental techniques. Once the system is identified, tuning rules can be used to find the PID gains which best control the system.

The PID controller used in this paper is a non-interacting series form of controller. It is of the form:

$$u(t) = kp\, e(t) + ki \int e(t)\, dt + kd\, \frac{d(Ttool(t))}{dt} \tag{1}$$

Where:
u(t) – the change in power commanded by the PID controller
kp – the proportional gain
e(t) – the current tool temperature minus the set point temperature
ki – the integrator gain
kd – the derivative gain
Ttool(t) – the current value of the tool temperature

The Relay Feedback Test is a frequency domain system identification method. It uses a continuous cycle of high and low step responses to yield an output from which the ultimate gain (*ku*) and ultimate period (*pu*) can be calculated. The output for this system is an oscillating wave. The shape, magnitude, and period of the output response change depending on the system type and its parameters. A typical input and output signal can be seen below in Figure 1.

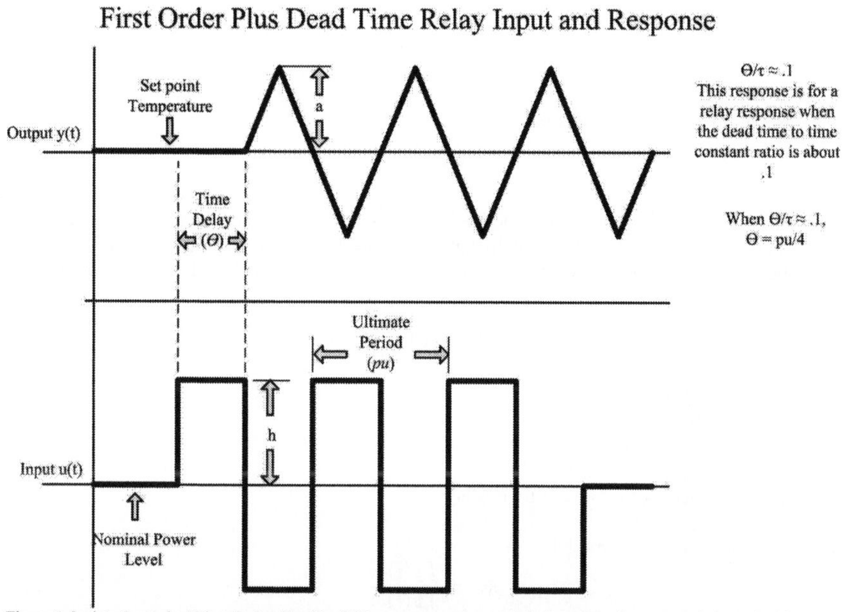

Figure 1 depicts the typical First Order Plus Dead Time response observed during a~~the~~ relay test, and the typical input. This response is only correct for a FOPDT with a time delay to time constant ratio of .1.

The relay feedback test was developed by Åström and Hägglund [4]. It is different than most other methods because it is a feedback method. The relay test provides a different input value to the system depending on the system response. Yu [5] identified some main advantages to the relay test, they are:

1) The useful process information given by the relay feedback test is close to the ultimate frequency.
2) The process is closed loop, which keeps it from being driven away from the nominal operating point.
3) This method is a more time efficient method for processes with a long time constant.
4) The experimental time is short, around two to four times the ultimate period.

In the case of friction stir welding, when the system temperature crosses the specified set point value, the relay controller adjusts the power. If the temperature is above the set point, the system is given a power input which is 5-10% below the nominal value for the set point temperature. When the temperature drops below the set point temperature, the power is set to 5-10% above the nominal value for the set point temperature.

The input value is being constantly adjusted in response to the output; this causes a wave form output response to be created. From this wave form, the ultimate gain and frequency may be calculated, and subsequently, the model parameters may be determined. In order to calculate the ultimate gain (*ku*) and ultimate frequency (*ωu*), *a*, *pu*, and *h* must be found. The value *a* is the amplitude of the wave signal and *h* is the amplitude of the square wave input signal. *Pu* is called the ultimate period; it is the period of the output wave signal. The ultimate gain is calculated as[5]:

$$Ku = \frac{4h}{\pi a} \quad \textbf{(2)}$$

The ultimate frequency is calculated as:

$$\omega u = \frac{2\pi}{pu} \quad \textbf{(3)}$$

With the parameters, *Ku*, ωu and *Pu*, the model parameters may be calculated. Experimental results and Ken Ross's work demonstrate that the process model for friction stir welding between the spindle and the tool thermocouple can be considered a First Order Plus Dead Time (FOPDT) process. The basic FOPDT model is:

$$y(s) = \frac{Km}{\tau s + 1} e^{-\theta s} * u(s) \quad \textbf{(4)}$$

Where:
y(s) – process output
u(s) – process input
Km – model or process gain
τ – model or process time constant
Θ – model or process dead time or time delay

Km can be calculated for a model as:

$$Km = \frac{\int y(t)dt}{\int u(t)dt} \quad \textbf{(5)}$$

Where *y(t)* and *u(t)* are the real time output and input signals respectively. The process time constant τ can be calculated with *Ku*, ωu, and *Km*[5].

$$\tau = \frac{\sqrt{(Km\, Ku)^2 - 1}}{\omega u} \quad \textbf{(6)}$$

The time delay Θ can be found by seeing how long it takes for the temperature to respond when the relay is first activated, but depending on the $\frac{\theta}{\tau}$ ratio, it can sometimes be calculated as $\theta = \frac{pu}{4}$. For the intent of this paper, it will be assumed that the time delay is one fourth of the ultimate period.

The tuning rules can now be used with the system models to calculate the PID gains. Tuning rules are available in books like Aidan O'Dwyer's, "Handbook of PI and PID Controller Tuning Rules." The tuning rules used for these welds were Chien 0% overshoot rule, and the minimum ITAE tuning rule. Chien's tuning rule is used for servo control, and the minimum ITAE tuning rule is used for regulator control. Servo tuning rules are used when a larger error is expected between the tool temperature and the set point temperature. This error may be due to the initial startup error, or due to changes in the set point. While these tuning allow large errors, they sacrifice set point tracking and disturbance rejection by using smaller integrator and proportional gains. The values for the Chien 0% overshoot tuning rule are calculated as followed [6]:

$$kp = \frac{0.6\,\tau}{Km\,\theta} \quad \textbf{(7)}$$

$$\tau i = \tau \quad \textbf{(8)}$$

$$\tau d = 0.5\,\theta \quad \textbf{(9)}$$

Since the form of PID controller being used requires ki and kd, the values can be converted as follows: $ki = \frac{kp}{\tau i}$, and kd is equal to $kd = kp * \tau d$.

Regulator tuning rules are to reject disturbances. The regulator tuning rules have the tendency to cause large overshoot, oscillation, and potentially severe oscillation and instability to even a small difference in the set point and tool temperature. Ideally these tuning rules should be running when the tool temperature is within a few degrees of the set point temperature. The PID gains are calculated using the minimum ITAE tuning rule as follows [6]:

$$kp = \frac{1.357}{Km}\left(\frac{\tau}{\theta}\right)^{0.947} \quad \textbf{(10)}$$

$$\tau i = \frac{\tau}{0.842}\left(\frac{\theta}{\tau}\right)^{0.738} \quad \textbf{(11)}$$

$$\tau d = 0.381\,\tau\left(\frac{\theta}{\tau}\right)^{0.995} \quad \textbf{(12)}$$

Methods

All welds for this paper were run on Brigham Young University's TTI model RM-2 friction stir welding machine. The machine code was modified to allow relay test welds to be run. A derivative of Ross's temperature control algorithm, and power control algorithm were also used. All temperature data was collected by an Omega K type thermocouple located within the tool. All data was recorded by the PLC and a data logging program.

Weld Setup

All relay and temperature control welds referred to in this paper were run in .25" thick 7075-T6 Aluminum. All welds plates were approximately 6" in width, and 48" in length. The backing plate used was .25" thick ASTM A-36 hot rolled steel. The weld plate and backing plate were held in place by 1" tall by 1/2" wide rectangular bars running lengthwise along the top outside edge of the aluminum plate. The rectangular bars were held in place by toe clamps, which were spaced roughly 8-10" apart.

Tool

The tool used in these experiments is a CS4 tool design made out of H-13 steel. The tool has a diameter of 25 mm, a pin length of 5.08 mm and a pin cone angle of 32.5 degrees. The tool was heat treated prior to use, and had a hole for a thermocouple machined in. The thermocouple hole is about .09" in diameter, and extends down to approximately .125" away from the end of the tool pin.

Relay Welds

All relay welds were run in a consistent manner. They all had a weld length of 22" with a feed rate of 6 IPM. The first 2 inches after the weld plunge were run in position control at 300 rpm. After the first two inches, a 10 inch power control section was run. The power value that was used was equal to the spindle power being applied to the work piece right before the controller switched from position control to power control.

At the end of the power control section, both the tool temperature, and spindle power were recorded and used for the relay. The tool temperature at the end of the power control section is used as the set point temperature, and the spindle power is used as a nominal power for the relay control. As the temperature drops below or rises above the set point temperature, the relay controller would respond. The relay controller used a 10% adjustment to the power. The relay section was run for 10 inches. At the end of the relay section, the tool would extract, and all data was saved to a data file for later access.

The formulas required to calculate the system parameters from the relay weld results are mentioned above in the background section. All relay calculations were made in Microsoft Excel. The data was imported, and calculations were made in a partially automated process. The process or system gain, Km was calculated using the midpoint integration method to integrate both the temperature data (output), and also the power data (input). All parameters were calculated using both an a value for cooling, and also an a value for heating.

It was noticed early on that the relay welds for friction stir welding are asymmetric. Load disturbances can cause asymmetry, and it is possible to adjust their effects by providing an asymmetric power input, but with friction stir welding, it is assumed that this asymmetry is related to the heating and cooling and is actually important to the process.

When the parameters are calculated, the system, either heating or cooling, which has the smallest τ value (the faster system), is used for tuning. This is used because a larger τ value will result in more aggressive PID tuning gains, which could cause the faster system to become unstable. PID gains were calculated for both the Chien 0% overshoot tuning rule, and the minimum ITAE tuning rule.

Temperature Control Welds

Temperature controlled welds, 22 inches in length, were run using a set of specified PID gains. These welds were run with a feed rate of 6 IPM. The first 2 inches were a position controlled section with a spindle speed of 300 rpm. After the 2 inch position controlled section, the PID controller took over, and attempted to bring the tool temperature to the set point temperature. If necessary, the set point temperature can be adjusted to reduce the error size between the tool temperature and the set point temperature, or shift the set point value.

Three relay test welds were run for the tool described above. A typical relay weld with this tool can be seen in Figure 2. The system parameters were calculated for the relay welds. They can be seen in Table 1.

Results and Discussion

Three relay test welds were run for the tool described above. A typical relay weld with this tool can be seen in Figure 2. The system parameters were calculated for the relay welds. They can be seen in Table 1.

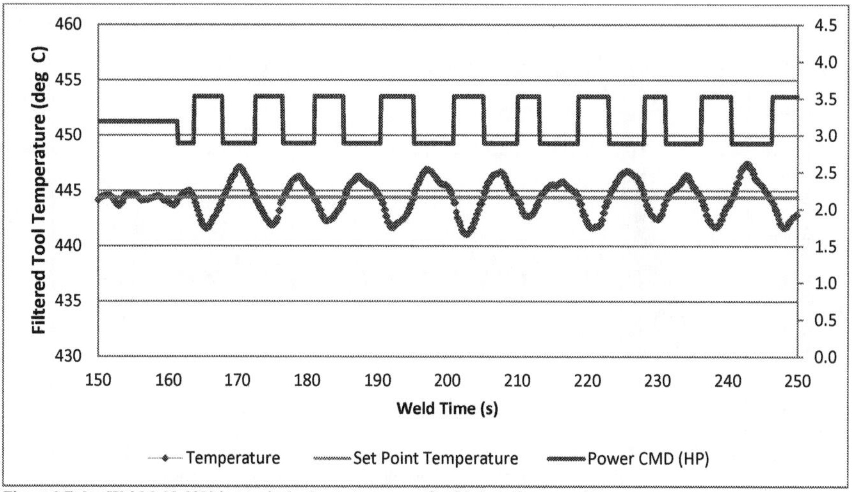

Figure 2 Relay Weld 8-23-0000 is a typical relay test response for friction stir processing.

Table 1 shows the system parameters calculated for the three relay welds which were run.

Weld Number	Type	Set Point Temperature	K	τ Heating	τ Cooling	θ
8-22-0001	Power Control Before Relay	446.85	139.21	10.24	32.56	4.82
8-23-0000	Power Control Before Relay	444.40	139.15	17.00	18.77	1.15
8-23-0001	Power Control Before Relay	436.82	141.69	10.74	14.53	1.02

A temperature controlled weld was run for each of the three relay welds. The PID gains used for each temperature controlled weld corresponded to the system parameters from one of the relay welds. The PID gains used an average of the heating and cooling constants for their calculation, rather than using either just the heating or cooling values. The minimum ITAE tuning rule was used.

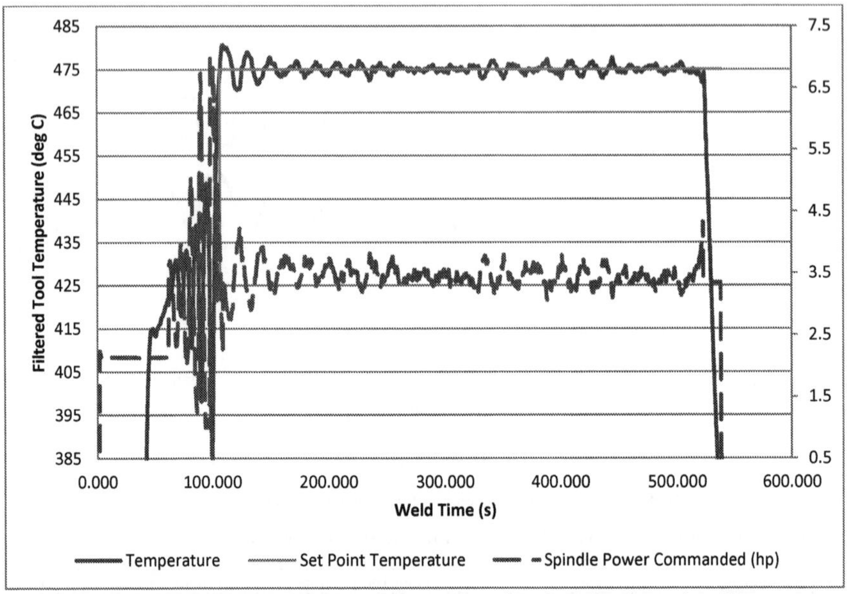

Figure 3 shows temperature weld 08-24-0000. It was the first weld run after relay tuning. It experienced integrator windup which caused oscillations to occur. These oscillations were removed by moving the set point temperature from 425 degrees Celsius to 475 degrees Celsius.

These three temperature control welds all demonstrated instability on weld startup. The instability can be seen in the form of oscillation. Weld 08-24-0000 was the weld first run (see Figure 4). There was only about 4 degrees of error between the tool temperature and the set point temperature. After startup, the oscillations quickly grew until the power limits prohibited further growth. The initial set point temperature was 425 degrees Celsius. After the oscillation began, the set point temperature was moved up 50 degrees to 475 degrees Celsius. At this temperature, the oscillations began to settle out and eventually disappeared.

Weld 08-24-0001 and weld 08-24-0002 were similar to weld 08-24-0000. Both required an increase in set point temperature to reduce the oscillations. Table 2 lists the PID gains, initial error, and final set point used in these three temperature controlled welds.

Table 2 includes the PID gains, initial temperature error and final set point values for the three temperature controlled welds which were tuned using the minimum ITAE regulator tuning rule.

PID Gains for Welds Run						
Weld Number	Corresponding Relay Weld	Initial Error	Final Set Point	kp	ki	kd
9-19-0002	8-22-0001	4 deg. C	475 deg. C	0.1205	0.0340	0.0703
9-19-0003	8-23-0001	7 deg. C	465 deg. C	0.1009	0.0295	0.0588
9-19-0004	8-23-0000	4 deg. C	445 deg. C	0.0670	0.0218	0.0390

It was seen that the minimum ITAE tuning rule didn't control the temperature controlled welds well from the beginning despite relatively small temperature errors when the controller was activated. It was decided that a servo tuning rule should be used for additional welds. The Chien 0% overshoot rule was chosen.

Three temperature controlled welds were run using the Chien 0% tuning rule. The PID gains were calculated off an average of the heating and cooling system parameters. All three were run with the set point set at 437 degrees Celsius. All three demonstrated good control. They started with a larger temperature error, but the controller was able to move the tool temperature to the set point without introducing unstable oscillations. Weld 9-19-0003 is typical of the response. It can be seen in Figure 5.

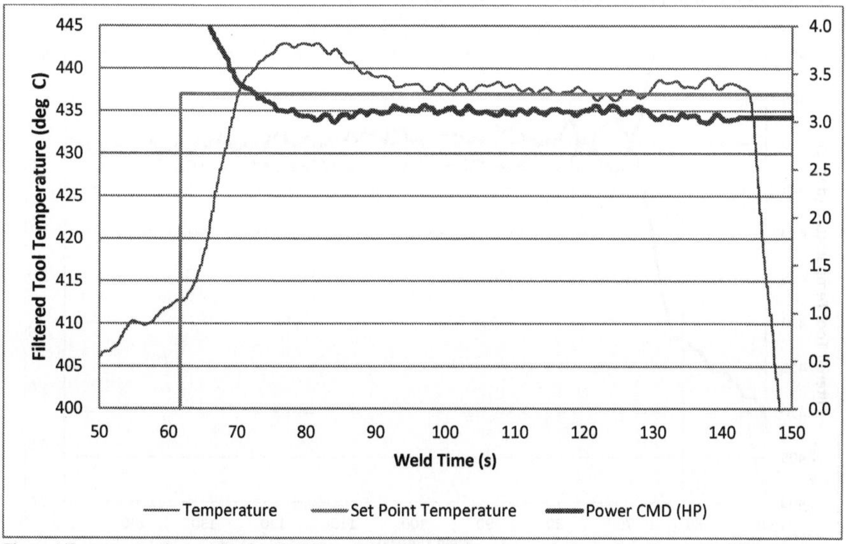

Figure 4 Temperature Control Weld 9-19-0003 used the Chien 0% overshoot rule for tuning. It was tuned based on an average of the heating and cooling system parameters.

In all three, the tool temperature stays steady around the set point after settling out. The PID gains for these three welds can be seen in Table 3.

Table 3 includes the PID gains, initial temperature error and final set point values for the three temperature controlled weld tuned with Chien's 0% overshoot servo tuning rule.

Weld Number	Corresponding Relay Weld	Initial Error	Final Set Point	kp	ki	kd
9-19-0002	8-22-0001	26.5 deg. C	437 deg. C	0.029	0.003	0.015
9-19-0003	8-23-0001	24 deg. C	437 deg. C	0.048	0.003	0.028
9-19-0004	8-23-0000	26 deg. C	437 deg. C	0.052	0.003	0.034

It was assumed that integrator windup was the cause of the oscillation observed in welds 08-24-0000, 08-24-0001, and 08-24-0002, rather than having to large of a proportional gain. To test this theory, another temperature controlled weld was run. It was given the gains of kp = 0.0926, ki = 0.00435, and kd = 0.0463. The proportional gain is approximately an average of the proportional gain for all three welds, being closest to the second, which had a gain of kp = 0.1009. The integrator gain was an order of magnitude smaller than the values used in the three welds, and the derivative gain is very close in value. On weld startup, there was a much larger temperature error (see Figure 6). The set point temperature was set to 437 degrees Celsius, and the tool temperature when the controller took over was 417 degrees Celsius. Despite the 19.3 degree Celsius temperature error, no oscillations were observed. In fact, the controller caused the tool temperature to climb to the set point value quickly. There was an overshoot of 6 degrees Celsius, but then it quickly returned to the set point value.

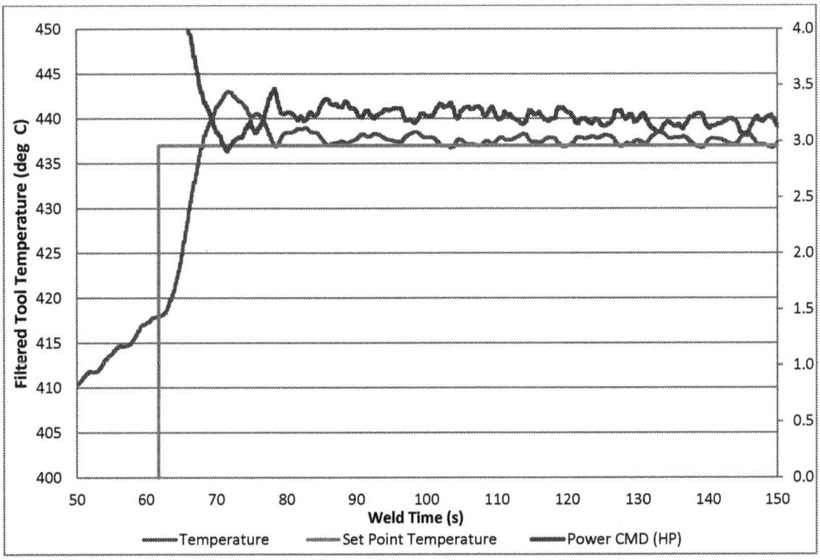

Figure 5 shows temperature weld 9-19-0006, which demonstrates that the welds run with disturbance rejection welds were unstable due to integrator windup, rather than too large of a proportional gain.

Conclusions

This paper demonstrates the use of the relay test with the friction stir welding process. It demonstrates that the relay test allows for good system identification tool for friction stir welding in aluminum alloys. The relay test can be run in a material and it will yield system parameters which can be used for controller tuning even without a prior knowledge of the system parameters. Overall, the relay test allows for system identification to aid in controller tuning. This allows the user to specify controller parameters which will result in stable temperature control, whether the desire is to maintain a steady temperature or if it is to allow step changes in temperature.

1. Ross, K.A., *Investigation and Implementation of a Robust Temperature Control Algorithm for Friction Stir Welding*, in *Mechanical Engineering*2012, Brigham Young University: Brigham Young University Press.
2. Cederqvist, L., et al., *Cascade control of the friction stir welding process to seal canisters for spent nuclear fuel.* Control Engineering Practice, 2011.
3. Mayfield, D.S., CD. *An Improved Temperature Control Algorithm for Friction Stir Processing.* 2010. Immendorfer Strand, Germany.
4. Åström, K.J. and T. Hägglund, *Automatic tuning of simple regulators with specifications on phase and amplitude margins.* Automatica, 1984. **20**(5): p. 645-651.
5. Yu, C.C., *Autotuning of PID Controllers: A Relay Feedback Approach*2012: Springer.
6. O'Dwyer, A., *Handbook of Pi And Pid Controller Tuning Rules*2006: Imperial College Press.

Friction Stir Welding and Proceeing VII
Edited by: Rajiv Mishra, Murray W. Mahoney, Yutaka Sato, Yuri Hovanski, and Ravi Verma
TMS (The Minerals, Metals & Materials Society), 2013

Advances in Temperature Control for FSP

Kenneth Ross[1] and Carl Sorensen[2]

[1]Manufacturing Technology, Inc.; 1702 West Washington Street, South Bend, IN 46628, USA

[2]Brigham Young University, 435 CTB, Provo, UT, USA

Keywords: temperature, power, control, friction stir

Abstract

A temperature control algorithm has been developed that contains an inner loop to control power and an outer loop to command power based on temperature feedback. The foundation of the control algorithm used in this work is the fact that spindle power leads tool temperature. This fact will be proven through analytical models and experimental data. Commanding spindle power to control temperature is a significant paradigm shift for some members of the friction stir processing (FSP) community.

The anticipated benefits of temperature control during FSP include: decreased variation in properties throughout the length of the weld, increased repeatability, increased tool life and extension of the application of FSP. This paper summarizes previous research, and investigates theory relevant to temperature control in FSP. Subsequently, the current state of temperature control technology and experimental data are presented.

Introduction

Controlling the weld temperature throughout the length of the weld is an important undertaking because weld properties (Nelson, et al. June, 2000) (Richards, et al. 2010) (Reynolds, Long and Tang 2007) (Sato, Urata and Kokawa 2002), such as fracture toughness and corrosion resistance, vary with weld temperature. If specified properties are desired throughout the weld, the weld temperature must be adjustable and in control throughout the length of the weld.

As the application of FSP in industry increases, more is demanded of the process. We believe that temperature control will increase the quality, repeatability and robustness of FSP. We anticipate the advent of temperature control will accelerate and expand the use of FSP in industry. This paper is an overview of research performed by Ross and Sorensen (K. A. Ross 2012) (Ross and Sorensen, Development and Implementation of a Robust Temperature Control Algorithm 2012).

Previous Work

The first efforts to control weld temperature used passive control techniques. Researchers considered that the weld temperature was proportional to weld pitch (spindle speed)/(travel speed) or various "pseudo heat indexes" which are functions of spindle speed and travel speed (Chimbli, Medlin and Arbegast 2007) (Kalya, et al. 2007). Passive control techniques assume that the process has reached a self-limiting equilibrium condition.

Passive control techniques are not adequate for temperature control because equilibrium conditions may not exist along the length of a weld. Causes of temperature changes along the length of the weld include: inadequate cooling of the tool or backing plate, changes in the thermal boundary condition and insufficient time to reach equilibrium. Passive control techniques are not versatile because they do not adjust for process disturbances.

Cederqvist et al. (Cederqvist, et al. 2009) noted a clear relationship between spindle power and tool temperature. It was suggested that this this relationship could be used to more accurately control the process.

Fehrenbacher et al. present the results of using a closed-loop temperature control system in the FSW of aluminum (Fehrenbacher, et al. 2010). Spindle speed was adjusted in response to temperature feedback. The results indicate that closed-loop control can reject small disturbances and reduce the error associated with larger disturbances introduced by changes in backing plate material.

Mayfield and Sorensen (Mayfield and Sorensen 2010) observe torque leads tool temperature for welds run at constant spindle speed. This is shown explicitly in experimental data presented. A common assumption in the FSP community has been that changes in temperature lead changes in torque. Data presented by Mayfield and Sorensen disproves this assumption.

In order to increase understanding of the system, a simplified thermal model of the stir zone was developed from a simplified thermal balance of the stir zone and is shown in Figure 1. This model suggests that the temperature of the stir zone can be controlled by the power input to the stir zone.

Power flowing into the stir zone is given by torque multiplied by the rotational velocity of the tool. Under constant spindle speed, a disturbance in torque causes a disturbance in power which changes the temperature of the stir zone. Spindle speed can be adjusted in response to disturbances in torque to maintain constant power. If thermal boundary conditions are constant, a constant spindle power results in a constant temperature.

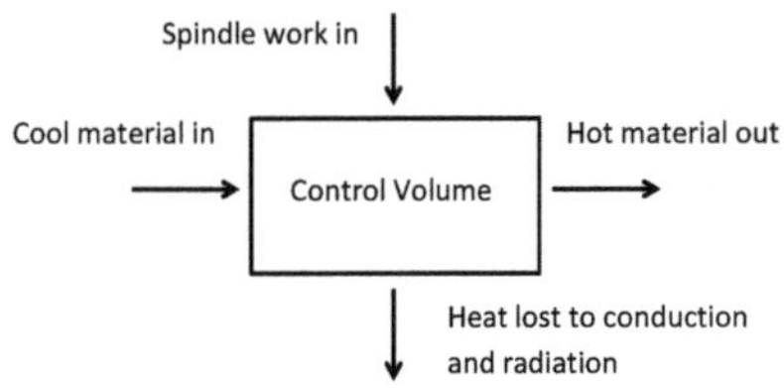

Figure 1: Thermal balance in an isothermal control volume representative of the stir zone presented by Mayfield and Sorensen

Mayfield and Sorensen present an improved temperature control algorithm (Mayfield 2010). The model proposed is a two-stage control model containing an inner and outer loop. The inner loop adjusts spindle speed to keep power constant, while the outer loop sets the desired power based on temperature. A simple feasibility test was performed by manually adjusting spindle speed in response to changes in torque. Decreased temperature variation was achieved.

It is expected that the temperature control algorithm presented by Mayfield and Sorensen will provide a faster response to disturbances than previously developed temperature control methods. The inner power control loop responds to thermal disturbances faster than they are registered by the temperature measurement system. Therefore, the advantage of using the improved temperature control algorithm increases as the time delay of the temperature measurement increases.

Cederqvist et al. (Cederqvist, Garpinger, et al. 2012) implemented the control scheme proposed by Mayfield and Sorensen for the welding of copper canisters containing nuclear waste. In 2012 they reported that a cascade control scheme kept the temperature within ±10 °C of desired temperature for their process.

Control of the Spindle Axis

Inner Loop

Power control is best achieved by controlling torque (Ross and Sorensen 2013) rather than spindle speed (patent pending). When spindle speed is used to control power, motor toque feedback is used as the control signal. Motor torque feedback is a poor control signal due to the large spikes in motor torque that occur when changes in spindle speed are commanded. When these torque spikes occur, the motor torque cannot be used to approximate spindle torque. When torque is adjusted to control power, spindle speed is used as the control signal. Uncertainty in spindle speed measurement is small. Motor torque can be used to approximate spindle toque when spindle torque is adjusted to keep power constant.

Outer Loop

Proportional-integral-derivative (PID) controllers are the most commonly used type of closed-loop feedback controller (Åström and Murray 2008). PID control uses the magnitude, integral and derivative of the error to control the process. The outer loop uses a PID algorithm to command power based on tool temperature feedback. PID gain tuning can be used to achieve a desired temperature response.

Identifying the Form of the System Transfer Function

Spindle power leads tool temperature. Thermal modeling and experimental results show that when a weld is run at constant power a steady state temperature is achieved. When the power is changed, there is a time delay, followed by a characteristic change in temperature resulting in a new steady state temperature. Thermal modeling and experimental results indicate that the temperature response of the tool can be approximated by a first-order transfer function with time delay.

The temperature response is dominated by thermocouple location. Experimental data and numerical modeling show that both time delay and time constants increase with distance from the thermocouple to the tool-workpiece interface. Second-order dynamics could not be seen in welds run in aluminum. Numerical solutions indicate that second-order dynamics become increasingly significant as distance between the tool-workpiece interface and the thermocouple increases.

Thermal Modeling

The 1-D heat diffusion equation is used to investigate the thermal response of the tool to changes in weld power. Figure 2 shows the calculated temperature rise of a weld due to an increase in weld power. Previous to time 0 the temperature and power are constant. At time 0 there is a step increase in weld power. The time between the step change in power and the time the temperature begins to change is the time delay, t_d. Thermocouple locations are defined by the distance from the thermocouple to the tool-workpiece interface.

Figure 2 shows that the temperature response of the tool is dominated by thermocouple location. When the thermocouple is located close to the tool-workpiece interface, time delay and time constants are small and only first-order dynamics are visible. As the thermocouple is located further from the tool-workpiece interface, time delay and time constant increase. Dynamics become increasingly second-order, immediately after the time delay, as distance increases.

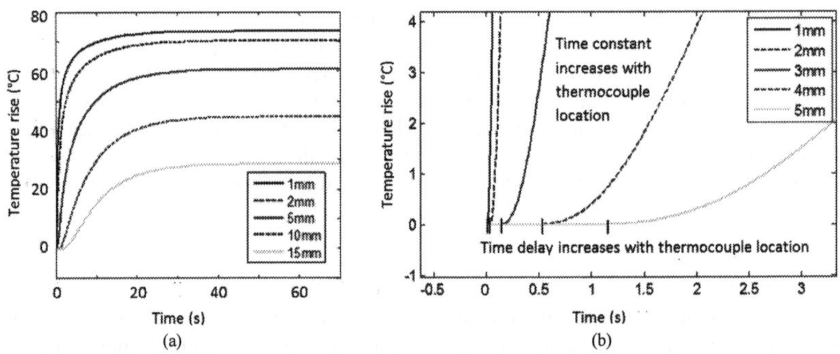

Figure 2: (a) Simulated temperature rise at various thermocouple locations using the heat diffusion equation with a linear initial temperature profile (b) Magnified view to show time delay

System identification was performed for all thermocouple locations shown in Figure 2 using the *System Identification Toolbox™* in MATLAB. For all thermocouple locations shown, first-order models with time delay had a fits above 96% and second-order models with time delay had fits greater than 97%. In each case the second-order fit was slightly higher. Because all the fits are above 95%, the heat diffusion equation suggests that tool dynamics can be modeled using a first-order system with time delay.

Experimental Results

Results for a power step from 2.24 kW (3.00 HP) to 2.54 kW (3.40 HP) are shown in Figure 4. This figure shows that as distance from the tool-workpiece interface to the thermocouple increases the time constant, τ, increases. Steady state temperature increases with thermocouple distance. Thermocouple locations are defined in Figure 3. The time constants for thermocouple locations 1, 2 and 3 are 25.9 s, 38.5 s and 58.3 s respectively. The steady-state temperature for thermocouple locations 1, 2 and 3 are 493°C, 467°C and 438°C respectively. This suggests that the time constant and steady state temperature are dominated by the thermocouple location. .

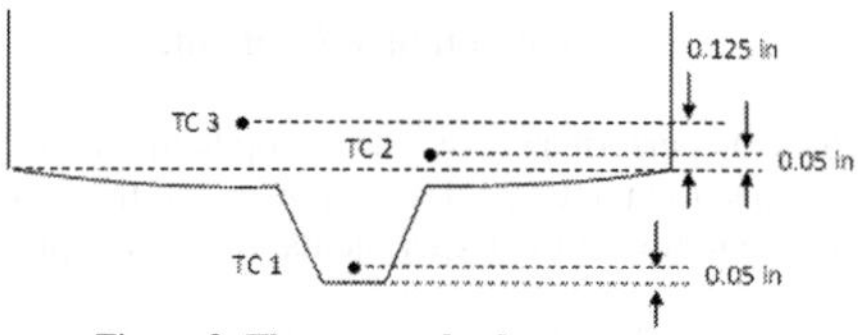

Figure 3: Thermocouple placement

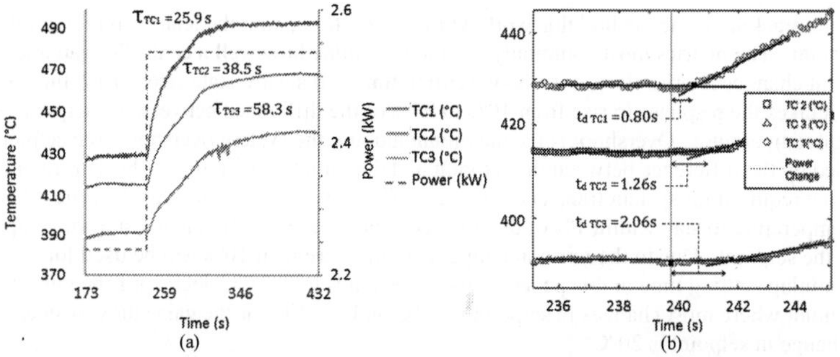

Figure 4: Power step from 2.24 kW (3.00 HP) to 2.54 kW (3.40 HP) temperature response

The time delays for thermocouple location 1, 2 and 3 are 0.80, 1.26 and 2.06 seconds respectively. The plot used to calculate the time delay is shown in Figure 4b. This figure indicates time delay is dominated by thermocouple placement.

A graph showing the first and second-order approximations and response data from TC1 is shown in Figure 5. The System Identification Toolbox^{TM} in MATLAB is used to obtain first and second-order approximations of the temperature response to changes in power. The goodness of fit is calculated as the percentage of output variation that can be explained by the model. Both first and second-order models had fits of 95.72%. Therefore the response can be classified as first-order.

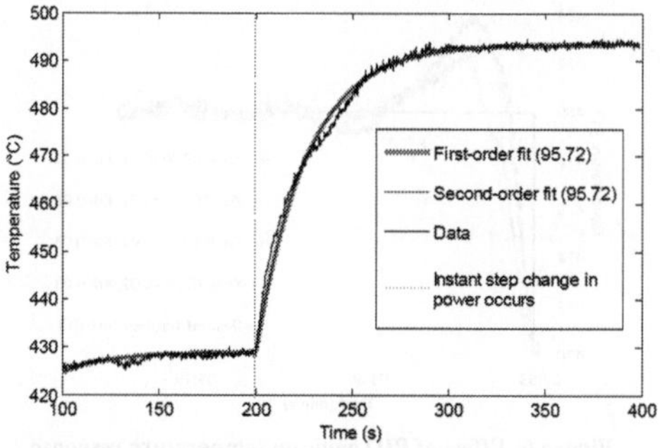

Figure 5: 1st and 2nd order fits of temperature response to step change in power from 2.24 kW (3.00 HP) to 2.54 kW (3.20 HP)

Validation of Controller

This section investigates the performance of a dual loop temperature control algorithm where the inner loop modulates torque to maintain a desired power and the outer loop modulates desired power to achieve a desired temperature. Experimental results are presented for welds run in both aluminum and steel.

Performance metrics

Setpoint tracking is the method this work will focus on for testing the temperature control algorithm. The metrics most commonly used to determine how well a controller can track setpoint changes are rise time, overshoot, settling time and steady-state error. Rise time is the time it takes the response to rise from 10% to 90% of the difference between the setpoint and current temperature. Overshoot is the maximum amount the system overshoots the setpoint divided by the difference between the setpoint and the initial temperature. The settling time is the time required for system transients to decay. This is typically defined as the time it takes for the temperature to stay within 1% of the difference between the setpoint and starting temperature from the setpoint. Due to the noise in temperature measurement 10% will be used for determining settling time in this paper. This corresponds to +/- 1°C degree for the gain study in aluminum where most changes in setpoint is 10°C and +/- 2°C for the gain study in steel where the change in setpoint is 20°C.

Aluminum

A gain study was performed by running welds in aluminum with various PID control gains. For each weld, the commanded temperature experienced a step change from 470 °C to 480 °C. The data collected indicates changes in PID gains affect the performance metrics in a manner constant with PID control theory. Therefore manual PID tuning can be used to achieve desired performance metrics. The temperature response and corresponding PID gains for various welds using in the gain study are shown in Figure 6. Welds shown in Figure 6 are run in 6.35 mm (0.25 in) thick AA 7075-T735, have a feed rate of 150 mm/min (6.0 IPM) and plunge 5.8 mm (0.23 in) into the material. Temperature measurements and control are taken from TC1, shown in Figure 3.

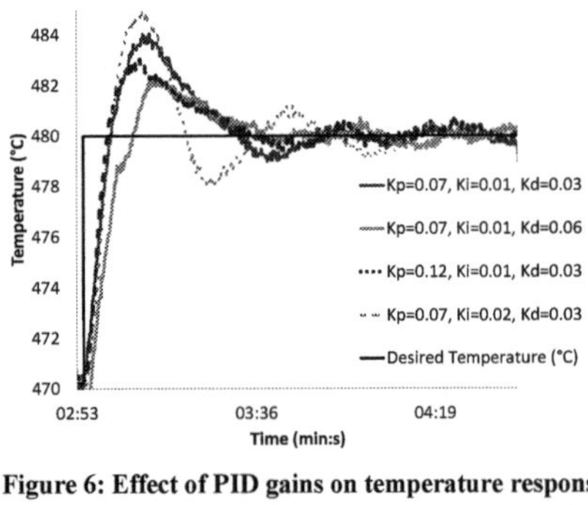

Figure 6: Effect of PID gains on temperature response

Data from a weld run in thicker material at a slower speed rate shows that temperature response to a step change is similar across varying plate thicknesses and feed rates for a given set of PID gains. A weld was tested using a 9.53 mm (0.375 in) plate and a feed rate of 101.6 mm/min (4.0 IPM) and is shown in Figure 7. The weld shown in Figure 7 uses a Kp of 0.07, a Ki of 0.01 and a Kd of 0.06. A weld with these gains is also shown in Figure 6.

The weld started in RPM control mode and reached a temperature over 51°C higher than the setpoint. When temperature control was engaged, the system responded with an overshoot of 29%, rise time 3.33 seconds, 10% settling time of 17.30 seconds and 1.0°C settling time of 29.5 s. The temperature response of this weld is similar to that of weld with identical gains shown in Figure 6, despite a different material thickness, feed rate and larger difference between the setpoint and measured temperature previous to the setpoint command. The similar temperature response of the two welds is significant because it indicates that a given tool with a given thermocouple location can use the same gains with varying thermal boundary conditions and obtain similar temperature responses.

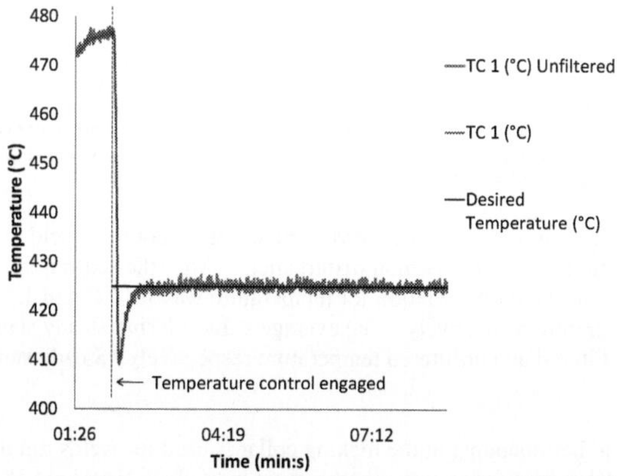

Figure 7: Kp=0.06, Ki=0.01, Kd=0.03 in 0.375" AL 7075 at 101.6 mm/min

Data collected from the weld shown Figure 7 indicates that steady state error is negligible. After the 1.0°C settling time, the temperature was held at 425°C with a standard deviation of 0.72°C for non-filtered temperature and 0.31°C for filtered. The average value for all temperature data collected after the 1.0°C settling time is 424.997°C for unfiltered temperature data and 425.020°C for filtered data. Both values can be expressed as 425.0°C.

Noise seen for air welds is similar to noise seen in temperature control welds. For air welds, the standard deviation in temperature is 0.775°C for non-filtered temperature and 0.217°C for filtered temperature. These values represent instrumentation noise and are approximately the same as the temperature deviations reported for temperature control in aluminum after the settling time. The performance of the temperature control algorithm may be limited by instrumentation noise.

<u>Steel</u>

Large temperature variation occurs when running welds in steel without temperature control.

Figure 8a shows a weld run with a constant spindle speed. The temperature varies by over 20°C in only 254 mm (10.0 in). The temperature control algorithm can maintain a temperature within a degree and a half of the setpoint in steels, as shown in Figure 8b.

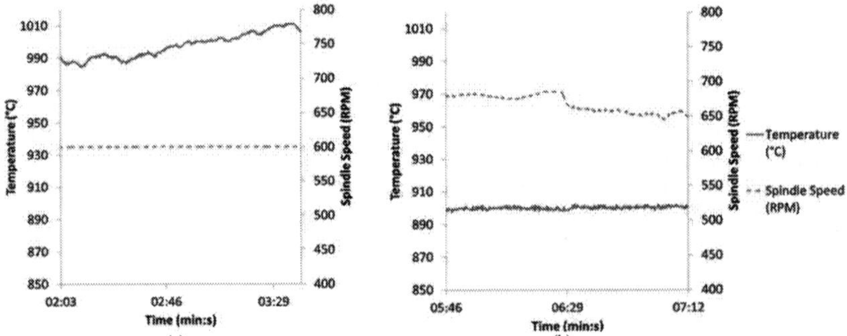

(a) (b)

Figure 8: (a) 254mm segment of constant RPM weld in steel demonstrating large temperature variation (b) 254mm segment of a temperature control weld in steel demonstrating the algorithms ability to hold temperature within a degree and a half of a given setpoint

While temperature remains constant, the RPM changes throughout the weld, indicating that the algorithm is successfully rejecting thermal disturbances. After the settling time for the weld shown in Figure 8b the standard deviation for temperature was 1.27°C and 1.42°C for filtered and unfiltered temperature respectively. The average values during steady state are 899.692°C and 899.726°C for filtered and unfiltered temperature respectively. Steady state error is negligible.

A PCBN tool with a thermocouple in the locking collar is used for welds run in steel. The time constant and time delay for temperature measurement in the PCBN tool are 38.76 and 2.64s respectively. These values are significantly higher than the thermal response of TC1 in aluminum welds. Therefore, less aggressive PID gains are used. Manual tuning techniques were used to obtain a rise time of 12.65 seconds, overshoot less than 20.63% and settling time of less than 89.20s.

Performance metrics in PCBN tools could be increased by isolating the thermocouple from the thermal mass of the locking collar and decreasing the distance between the thermocouple and tool-workpiece interface.

Conclusions

Spindle power leads tool temperature. Experimental results show that when a weld is run at constant power, a steady state temperature is achieved. When the power is changed, there is a time delay followed by a characteristic change in temperature resulting in a new steady state temperature.

First-order with time delay is an appropriate model for temperature response. Experimental and numerical results indicate that the temperature co the tool can be approximated by a first order function with time delay.

The temperature response of the tool is dominated by thermocouple location. Experimental results and numerical modeling show that both time delay and time constant increase with distance from the thermocouple to the tool-workpiece interface. Second-order dynamics could not be seen in welds run in aluminum. Using the PCBN tool, the second-order model had a slightly higher fit than the first-order model. Numerical solutions indicate that second-order dynamics become increasingly significant as distance between the tool-workpiece interface increases.

Standard PID gain tuning can be used to achieve a desired temperature response. Experimental data shows that changes in PID gains affect the temperature response as expected for welds run in aluminum and steel.

The algorithm has negligible steady state error. For a weld run in aluminum, after the settling time, the standard deviation from the setpoint was 0.717°C for non-filtered temperature and 0.312°C for filtered temperature. The difference between the average temperature and the setpoint is 0.003°C for unfiltered temperature data and 0.020°C for filtered data. Using unfiltered temperature data, the difference between the setpoint and the average temperature at steady state is 0.003°C for aluminum and 0.274°C for steel. The steady state error for aluminum is $7.06e^{-4}$% of the setpoint and 0.0304% of the setpoint of steel.

Noise seen for air welds is similar to noise seen in temperature control welds. For air welds used to calculate spindle inertia, the standard deviation in temperature was 0.775°C for non-filtered temperature and 0.217°C for filtered temperature. These values represent instrumentation noise and are approximately the same as the temperature deviations reported for temperature control in aluminum after the settling time. The performance of the temperature control algorithm may be limited by instrumentation noise.

References

Åström, Karl J., and Richard M. Murray. *Feedback Systems: An Introduction for Scientists and Engineers.* Princeton: Princeton University Press, 2008.

Cederqvist, Lars, Carl D Sorensen, Anthony P Reynolds, and T Öberg. "Improved process stability during friction stir welding of 5 cm thick copper canisters through shoulder geometry and parameter studies." *Science and Technology of Welding and Joining*, 2009: 178-184.

Cederqvist, Lars, O Garpinger, T Hägglund, and A Robertsson. "Cascade control of the friction stir welding process to seal canisters for spend nuclear fuel." *Control Engineering Practice* 20 (2012): 35-48.

Chimbli, S.K., D.J. Medlin, and W.J. Arbegast. "Minimizing Lack of Consolidation Defects in Friction Stir Welds." *Friction Stir Welding IV.* Warrendale, PA: The Minerals, Metals and Materials Society, 2007. 135-142.

Fehrenbacher, Axel, Neil A Duffie, Nicola J Ferrier, Michael R Zinn, and Frank E Pfefferkorn. "Temperature Measurement and Closed-Loop Control in Friction Stir Welding." *8th International Friction Stir Welding Symposium.* Timmendorfer Strand, Germany, 2010.

Kalya, P., K. Krishnamurthy, R.S. Mishra, and J. Baumann. "Specific Energy and Temperature Mechanistic Models for Friction Stir Processing of AL-F357." *Friction Stir Welding IV.* Warrendale, PA: The Minerals, Metals and Materials Society, 2007. 113-125.

Mayfield, David W, and Carl D Sorensen. "An Improved Temperature Control Algorithm for Friction Stir Processing." *8th International Friction Stir Welding Symposium.* Timmendorfer Strand, Germany, 2010.

Nelson, T.W., C.D Sorensen, R.J. Steel, and W.A. Arbegast. "Effects of Active and Passive Cooling in Friction Stir Welding." *Aeromat 2000.* Seattle, WA., June, 2000.

Reynolds, A P, T Long, and W Tang. "Process Response Parameter Relationships in Aluminium Alloy Friction Stir Welds." *Science and Technology of Welding and Joining* 12, no. 4 (2007): 311-317.

Richards, D G, P B Prangnell, P J Withers, S W Williams, T Nagy, and S Morgan. "Efficacy of Active Cooling for Controlling Residual Stresses in Friction Stir Welds." *Science and Technology of Welding and Joining* 15, no. 2 (2010): 156-165.

Ross, Kenneth A, and Carl D Sorensen. "Development and Implementation of a Robust Temperature Control Algorithm." *9th International Symposium on Friction Stir Welding.* Huntsville, Alabama, USA, 2012.

—. "Paradigm Shift in Control of the Spindle Axis." *Friction Stir Welding and Processing VII.* San Antonio, TX: John Wiley & Sons, Inc., 2013.

Ross, Kenneth Alec. "Investigation and Implementation of a Robust Temperature Control Algorithm for Friction Stir Welding." MS Thesis, Mechanical Engineering, Brigham Young University, Provo, UT, 2012.

Sato, Y.S., M. Urata, and H. Kokawa. "Parameters controlling microstructure and hardness during friction-stir welding of precipitation-hardened aluminum alloy 6063." *Metallurgical and Materials Transations A* 33, no. 3 (March 2002): 625-635.

Friction Stir Welding and Proceesing VII
Edited by: Rajiv Mishra, Murray W. Mahoney, Yutaka Sato, Yuri Hovanski, and Ravi Verma
TMS (The Minerals, Metals & Materials Society), 2013

Analysis of Tool Feedback Forces and Material Flow during Friction Stir Welding

Enkhsaikhan Boldsaikhan[1], Michael McCoy[1]

[1]Wichita State University
1845 Fairmount Street, Wichita, KS 67260, USA

Keywords: Friction stir welding, tool feedback forces, prevailing pressure field, revolving pressure field

Abstract

An innovative heuristic force model has been developed to characterize material flow induced by the Friction Stir Welding (FSW) process. When the rotating weld tool travels inside the plasticized material, the leading interfacial surfaces of the tool experience the prevailing pressure field (PPF) due to the tool travel motion. Furthermore, a plasticized shear layer of material is periodically extruded from the leading side to the trailing side of the weld tool by the rotational and travel motions of the weld tool. This periodic material flow introduces revolving pressure field (RPF) around the weld tool. Due to PPF and RPF, the tool feedback forces grow and evolve during the FSW process. A relationship between the pressure fields and the tool feedback forces was identified by the heuristic force model. The model has been simulated and the actual weld data was analyzed with regards to the simulation results.

Introduction

The present study introduces a heuristic force model to characterize tool feedback forces with regards to the material flow induced by the friction stir welding process. The model uses the tool feedback forces, such as the drag force (F_X or X force) and the transverse force (F_Y or Y force), based on the assumption that they contain crucial information about the material flow dynamics.

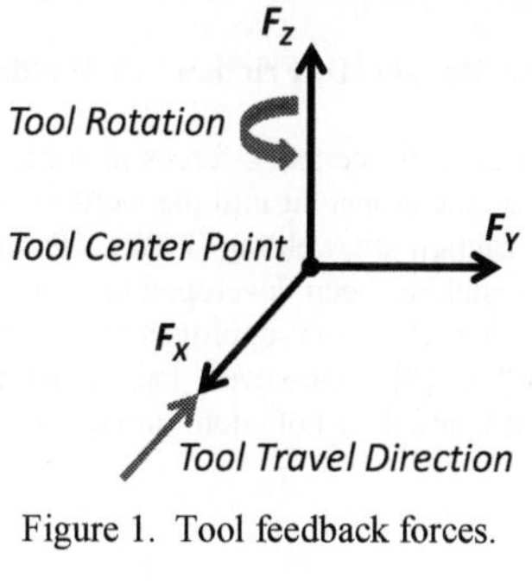

Figure 1. Tool feedback forces.

Feedback forces from the weld tool are measurements of the material resistance forces induced by the rotational and travel motions of the weld tool. The feedback forces including F_X, F_Y, and F_Z, in the directions shown in Figure 1 can be acquired through pressure transducers installed on the weld head assembly of the weld tool.

When the rotating tool travels inside the plasticized material during the welding process, the leading surfaces of the tool experience the prevailing pressure field (PPF) attributed to the material resistance in response to the tool travel motion. Furthermore, the rotational and travel motions of the weld tool induce a plasticized shear layer of material to periodically move from the leading side to the trailing side of the weld tool. The periodic nature of the material flow was observed by many researchers [1-8]. Colligan [1] showed highly cyclical plasticized material flow around the weld tool from the leading face to the trailing face using steel markers placed in butt-weld joints. London [2] and Reynolds [3] have verified this experiment using different marker materials. Chen et al [4], Schmidt et al [5], Schneider et al [6], and Arbegast [7] showed that a cyclic shear material layer is periodically deposited at the trailing side of the moving weld tool. The periodic nature of the material flow during friction stir welding (FSW) was also indicated by the tool feedback forces as demonstrated in [8]. All these research results reported are related to friction stir welding of aluminum alloys, and lead to an assumption that the material flow during friction stir welding is highly periodic and the period of that process is defined by the tool rotation speed.

In this study, it was hypothesized that the periodic material flow introduces a revolving pressure field (RPF) that is resulted from instantaneous pressure differences between two opposing sides of the weld tool. The revolving pressure field (RPF) revolves around the weld tool when a plasticized shear layer of material is periodically extruded from the leading side to the trailing side of the weld tool in the direction of the tool rotation. The prevailing pressure field and the revolving pressure field play important roles for defining the dynamics of the tool feedback forces and may provide an important insight into the process nature.

In the next section, a two dimensional (2D) friction stir welding (FSW) model is introduced and then its simulation settings are presented. In the results and discussion section, the simulation results of the 2D model are discussed and compared with actual weld data.

Two Dimensional (2D) Friction Stir Welding Model

Understanding the dynamics of the tool feedback forces in conjunction with the dynamics at the tool-material interface may give crucial insight into the welding process and help to facilitate in-process quality monitoring for friction stir welding (FSW). To date, due to the highly nonlinear process nature, no significant model has been developed to characterize the dynamics carried out at the tool-material interface. A preliminary approach to modeling the dynamics at the tool-material interface was reported in [8]. However, the model proposed in [8] uses only the shear/tangential stress component, and does not count impact of the radial pressure component.

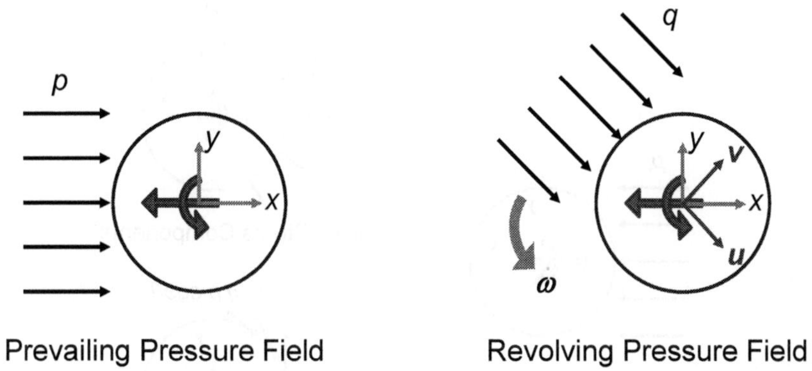

Figure 2. Prevailing pressure field (PPF) and revolving pressure field (RPF). Note that tool XY coordinate system is stationary and the tool UV coordinate system rotates around the tool along with the RPF at the revolving speed of ω.

An innovative 2D FSW model proposed herein uses PPF and RPF for representing the nonlinear dynamics that takes place at the tool-material interface. A schematic view of the 2D model is illustrated in Figure 2. For the sake of simplicity, the 2D model uses a circular weld tool with the unit radius, uniform PPF, uniform RPF, and a constant rotation speed, ω, of RPF. It is assumed that the simplified 2D model provides approximate representation of the nonlinear nature of the actual FSW process.

In Figure 2, the leading surface of the tool experiences uniform PPF, p, attributed to the material resistance in response to the tool travel motion. In addition, uniform RPF, q, revolves around the weld tool at a rotation speed of ω ($\omega \leq tool\_spindle\_speed$) as an indication of a plasticized shear layer of material periodically extruded from the leading side to the trailing side of the weld tool in the direction of the tool rotation (Figure 2).

Since the tool used in the 2D model is circular, a pressure field imposes radial stress component and the shear stress component on the tool surface as illustrated in Figures 3 and 4. It is intuitive to assume that the radial and shear stress components are proportional to each other and the shear coefficient of the shear component is represented by η ($\eta \geq 0$) in Figures 3 and 4. Also, the shear components are oriented in the opposite direction of the tool rotation due to the material resistance. The radial stress components of PPF and RPF are $p \cos\theta$ and $q \cos\varphi$, respectively, and the shear stress components of PPF and RPF are $\eta p \cos\theta$ and $\eta q \cos\varphi$ in the respective order. The angular variable θ is used for integrating PPF in the tool XY coordinate system, and the angular variable φ is used for integrating RPF in the tool UV coordinate system. Note that tool XY coordinate system is stationary and the tool UV coordinate system rotates around the tool along with RPF as shown in Figure 2.

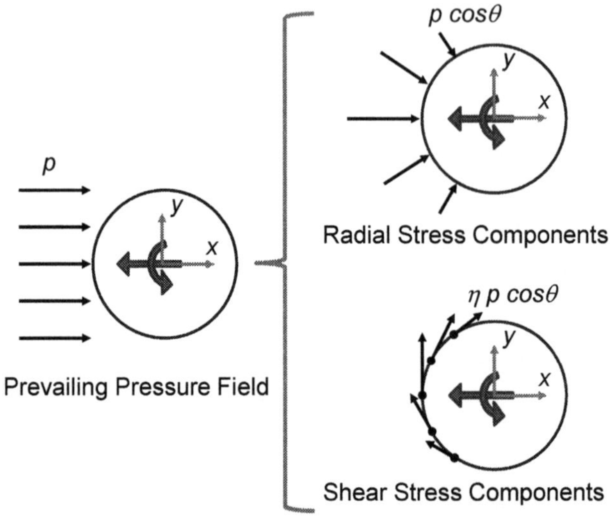

Figure 3. Radial and shear stress components of PPF.

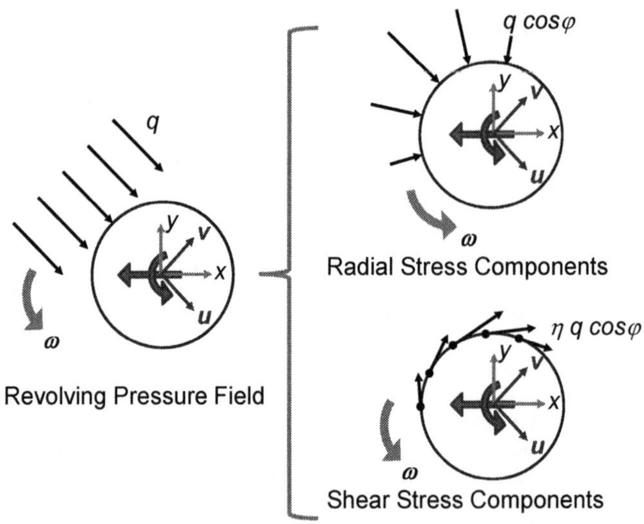

Figure 4. Radial and shear stress components of RPF.

Furthermore, analytical equations for the drag force (F_X or X force) and the transverse force (F_Y or Y force) of the weld tool are integrated from the uniform pressure field components according to Equations 1, 2 and 3. Note that Equations 1, 2, and 3 are for the unit circle tool.

$$\begin{cases} f_x = \int_{-\pi/2}^{\pi/2} p\cos^2\theta\, d\theta + \int_{-\pi/2}^{\pi/2} \eta\, p\cos\theta\sin\theta\, d\theta = \frac{\pi p}{2} \\ f_y = \int_{-\pi/2}^{\pi/2} \eta\, p\cos^2\theta\, d\theta + \int_{-\pi/2}^{\pi/2} p\cos\theta\sin\theta\, d\theta = \frac{\pi\eta p}{2} \end{cases} \quad (1)$$

Equation 1 demonstrates analytical calculations of the XY force components of uniform PPF.

$$\begin{cases} f_u = \int_{-\pi/2}^{\pi/2} q\cos^2\varphi\, d\varphi + \int_{-\pi/2}^{\pi/2} \eta\, q\cos\varphi\sin\varphi\, d\varphi = \frac{\pi q}{2} \\ f_v = \int_{-\pi/2}^{\pi/2} \eta\, q\cos^2\varphi\, d\varphi + \int_{-\pi/2}^{\pi/2} q\cos\varphi\sin\varphi\, d\varphi = \frac{\pi\eta q}{2} \end{cases} \quad (2)$$

Equation 2 provides analytical calculations of the UV force components of uniform RPF as shown in Figure 4.

$$\begin{pmatrix} F_x \\ F_y \end{pmatrix} = \begin{pmatrix} f_x \\ f_y \end{pmatrix} + \begin{pmatrix} \cos\omega t & -\sin\omega t \\ \sin\omega t & \cos\omega t \end{pmatrix} \begin{pmatrix} f_u \\ f_v \end{pmatrix} \quad (3)$$

Equation 3 is the matrix form of the 2D FSW model that includes the 2D tool feedback forces integrated from the uniform pressure field components. The parameter t is time in Equation 3. The 2D model takes four parameters, which are the prevailing pressure p, the revolving pressure q, the shear coefficient η, and the rotation speed ω.

- The prevailing pressure field (PPF) is attributed to the travel motion of the weld tool. The assumption is that higher travel speed produces higher prevailing pressure on the leading surface of the weld tool.
- The revolving pressure field (RPF) is related to how much material is extruded from the leading side to the trailing side of the weld tool. The revolving pressure q is assumed to be higher if the advance per revolution (APR) of the weld tool is higher.
- The shear coefficient η ($\eta \geq 0$) is a proportional coefficient (slope) of the relationship between the radial pressure and the shear stress. The shear coefficient is defined by the material types and thermo-mechanical conditions at the tool-material interface. It is assumed that greater sticking condition at the tool-material interface corresponds to greater shear coefficient. Further research is needed to investigate stick and slip conditions associated with FSW.
- In general, the rotation speed ω of RPF is equal to or less than the tool spindle speed. In this study, it was assumed that the rotation speed ω is equal to the tool spindle speed.

Based on the assumptions listed above the parameter values listed in Table I were arbitrarily selected and used for simulation of the 2D FSW model shown in Equation 3. The simulation settings are also included in Table I.

Table I. Selected arbitrary parameter values and simulation settings for 2D FSW model.

Parameters	Values
p	400 psi, 500 psi, 600 psi
q	100 psi, 200 psi, and 300 psi
η	0, 0.5, and 1
ω	1200 rpm
Start time	0 sec
End time	0.048 sec (one tool-rotation period)
Sampling rate	250 Hz

Results and Discussion

The periodic nature of the FSW process can also be observed from the XY force plots of the tool feedback forces. They usually exhibit cyclic trajectories resulted from the fluctuation of the tool feedback forces [8]. Similar cyclic patterns can be observed from the XY force plots generated by the simulation of the 2D FSW model (Figures 5, 6, and 7). The simulation parameters are listed in Table I.

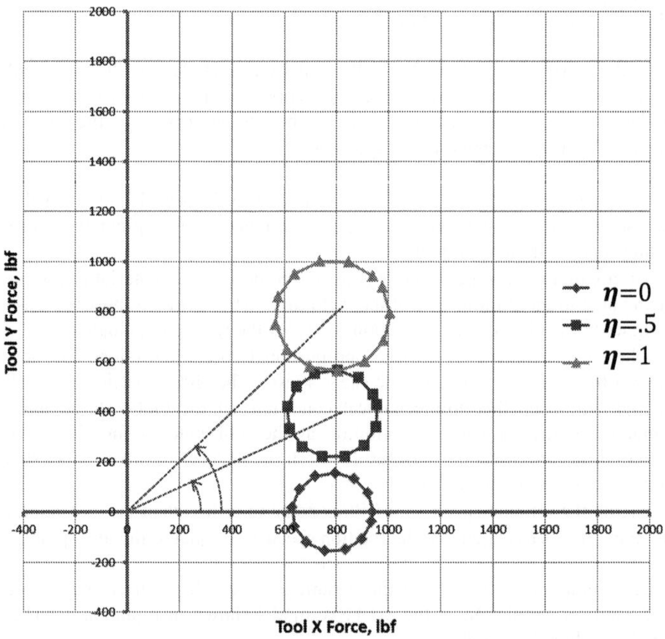

Figure 5. Effect of shear coefficient on XY force trajectory. Note: $p = 500$ psi, $q = 100$ psi.

In Figure 5, the shear coefficient η was varied and the prevailing pressure p and the revolving pressure q were kept constant during the simulation. It looks like the force trajectory rotates more in the counter clockwise direction around the coordinate origin while the shear coefficient is increased from 0 to 1. Note that the values of the shear coefficient were arbitrarily selected to investigate the impact of the shear coefficient on the tool feedback forces. It is assumed that greater shear coefficient corresponds to greater sticking condition at the tool-material interface. In addition, a slight increase in the size of the force trajectory was observed after each increment in the shear coefficient.

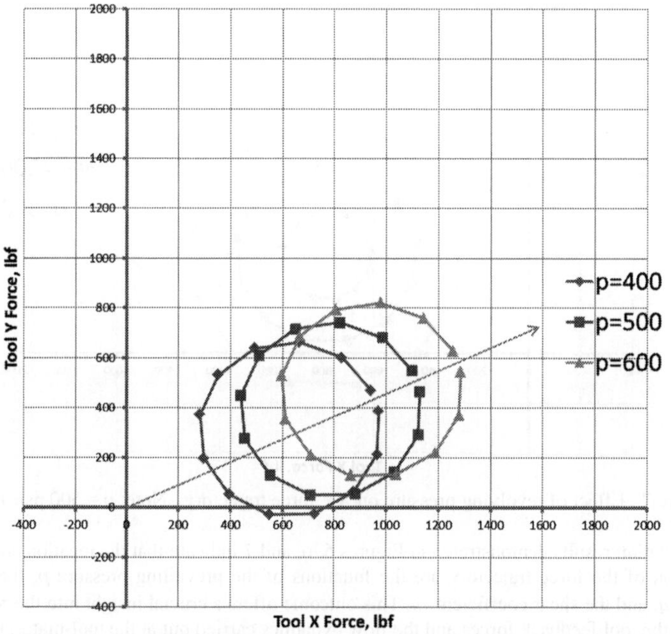

Figure 6. Effect of prevailing pressure on XY force trajectory. Note: $q = 200$ psi, $\eta = 0.5$.

In Figure 6, the prevailing pressure p was varied and the revolving pressure q and the shear coefficient η were kept constant during the simulation. The force trajectory moves away from the coordinate origin along the radial line when the prevailing pressure is increased from 400 psi to 600 psi. The assumption is that higher prevailing pressure on the leading surface of the weld tool corresponds to higher travel speed.

In Figure 7, the revolving pressure q was varied and the prevailing pressure p and the shear coefficient η were kept constant during the simulation. In this case, only the size of the force trajectory increases when the revolving pressure q is increased from 100 psi to 300 psi. The revolving pressure q is assumed to increase when the tool advance per revolution (APR) is increased.

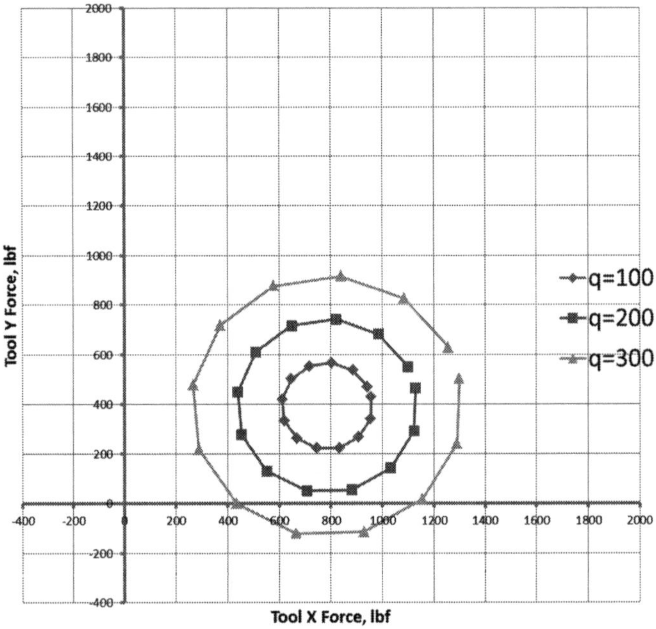

Figure 7. Effect of revolving pressure on XY force trajectory. Note: $p = 500$ psi, $\eta = 0.5$.

The simulation results demonstrated in Figures 5, 6, and 7 indicate that the location, size, and the orientation of the force trajectory are the functions of the prevailing pressure p, the revolving pressure q, and the shear coefficient η. This outcome offers a crucial insight into the relationship between the tool feedback forces and the flow dynamics carried out at the tool-material interface. The force trajectories from actual welds can be evaluated using the heuristic 2D FSW model assuming that the model is an ideal case of the FSW process.

In Figure 8, force trajectories from two friction stir joints generated by Buller [9] are illustrated. Two tools with slightly different probe features were used for making the joints. The tools are labeled as Tool 1 and Tool 2. The geometries of the two tools are almost identical except Tool 1 has more aggressive probe flats than Tool 2.

The angular location of the Tool-2 trajectory is less than the angular location of the Tool-1 trajectory as shown in Figure 8. The simulation results of the 2D FSW model indicate that the shear stress components of Tool 1 are higher than that of Tool 2 according to Figure 5. This indication is in a good agreement with the tool features of the weld tools whereas Tool 1 had more aggressive probe flats than Tool 2.

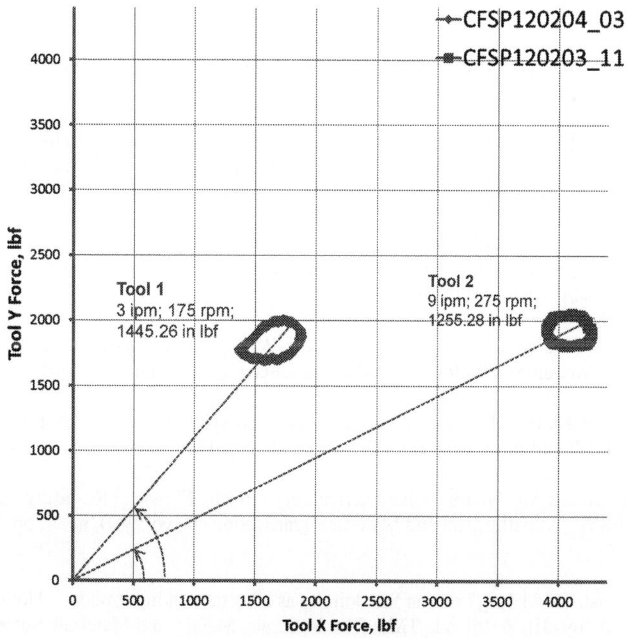

Figure 8. Actual weld data [9].

Furthermore, the Tool-2 trajectory is located further away from the coordinate origin compared to the Tool-1 trajectory. This can be explained by higher prevailing pressure p caused by higher travel speed in accordance with Figure 6. In fact, the Tool-1 joint was produced at the travel speed of 3 in/min while the Tool-2 joint was made at the travel speed of 9 in/min. In addition, the area of the Tool 2 trajectory is somewhat less than that of Tool 1, which may suggest that RPF of Tool 1 might be greater than RPF of Tool 2 in accordance with Figure 7.

In conclusion, further investigation is necessary to characterize the location, size, shape, and orientation of the force trajectories in terms of different process parameters and also with respect to weld quality in order for developing an advanced in-process monitoring tool for FSW.

Acknowledgements

Funding for this research was made available through the Wichita State University (WSU)

Research Site of the Center for Friction Stir Processing (CFSP), a National Science Foundation (NSF) Industry / University Collaborative Research Center (IUCRC).

References

1. K. Colligan, "Material Flow Behavior during Friction Stir Welding of Aluminum," Welding Journal, July 1999, p. 229.

2. B. London, M. Mahoney, W. Bingel, M. Calabrese, R.H. Bossi, and D. Waldron, "Material Flow in Friction Stir Welding Monitored with Al-SiC and Al-W composite markers," Symposium on Friction Stir Welding and Processing II, Warrendale, PA, USA, TMS, 2003.

3. A.P. Reynolds, T.U. Seidel, and M. Simonsen, "Visualization of Material Flow in an Autogenous Friction Stir Weld," 1st International Symposium on Friction Stir Welding, Thousand Oaks, CA, USA, TWI, 1999.

4. Z.W.Chen, T. Pasang, Y. Qi and R. Perris, "Tool-Workpiece Interface and Shear Layer Formed during Friction Stir Welding," TWI Conference paper, Montreal, Canada, October 2006.

5. H. Schmidt, and J. Hattel, "CFD modeling of the shear layer around the tool probe in friction stir welding," FSW and Processing III, K.V.Jata et. al., ed., TMS, 2005.

6. J.A.Schneider and A.C.Nunes, "Characterization of Plastic Flow and Resulting Micro-textures in a FSW", Metallurgical and Materials Transactions B, vol. 35B, no. 4, pp. 777-783, 2004.

7. W.J. Arbegast, "Modeling Friction Stir Joining as a Metalworking Process", Hot Deformation of Aluminum Alloys III, Z. Jin, ed., TMS (The Minerals, Metals, and Materials Society), 2003

8. E. Boldsaikhan, D.A. Burford and P.J.G. Britos, (2011) Effect of Plasticized Material Flow on the Tool Feedback Forces during Friction Stir Welding, in Friction Stir Welding and Processing VI (eds. R. Mishra, M. W. Mahoney, Y. Sato, Y. Hovanski and R. Verma), John Wiley & Sons, Inc., Hoboken, NJ, USA. doi: 10.1002/9781118062302.ch39.

9. J. Buller, Masters Project Report, Aerospace Engineering Department, Wichita State University, Wichita, Kansas, USA, 2012.

Friction Stir Welding and Proceesing VII
Edited by: Rajiv Mishra, Murray W. Mahoney, Yutaka Sato, Yuri Hovanski, and Ravi Verma
TMS (The Minerals, Metals & Materials Society), 2013

Paradigm Shift in Control of the Spindle Axis

Kenneth Ross[1] and Carl Sorensen[2]

[1]Manufacturing Technology, Inc.; 1702 West Washington Street, South Bend, IN 46628, USA

[2]Brigham Young University, 435 CTB, Provo, UT, USA

Keywords: temperature, power, control, friction stir

Abstract

Previous work has demonstrated the value of a cascaded control loop to control tool temperature, with an inner high-speed loop controlling spindle power, and an outer, low-speed loop controlling temperature by adjusting the power setpoint. Such control schemes have demonstrated the ability to control process temperature.

Traditional power control schemes produce a relatively noisy power signal, which potentially decreases the stability of the control scheme and leads to unnecessary stress on the FSP tool and equipment. Rethinking the fundamental physics of power control has led to a revised power control loop. The revised loop has much less noise and provides improved temperature control of the process.

This paper presents the revised power control method, along with the model supporting it. It also presents both analytical and experimental demonstration that the revised method is superior to the original method.

Previous Work

Mayfield and Sorensen (Mayfield and Sorensen 2010) present an improved temperature control algorithm. The model proposed is a two-stage control model containing an inner and outer loop. The inner loop adjusts spindle speed to keep power constant, while the outer loop sets the desired power based on temperature.

A simplified component block diagram of the improved temperature control algorithm is shown in Figure 1 where T is temperature ω is spindle speed, M is torque and P is power. Power is calculated using Equation (1). The plant is the combination of the spindle motor and the stir zone. For the outer loop the reference is desired temperature, the controlled variable is temperature and the manipulated variable is power. For the inner loop the reference is commanded power and the controlled variable is power.

$$P = \omega M \qquad (1)$$

Mayfield and Sorensen do not specify controllers for the inner and outer loop. They do specify that the inner loop adjusts spindle speed to maintain constant power. The following sections present development of controllers for the inner and outer loops.

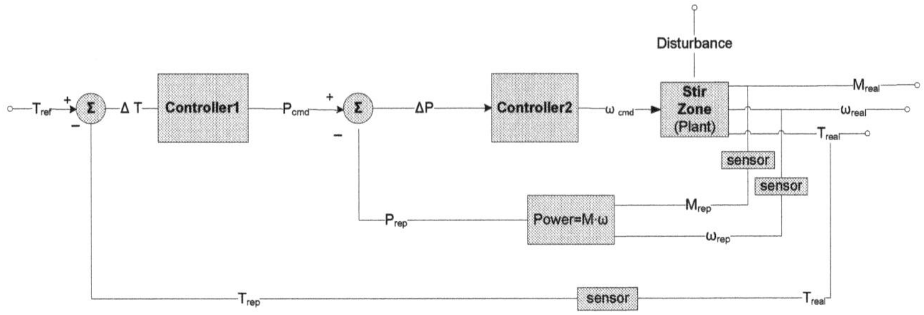

Figure 1: Simplified component block diagram for the temperature control algorithm presented by Mayfield and Sorensen

Ross and Sorensen (Ross and Sorensen 2011) investigated control for inner loop by modulating spindle speed as shown in Figure 1. Spindle motor torque is used as the control signal. Filtering was used to prevent the system from going unstable due noise in the torque signal. Therefore and all torque data used for control and presented was filtered data.

Figure 2 shows process data from a power control weld presented by Ross and Sorensen with the addition of unfiltered power. When power control is enabled, the torque is high and the RPM is low. As the weld progresses the plate heats and softens causing a decrease in torque. As torque decreases, the RPM increases maintaining constant power. Large spikes in power persist throughout the weld. Thesis power spikes occur because the spindle motor attempts to accelerate the spindle instantaneously to achieve the commanded RPM.

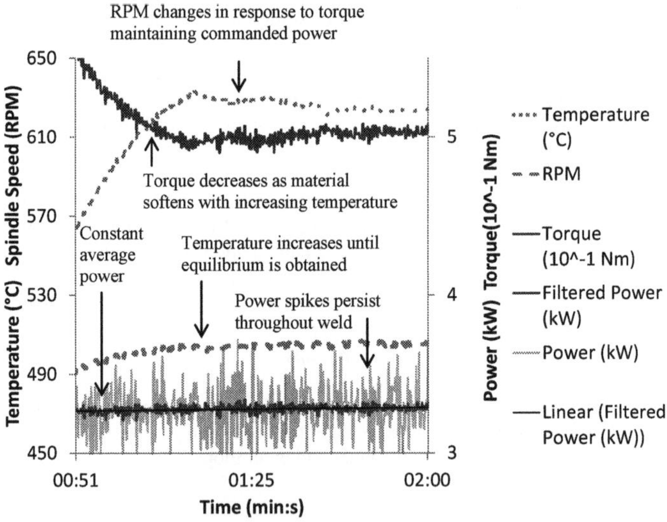

Figure 2: Power control by adjusting spindle speed results with filtered and unfiltered power shown

The linear fit shows that the average power is held constant throughout the weld. The error between the linear fit of the reported power and power desired is 0.05%. Adjusting spindle speed to control power results in large power spikes throughout the weld. The average of these power spikes is the desired power value. Ross and Sorensen were displeased with noise and uncertainty in torque and power values and stated that further investigation be done to improve power control

Cederqvist et al. have written various publications on the use of the scheme proposed by Mayfield and Sorensen in the FSW of thick section copper canisters for nuclear waste storage. In their most recent publication (Cederqvist, et al. 2012) their implementation of the control scheme proposed by Mayfield and Sorensen kept the temperature within ±10 °C of desired temperature for their process.

Ross and Sorensen develop a revised power control method. This revised method is foundation of the robust temperature control algorithm reported by Ross and Sorensen (K. A. Ross 2012) (Ross and Sorensen 2012). This paper presents the revised power control method, along with the model supporting it. It also presents both analytical and experimental demonstration that the revised method is superior to the original method.

Power Control by Adjusting Torque

This control scheme uses the relationship between torque and power, Equation (1), to calculate the torque required to achieve the desired power (patent pending). The spindle motor is a variable frequency AC induction motor. When running under torque control the motor driver commands and maintains a constant torque using flux vector control. The desired torque is obtained by controlling the current fed to the induction motor. This control method is shown in Figure 3.

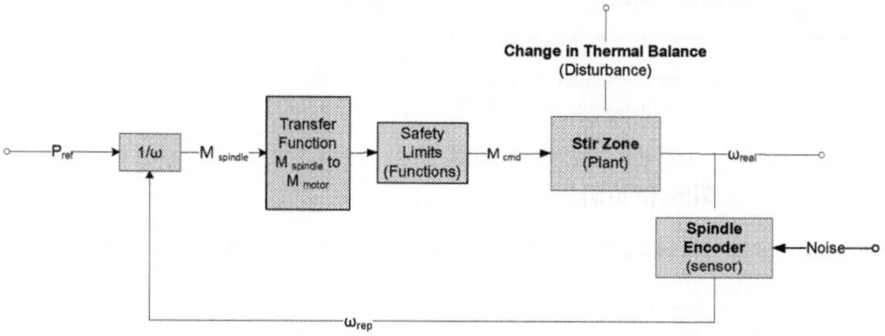

Figure 3: Power control by adjusting torque block diagram

It is important to note that torque control without spindle speed feedback is unstable. A constant torque can only be maintained for a short time. The torque can only be controlled when the load supports the torque. Torque that is greater than the natural process torque leads to greatly decreasing loads causing an exponential increase in spindle speed. This is because when both spindle speed and torque are increasing, the power increases and the material softens. Conversely, if the torque is lower than the natural process torque, the spindle speed will decrease exponentially as the material cools and hardens. Process variation causes the commanded torque to be either too low or too high to maintain equilibrium causing the spindle to rapidly decelerate until it stops or accelerate until machine safety limits are triggered.

Power control achieved by adjusting torque in response to changes in spindle speed is a stable process. Torque increases in response to decreasing spindle speed to maintain a constant power. Torque decreases in response to increasing spindle speed to maintain a constant power. Therefore power control achieved by adjusting torque in response to spindle speed is a stable process.

Results from a weld where torque is adjusted to maintain constant power is shown in Figure 4. As the workpiece is heated, the material softens. The torque and RPM signals are mirror images of each other. The spindle speed increases as the material softens. The torque decreases with increasing spindle speed to maintain the constant power. Figure 4 shows that power control achieved by torque control is a stable process.

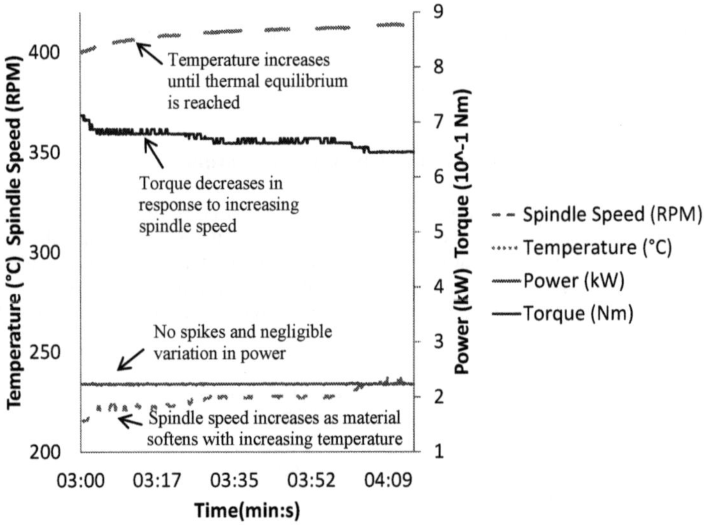

Figure 4: Constant power weld at 2.238 kW where torque is adjusted to keep power constant

Torque Resolution

When spindle speed is adjusted to control power, the commanded spindle speed is determined by the torque reported by the spindle controller. When torque is adjusted to control spindle speed, a commanded torque signal is sent to the spindle controller. The difference in resolution between reported and commanded torque affects the design of the power control loop.

Torque measurements used in this work are reported by the spindle motor controller. The measurement is reported as a fraction of half the motor rated torque via a +/-10 Vdc +/-9 bit signal. Therefore the resolution for torque measurement is 5.984 N-m. The commanded torque is sent as a fraction of the motor rated torque to the spindle controller via a +/-10 Vdc +/-11 bit signal. Therefore the resolution for commanded torque is 0.748 N-m. Because changes in torque during temperature and power control are small spindle torque will be assumed to be equal to commanded torque.

Torque Control vs. Spindle Speed Control

A dynamics analysis was performed (K. A. Ross 2012) to compare adjusting spindle speed to adjusting torque to maintain constant power. A free body diagram of the spindle motor and spindle is given in Figure 5 where M_{mtr} is the motor torque, M_{spn} is spindle torque, D_{mtr} is the diameter of the motor pulley, D_{spn} is the diameter of the spindle pulley, M_b is torque lost due to bearings, ω_{spn} is the rotational velocity of the spindle, and J is the mass moment of inertia of associated with the spindle. The gear ratio, R, is given by:

$$R = \frac{D_{spn}}{D_{mtr}} \tag{2}$$

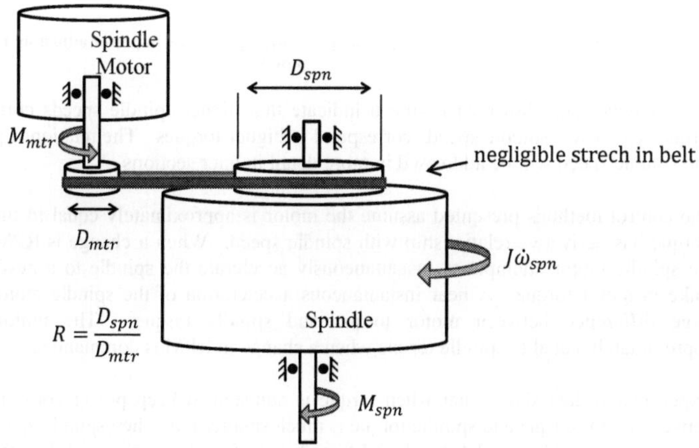

Figure 5: Free-body diagram of spindle and spindle motor

The model derived from Figure 5 shows that torque has a derivative relationship with spindle speed and spindle speed has an integral relationship with torque. Experimental results for step changes in spindle speed and torque validate that torque has a derivative relationship with spindle speed, and spindle speed has an integral relationship with torque. Figure 6a shows the effective motor torque response to a step change in RPM. At the instant the change in desired spindle speed is made, the motor attempts to instantaneously accelerate the spindle in order to achieve the desired spindle speed. This results in a large spike in motor torque.

Figure 6b shows the spindle speed response to a step increase in power where power control was obtained by adjusting torque. When the change in power is commanded, the torque instantly increases slightly to obtain the desired power value. Due to the increased power the material begins to soften causing the spindle speed to increase and the motor torque to decrease. The spindle speed response to torque contains no discontinuities.

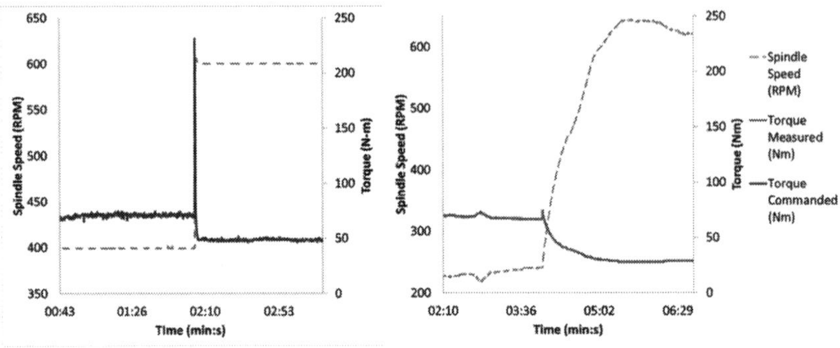

(a) (b)

Figure 6: (a) Step increase in spindle speed (b) step increase in power adjusting torque to control power

Experimental data shown in Figure 6 indicate that higher spindle speeds correspond to lower torques and lower spindle speeds correspond to higher torques. The relationship between torque and spindle speed will be addressed in more detail in later sections.

The control methods presented assume the motor is approximately equal to the spindle torque. Torque has derivative relationship with spindle speed. When a change is RPM is commanded, the spindle motor attempts to instantaneously accelerate the spindle to a new RPM causing a spike in motor torque. A near instantaneous acceleration of the spindle motor would cause a large difference between motor torque and spindle torque. The motor torque is not approximately equal to spindle torque when a change in RPM is commanded.

Experimental data shows that when torque is adjusted to keep power constant, the difference between motor torque and spindle torque is much smaller that when spindle speed is used to keep power constant. The model developed from Figure 5 and experimental data from welds shown in Figure 6 are used to compare the error between motor torque and spindle torque.

When a step change in power is commanded, Figure 6a, the torque error during the spindle spike is 203.0 Nm (149.7ft-lb) which corresponds to a 406% error in power at the high power level and a 565% error at the low power level. For power control by adjusting torque, Figure 6b, the error associated with the spindle acceleration after the change in power, where acceleration is greatest, is 7.4 Nm (5.46ft-lb) which corresponds to 10% error in power.

Adjusting spindle speed to keep power constant is undesirable because reported torque is a poor control signal. Torque spikes cause the difference between motor and spindle torque to always be at least 400%. The reported torque signal has low resolution, adding to the error.

The main advantage of using torque to control power is the avoidance of artificial torque spikes caused by attempting to change the RPM instantaneously. Using torque control to control power results in a smooth power signal with low uncertainty. The difference between the motor torque and actual spindle torque is proportional to the acceleration of the spindle. Low uncertainty exists because under torque control, power and spindle speed changes slowly. Power control by adjusting torque provides a more accurate torque measurement due to the resolution difference between the commanded and reported torque signals.

Use of torque control in temperature control algorithms

Torque, spindle speed power and temperature response to step increases in desired temperature are shown in Figure 7. When a higher temperature is commanded, the torque is increased to increase the power. This causes an increase in temperature and spindle speed. As the material softens the spindle speed continues to increase and the torque must be reduced to maintain constant power. When the steady state at the higher temperature value is reached, the power and spindle speed are higher and the torque is lower than at the previous steady state temperature.

The converse is true if a lower temperature is desired. When torque decreases, the power drops and material cools. This causes increase in torque that is offset by a larger reduction in spindle speed. The steady state at a lower temperature is a higher torque, lower spindle speed and lower power. The non-intuitive finding is that in order to achieve a desired temperature, the torque must initially increase in the direction opposite of its steady state value.

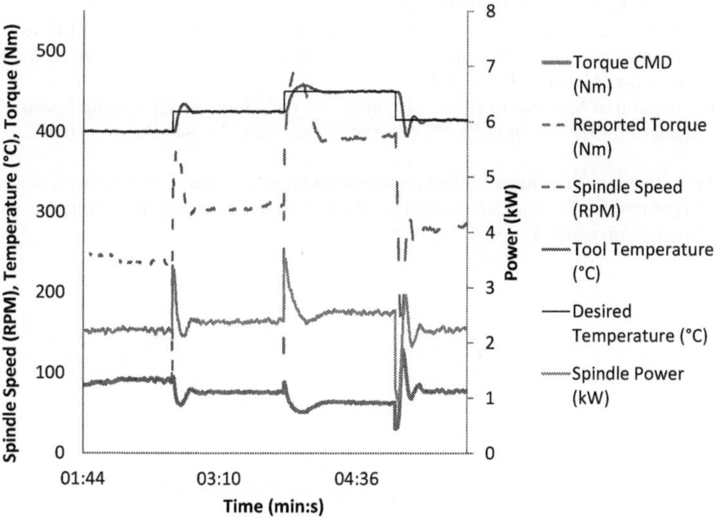

Figure 7: Torque, spindle speed power and temperature response to step changes in desired temperature

The reported torque and the commanded torque are approximately the same except for the instant a new temperature setpoint is commanded. Immediately after that step, there is one data point in reported torque that is extremely high. There is no physical justification for the single point spike in reported torque. Therefore it will be treated as an error and ignored until the anomaly is further investigated.

Conclusions

Power control is best achieved by controlling torque rather than spindle speed. When spindle speed is used to control power, motor toque feedback is used as the control signal. Motor torque feedback is a poor control signal due to the large spikes in motor torque that occur when changes in spindle speed are commanded. When these torque spikes occur, the motor torque cannot be used to approximate spindle torque. When torque is adjusted to control power, spindle speed is used as the control signal. Uncertainty in spindle speed measurement is small. Motor torque can be used to approximate spindle toque when spindle torque is adjusted to keep power constant.

References

Cederqvist, Lars, O Garpinger, T Hägglund, and A Robertsson. "Cascade control of the friction stir welding process to seal canisters for spend nuclear fuel." *Control Engineering Practice* 20 (2012): 35-48.

Mayfield, David W, and Carl D Sorensen. "An Improved Temperature Control Algorithm for Friction Stir Processing." *8th International Friction Stir Welding Symposium.* Timmendorfer Strand, Germany, 2010.

Ross, Kenneth A, and Carl D Sorensen. "Development and Implementation of a Robust Temperature Control Algorithm." *9th International Symposium on Friction Stir Welding.* Huntsville, Alabama, USA, 2012.

—. "Investigation of Methods to Control Friction Stir Weld Power with Spindle Speed Changes." *TMS: Friction Stir Welding and Processing VI.* San Diego: Wiley, 2011. 345-352.

Ross, Kenneth Alec. "Investigation and Implementation of a Robust Temperature Control Algorithm for Friction Stir Welding." MS Thesis, Mechanical Engineering, Brigham Young University, Provo, UT, 2012.

Friction Stir Welding and Proceesing VII
Edited by: Rajiv Mishra, Murray W. Mahoney, Yutaka Sato, Yuri Hovanski, and Ravi Verma
TMS (The Minerals, Metals & Materials Society), 2013

A Coupled Thermal/Material Flow Model of Friction Stir Welding Applied to Sc-Modified Aluminum Alloys

C. Hamilton[1], M. Kopyściański[2], O. Senkov[3], S. Dymek[2]

[1]Department of Mechanical and Manufacturing Engineering
Miami University, Oxford, OH

[2]Faculty of Metals Engineering and Industrial Computer Science
AGH University of Science and Technology, Kraków, Poland

[3]UES, Inc., 4401 Dayton-Xenia Rd., Dayton, OH 45432-1894, USA

Keywords: friction stir welding, onion rings, thermal analysis, aluminum, scandium

Abstract

A coupled thermal/material flow model of friction stir welding was developed and applied to the joining of Sc-modified aluminum alloy (7042-T6) extrusions. The model reveals that surface material is pulled from the retreating side into the weld zone where it is interleaved with in-situ material. Due to frictional contact with the shoulder, the surface material is hotter than the in-situ material, so that the final weld microstructure is composed of bands of material with different temperature histories. For this alloy and the associated FSW heating rates, secondary phase dissolution/precipitation temperatures are in proximity to the welding temperatures. Therefore, depending on the surface and in-situ material temperatures in relation to these transformation temperatures, disparate precipitate distributions can develop in the bands of material comprising the weld nugget. Based on the numerical simulation and on thermal analysis data from differential scanning calorimetry, a mechanism for the formation of onion rings within the weld zone is presented.

Introduction

Over the last twenty years, numerous investigations have sought to characterize the principles of friction stir welding (FSW) and to model the material flow behavior, the temperature distribution and the microstructural evolution within the weld. The review papers of Nandan et al.[1] and Threadgill et al.[2] provide an excellent account of past and current FSW research. Early numerical simulations focused on the temperature distribution during welding and studied its potential influence on the weld microstructure and precipitation kinetics. More recently, researchers have modeled both the material flow behavior and temperature characteristics during FSW despite the complex material flow associated with the process. The current investigation presents a coupled thermal/flow model of friction stir welding applied to Sc-modified Al–Zn–Mg–Cu extrusions (Al alloy 7042–T6).

Additions of scandium (Sc) and zirconium (Zr) stabilize the microstructure at temperatures greater than 150°C and increase strength of 7000 series alloys through the formation of coherent nanometer-sized $Al_3(Sc,Zr)$ particles.[3,4] These $Al_3(Sc,Zr)$ particles also stabilize the microstructure formed during hot working operations and inhibit recrystallization during heat treatment, thus potentially enhancing the residual properties after joining operations such as FSW.[5,6] These additions also affect the kinetics of precipitation and growth of the primary strengthening precipitates (GP zones, η'), thus modifying heat treatment conditions for enhancing the mechanical properties of these alloys.[7] The numerical simulation proposed here gives insight into the material flow and temperature distribution of the weld zone during the joining of 7042-T6 extrusions. Combined with thermal analysis data from differential scanning calorimetry (DSC), the precipitation behavior within the weld is discussed in terms of the volume fraction of metastable (GP zones and η′) and the equilibrium[η (MgZn $_2$) and/or T ($Al_2Mg_3Zn_3$)] strengthening particles found in the 7042 aluminum alloy.[8] It is assumed that FSW does not change the size and volume fraction of the Al_3(Sc,Zr) precipitates due to their high thermal stability.[9]

DSC is a powerful technique for the investigation of precipitation and dissolution processes in Al alloys.[7,10,11] Results of the DSC thermal analysis of the FSW regions of the 7042-T6 Al alloy, together with a developed coupled thermal/material flow model of FSW, were used to propose a mechanism of onion ring formation of within the weld zone. The model reveals that surface material is pulled from the retreating side into the weld zone where it is interleaved with in-situ material.

Experimental Procedure

The chemical composition of the 7042-T6 Al alloy used in this work is given in Table 1. For this investigation, a 76 mm diameter 7042 billet was produced by direct chill casting and then hot extruded into a bar with a rectangular cross-section of 50.4 mm x 6.35 mm. Following extrusion, the bar was heat treated to a T6 temper through the following schedule: (1) solution heat treat at 460°C for one hour followed by an additional hour at 480°C, (2) rapid quench in water to room temperature and (3) age at 120°C for 19 hours.

Table 1 - Chemical Composition of 7042

Element	Weight Percent	
	7042, this work	7042, nominal
Zn	7.11	6.5 – 7.9
Mg	2.14	2.0 – 2.8
Cu	1.56	1.3 – 1.9
Mn	0.25	0.2 – 0.4
Zr	0.17	0.11 – 0.20
Sc	0.38	0.18 – 0.50
Cr	< 0.05	≤0.05
Ti	–	–
Other, Total	0.35	≤0.55
Al	Balance	

After heat treatment, the bar was sectioned, and six longitudinal friction stir welds produced. The diameter of the FSW tool shoulder was 17.8 mm, the pin diameter tapered linearly from 10.3 mm at the tool shoulder to 7.7 mm at the tip, and the pin depth was 6.1 mm. With a weld velocity of 2.1 mm s^{-1} and an applied force of 22 kN, welds were produced at the following pin rotation speeds (PRS): 175, 225, 250, 300, 350 and 400 rev min^{-1}. The temperature profile across the weld surface was recorded using a Mikron M7815 Infrared Thermal Imaging Camera. These data were used to verify the temperature predictions of the simulation developed during this investigation. The uncertainty in these measurements was ± 2%.

Small samples (approximately 20 to 50 mg) were extracted from the T6-tempered baseline material and from the weld center of each welded sample for thermal analysis. The samples were sealed in Al pans and analyzed in a Perkin Elmer Jade differential scanning calorimeter (argon atmosphere). Samples were heated from room temperature to 400°C at a constant heating rate that ranged from 10°C min^{-1} to 100°C min^{-1}. Polarized optical microscopy was used to study the microstructure of the welds. To enhance the appearance of precipitate distributions and grains, the surfaces of the weld samples were polished and anodized in an electrolitic solution of 1.8% fluoroboric acid in water at room temperature and electric current of 0.15 A. Anodizing time was 2.5 to 3 minutes.

Results and Discussion

A coupled thermal/flow simulation of friction stir welding was developed utilizing Comsol multi-physics software. The model defines a flow-capable region between the advancing and retreating sides in which the temperature and action of the tool plasticizes the aluminum workpiece, and material flow occurs. The inlet velocity into the flow region is the weld velocity, u_{weld}. The boundary condition for the velocity field between the flow region and the advancing side, the retreating side and the backing spar is no slip, and these interfaces have mirrored meshes to ensure continuity in the thermal distribution. The velocity boundary condition for the surface of the flow region is:

$$u = u_{weld}, v = 0, w = 0 \tag{1}$$

where u, v and w are the velocity magnitudes in the x, y and z-directions, respectively. There are effectively four boundaries between the tool and the flow-capable region: 1) the tool shoulder, 2) the fillet radius between the shoulder and pin, 3) the pin side and 4) the pin bottom. The velocity field boundary conditions for the tool shoulder, the fillet radius and pin bottom are the same and are given in the following equation:

$$u = \omega y, v = -\omega x, w = 0 \tag{2}$$

where ω is the angular velocity of the tool. The pin is threaded with a pitch value, p, of 1.27 mm; therefore, the boundary condition for the pin side can account for the vertical displacement of material per tool revolution due to the threads and is given by:

$$u = \omega y, v = -\omega x, w = -p\left(\frac{\omega}{2\pi}\right) \tag{3}$$

To solve for the velocity field within the flow region, the viscosity of the region, μ, must be known. The viscosity is determined from the flow stress, σ_e, and the effective strain rate, $\dot{\varepsilon}$, through the following relationship:

$$\mu = \frac{\sigma_e}{3\dot{\varepsilon}} \tag{4}$$

The maximum strain rates within the flow region occur adjacent to the weld tool where the velocity gradients are greatest. The strain rates decrease rapidly away from the tool since the material flow velocities also decrease quickly away from the tool. In their work on aluminum, Frigaard et al.[12] estimated the maximum effective strain rate under the tool shoulder to be 20 s^{-1}, while Nandan et al.[13] calculated the maximum strain rate as 100 s^{-1} near the tool shoulder and as 30 s^{-1} approximately 4 mm below the shoulder. More recently, Arora et al.[14] computed the maximum strain rate in aluminum 2524 as 9 s^{-1} for their FSW parameters and tools.

In this study, the calculation of viscosity and flow stress is simplified by assuming a constant, maximum value for the effective strain rate calculated at the tool shoulder that is then applied to the entire flow-capable region. This approach overestimates the strain rates away from the tool shoulder and pin, but still accurately captures the material flow behavior in the vicinity of the tool. As noted by Colgrove [15], the tool surface dominates both heat generation and process kinematics; therefore, using a constant, maximum strain rate value derived from the tool shoulder provides a good approximation of the phenomena occurring in the entire weld region. To determine this value, the velocity components u and v are assumed to decrease linearly from the tool shoulder to the bottom of the workpiece as suggested by Heurtier et al.[16] and that the velocity component w remains constant. The velocity field in the area under the tool shoulder and adjacent to the pin then becomes (with $z = 0$ located at the workpiece bottom):

$$u = \omega y\left(\frac{z}{h}\right) - u_{weld}, v = -\omega x\left(\frac{z}{h}\right), w = -p\left(\frac{\omega}{2\pi}\right) \quad (5)$$

where h is the workpiece thickness. Nandan et al.[13] provide the expanded form of the effective strain rate as a function of the velocity gradients. Applying this to Eq. 5 and evaluating at the tool shoulder radius, $r_{shoulder}$, yields the following expression for the maximum effective strain rate:

$$\dot{\varepsilon} = \frac{r_{shoulder}\sqrt{6}}{3h}\omega \quad (6)$$

For the process parameters utilized in this investigation, the maximum effective strain rate varies from 20.9 s^{-1} at 175 rev min^{-1} to 47.9 s^{-1} at 400 rev min^{-1}.

Sheppard and Wright[17] proposed the familiar formulation for the flow stress utilizing the Zener-Hollomon parameter that captures the temperature influence on the effective strain rate. The constants for this formulation, Q, A, α and n, are presented in Table 2 and were taken from Colegrove et al.[15] for aluminum 7449. During the simulation, the Zener-Hollomon parameter is recalculated for each iteration based upon the predicted welding temperature. As such, the model captures the temperature dependence of the effective strain rate, the flow stress and the viscosity during friction stir welding.

Table 2 – Material constants for the Sheppard and Wright flow stress equation and Zener-Hollomon parameter.

Material Constant	Value	Units
Q	134158.4	J mol^{-1}
A	1.26×10^8	s^{-1}
α	0.03055	MPa^{-1}
n	3.24644	n/a

Within the flow-capable region, the heat transfer and material flow behavior are coupled. The thermal properties, i.e. the thermal conductivity, k, and the specific heat capacity, c_p, within the flow region are identical to those within the retreating and advancing sides of the aluminum workpiece. These thermal properties, as well as those of the tool and backing spar, and their temperature dependence can be found in [18]. The tool/workpiece slip behavior and heat flux calculations for the tool shoulder and pin bottom, the primary sources for heat generation during FSW, are also presented in this reference.

A thermal insulation constraint is applied at each interface of the flow-capable region with the non-flow areas of the model. These constraints assure temperature continuity across the flow-capable boundaries into the other areas of the model. Thermal insulation constraints are also applied to the tool shoulder/workpiece and pin bottom/workpiece interfaces. For the boundaries exposed to ambient conditions, i.e. workpiece top, workpiece side and tool side, the convective heat transfer coefficient is set to 15 $W/m^2{\cdot}K$ to approximate free convection on these surfaces. As suggested in [18] and [19], convection coefficients of 200 $W/m^2{\cdot}K$ and 250 $W/m^2{\cdot}K$ are applied to the tool top and spar bottom, respectively. For the underside of the workpiece and the sides of the backing spar, a convective coefficient of 100 $W/m^2{\cdot}K$ is used to represent the dissipation of heat into the backing plate. Heat dissipation due to radiation is ignored in this model.

Figure 1 from the simulation shows the three-dimensional flow behavior of surface material "tracers" (representatively shown for the 400 rev min^{-1} condition). The color shading indicates the tracer temperature during the material flow. Ahead of the advancing tool, surface material is swept toward the retreating side of the weld. From here the tracers follow three primary paths: 1) tracers are carried on the surface along the tool shoulder toward the advancing side and deposited behind the tool, 2) tracers are trapped by the rotating tool and remain on the surface under the tool shoulder and 3) tracers are pulled by the pin/shoulder from the retreating side of the surface down into the weld zone.

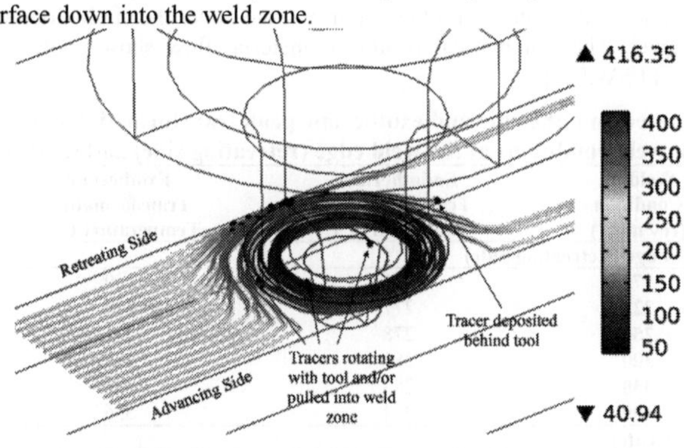

Figure 1 - Flow and temperature (°C) behavior of surface material tracers during FSW

These numerical results of material flow are consistent with the experimental observations of other researchers. Hamilton et al.[20], studying the friction stir welding of 6101-T6 extrusions plated with tin, observed the unique presence of tin within the weld nugget and the lack of tin within the TMAZ and concluded that the nugget forms as surface material is extruded from the retreating side into the weld region. Colligan[21] in his study of 6061-T6 concluded that surface material extrudes from the retreating side of the pin and deposits in the wake of the tool. Guerra et al.[22] in their study of 6061 similarly hypothesized that material from the front of the retreating side of the pin extrudes between deformed surface material rotating with the tool and parent material into the area behind the pin.

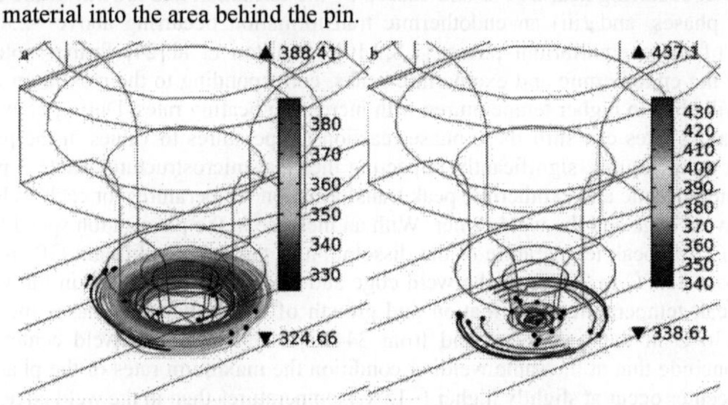

Figure 2 - Flow and temperature (°C) behavior of: a) mid-plane tracers and b) bottom plane tracers

Figure 2a shows the simulated three-dimensional material flow behavior of mid-plane (relative to the workpiece) material tracers (also representatively shown for the 400 rev min^{-1} condition). Here, "in-situ" tracers are primarily trapped by the rotating pin and either remain at the mid-plane level rotating with the pin or are pulled toward the bottom of the workpiece.

Similarly, Figure 2b shows the simulated flow behavior of bottom-plane material tracers and reveals that these tracers also rotate with the pin or migrate under the pin bottom and rotate with the tool at this level. These numerical results of material flow show good agreement with Reynolds' model of FSW.[23]

Table 3 - Estimated endothermic and exothermic peak transformation temperatures for different weld conditions at the weld edge (retreating side) and weld center.

Welding Condition (rev min^{-1})	Endothermic Transformation Temperature (°C)	Exothermic Transformation Temperature (°C)
Weld Edge (Retreating Side)		
175	275	329
225	277	330
250	278	331
300	281	335
350	283	336
400	284	336
Weld Center		
175	290	344
225	292	345
250	293	346
300	296	350
350	298	351
400	298	352

In their differential thermal analysis, Hamilton et al.[24] demonstrated that three phase transformations occur in the 7042-T6 alloy during heating at 10°C/min between room temperature and 400°C. These are (i) an endothermic transformation occurring near 165°C and caused by the dissolution of the non-equilibrium GP zones and η′ particles, (ii) an exothermic transformation occurring near 216°C and caused by the formation and growth of the equilibrium η and/or T phases, and (iii) an endothermic transformation occurring above ~250°C due to dissolution of these equilibrium phases.[7,8,9,10] Hamilton et al.[24] further noted that the positions of the endothermic and exothermic peaks, corresponding to the maximum rates of the reactions, will shift to higher temperatures with increasing heating rates. During FSW, therefore, the high heating rates can shift these phase reaction temperatures to values in the proximity of the welding temperatures, significantly impacting the final microstructure. Table 3 presents the estimated endothermic and exothermic peak transformation temperatures for each weld condition at both the weld edge and the weld center. With an increase in the pin rotation speed from 175 to 400 rev min^{-1}, the peak temperature of the dissolution of the non-equilibrium GP and η′ phases increases from 275°C to 284°C in the weld edge and from 290°C to 298°C in the weld center, while the peak temperature of formation and growth of the η and/or T phases increases from 329°C to 336°C in the weld edge and from 344°C to 352°C in the weld center. One may therefore conclude that at the same welding condition the maximum rates of the phase reactions in the weld center occur at slightly higher (~15°C) temperatures than in the weld edge.

Figure 3a presents the material flow behavior of surface material tracers during the 250 rev min^{-1} condition with color shading indicating the temperature history. Surface tracers under the shoulder and near the weld edge, can reach temperatures of 345°C, temperatures just greater than or approximately equal to the weld edge exothermic reaction temperature (331°C) for this weld condition. As such, the material represented by these tracers becomes enriched in the equilibrium η and/or T phase particles before being pulled by the pin from the retreating side into the weld zone. These η/T particles are much coarser than non-equilibrium GP zones and η′ particles. Upon cooling, the equilibrium phase particles are retained, forming bands of material that appear "particle-rich" under light microscopy within the weld zone.

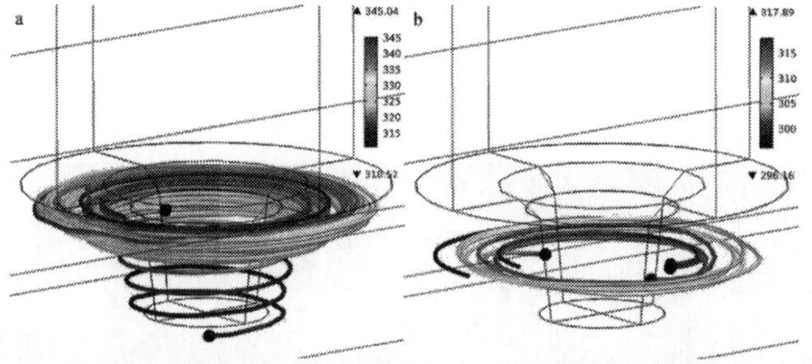

Figure 3 - Flow and temperature (°C) behavior of: a) surface material tracers pulled into the weld zone and b) mid-plane tracers that rotate around the pin (250 rev min^{-1} shown)

In contrast, the in-situ, mid-plane material tracers of the weld zone (Figure 3b) only reach temperatures of ~320°C near the pin surface and close to the weld center during the 250 rev min^{-1} condition. These temperatures are greater than the weld center endothermic reaction temperature (293°C), but certainly less than the weld center exothermic transformation temperature (346°C). As such, the in-situ weld material becomes depleted in the non-equilibrium η′ phase (due to dissolution of GP zones and η′ particles) and also becomes supersaturated with solute elements upon cooling since the precipitation of the η and/or T phases never takes place. Upon cooling, re-precipitation of fine GP zones occurs within the in-situ weld material, forming bands of material that appear "particle-poor" under light microscopy within the weld zone. The final weld microstructure, therefore, is comprised of coarse-particle-rich bands of material pulled from the surface's retreating side and interleaved with coarse-particle-poor bands of the in-situ weld material, as revealed in the SEM image of the 250 rev min^{-1} weld presented in Figure 4a.

The potential for onion ring formation in the 7042-T6 alloy, therefore, depends on the surface and in-situ material temperatures relative to the endothermic and exothermic reaction temperatures in these regions for the given weld conditions. If both the surface and in-situ material temperatures are less or both are greater than the exothermic temperature, then the appearance of onion rings is minimized. In the first case, no coarse particles will form in both regions, while in the second case, similar coarse particles will form in both regions; thus low optical contrast between the regions will develop. If, however, one of the surface and in-situ material temperatures is below and another is above the peak exothermic transformation temperature, then onion ring formation is maximized since each material zone will exhibit unique precipitation characteristics.

For welding at 175 rev min^{-1}, the in-situ material's temperature is ~ 60°C and the surface material's temperature is ~20°C below the respective exothermic reaction temperatures. This suggests that no visible coarse η/T particles would form in either region and thus onion rings would not develop for this weld condition. Indeed, the metallographic image of the 175 rev min^{-1} weld zone presented in Figure 4b reveals a primarily uniform microstructure. However, a faint onion pattern can still be seen, which probably indicates that precipitation of the η/T particles has already started at the surface region. At 225 rev min^{-1}, the in-situ material temperature is greater than the endothermic reaction temperature but about 45°C smaller than the peak exothermic temperature. On the other hand, the surface material temperature corresponds to the peak exothermic reaction temperature. The in-situ material will again be free of the η/T particles. However, the rate of formation of these particles now reaches a maximum value in the surface material before it is pulled from the retreating side down into the nugget. These stable secondary phases are retained in the microstructure upon cooling, thus creating the potential for onion ring formation due to the difference in precipitation behavior between the material zones. Examination of the metallographic image of the 225 rev min^{-1} weld reveals the emergence of a more prominent onion ring pattern.

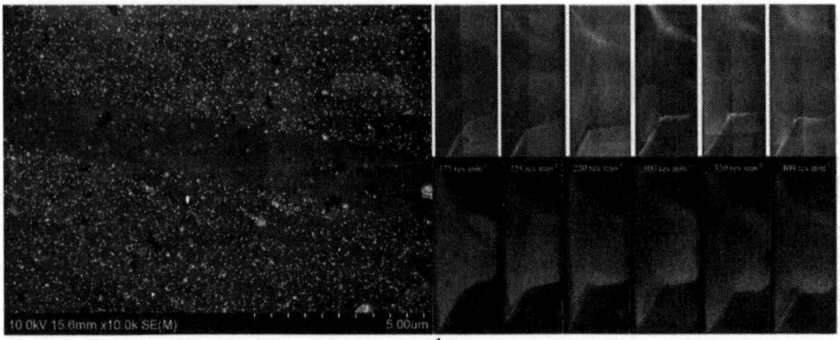

Figure 4 – a) SEM image of the 250 rev min^{-1} weld and b) metallographic images of the weld zone for each weld condition (top images were anodized and observed under polarized light to enhance the onion appearance)

The onion rings are most pronounced for the 250 and 300 rev min^{-1} welds. As previously discussed, at 250 rev min^{-1} the surface material slightly exceeds the peak exothermic transformation temperature, while the in-situ material reaches a temperature in between the respective endothermic and exothermic temperatures. Before being introduced into the weld zone, the surface material will experience a strong precipitation of stable η and/or T phases, and the in-situ material will dissolve fine GP and η′, which will re-precipitate upon cooling and holding at room temperature. The resulting microstructure will have defined bands of particle-rich surface material and particle-poor in-situ material comprising the weld. Similarly at 300 rev min^{-1}, the surface material now exceeds the exothermic temperature by 52°C, and the in-situ material temperature is effectively equal to the peak exothermic temperature. As such, η and/or T phases in the surface material will coarsen and begin to dissolve while they approach the maximum rate of precipitation in the in-situ material. Suggesting that the volume fraction of these phases is the same for these two conditions, the surface material would have a smaller particle number density and a larger average particle size than the in-situ material. Mixing the bands of surface material containing coarser η/T particles with the bands of in-situ material containing finer η/T particles again imparts a strong onion ring pattern due to the optical contrast in precipitation.

For pin rotation speeds greater than 300 rev min^{-1}, both the in-situ and surface material temperatures exceed the respective peak exothermic phase transformation temperatures. For the surface material, whose temperature now exceeds 400°C, the trend will be toward complete dissolution of the η and/or T phases and GP zone formation upon cooling and holding at room temperature. The in-situ material will experience limited dissolution of the η and/or T phases as well, but since the temperature lags behind that of the surface material, some overaged η and/or

T phases can be retained at room temperature. As a result, the onion ring pattern will become fainter for pin rotation speeds beyond 300 rev min^{-1}. The microstructural images for the 350 and 400 rev min^{-1} welds in Figure 4b show faint onion patterns, but they are certainly less distinct than those in the 250 and 300 rev min^{-1} welds.

Conclusions

A coupled thermal/material flow model of friction stir welding was developed and utilized to simulate the joining of Sc-modified aluminum extrusions (7042-T6). Within the weld zone, the model demonstrates that surface material approaching the rotating tool is swept to the retreating side. The material then either 1) rotates with the tool under the shoulder and is deposited behind the tool toward the advancing side or 2) rotates under the shoulder and is captured by the tool or 3) rotates under the shoulder and is pulled into the weld zone by the threaded pin. Mid-plane and bottom plane material rotate with the tool and show some downward migration. The extrusion of "hotter" surface material into the weld nugget where it interleaves with "cooler" in-situ material gives rise to the formation of onion rings in the 7042-T6 alloy. If the surface material temperature is greater than the exothermic transformation temperature of the alloy, then the surface becomes enriched in stable η and/or T phases before being pulled into the weld nugget. If, at the same time, the in-situ material temperature is greater than the endothermic reaction temperature, but less than the exothermic reaction temperature, the in-situ material becomes depleted in the non-equilibrium η′ phase and supersaturated. Upon cooling, GP zones form in the in-situ material, such that bands of particle-rich surface material interleaved with particle-poor in-situ material comprise the weld nugget and impart the characteristic onion ring appearance.

Acknowledgements

The authors would like to acknowledge the Polish Ministry of Science and Higher Education (Grant No. N507 446337) and UES, Inc. for their support of this work. ONS acknowledges financial support through the USAFRL Contract No. FA8650-10-D-52226.

References

[1]R. Nandan, T. DebRoy, H. K. D. H. Bhadeshia, "Recent Advances in Friction-Stir Welding – Process, Weldment Structure and Properties," *Progress in Materials Science,* 53 (2008) 980 – 1023.

[2]P. L. Threadgill, A. J. Leonard, H. R. Shercliff, P. J. Withers, "Friction Stir Welding of Aluminum Alloys," *International Materials Reviews*, 54 (2009) 49 – 93.

[3]S. V. Senkova, O. N. Senkov, D. B. Miracle, "Cryogenic and Elevated Temperature Strengths of an Al-Zn-Mg-Cu Alloy Modified with Sc and Zr," *Metallurgical and Materials Transactions A*, 57 (2006) 3569 – 3575.

[4]R. B. Bhat, J. D. Schloz, S. V. Senkova, O. N. Senkov, "Microstructure and Properties of Cast Ingots of Al-Zn-Mg-Cu Alloys Modified with Sc and Zr," *Metallurgical and Materials Transactions A*, 36 (2005) 2115 – 2126.

[5]P. S. De and R. S. Mishra, "Friction Stir Welding of Precipitation Strengthened Aluminium Alloys: Scopes and Challenges," *Science and Technology of Welding and Joining*, 16 (2011) 343 – 347.

[6] C. Hamilton, S. Dymek, O. Senkov, in: Proceedings of the 138th TMS Annual Meeting & Exhibition, San Francisco, CA, February 15 – 19, 2009.

[7]O. N. Senkov, M. R. Shagiev, S. V. Senkova, "Effect of Sc on Aging Kinetics in a Direct Chill Cast Al-Zn-Mg-Cu Alloy," *Metallurgical and Materials Transactions A*, 39 (2008) 1034-1053.

[8]I. J. Polmear, *Light Alloys: From Traditional Alloys to Nanocrystals,* Fourth Edition, pp. 131 – 160, Elsevier Butterworth-Heinemann, London, 2006.

[9]O. N. Senkov, M. R. Shagiev, S. V. Senkova, D. B. Miracle, "Precipitation of Al_3(Sc,Zr) Particles in an Al–Zn–Mg–Cu–Sc–Zr Alloy During Conventional Solution Heat Treatment and Its Effect on Tensile Properties," *Acta Materialia*, 56 (2008) 3723-3738.

[10]M. J. Starink, "Analysis of Aluminium Based Alloys by Calorimetry: Quantitative Analysis of Reactions and Reaction Kinetics," *International Materials Reviews*, 49 (2004) 191-226.

[11] C. Hamilton, S. Dymek, O. Senkov, in: Proceedings of the 140th TMS Annual Meeting & Exhibition, San Diego, CA, February 27 – March 3, 2011.

[12]O. Frigaard, O. Grong and O. T. Midling, "A Process Model for Friction Stir Welding of Age Hardening Aluminum Alloys," *Metallurgical and Materials Transactions A*, 32 (2001) 1189 – 1200.

[13]R. Nandan, G. G. Roy, T. DebRoy, "Numerical Simulation of Three-Dimensional Heat Transfer and Plastic Deformation during Friction Stir Welding," *Metallurgical and Materials Transactions A*, 37 (2006) 1247 – 1259.

[14]A. Arora, Z. Zhang, A. De, T. DebRoy, "Strains and Strain Rates during Friction Stir Welding," *Scripta Materialia*, 61 (2009) 863 – 866.

[15] P. A. Colegrove, H. R. Shercliff, R. Zettler, "Model for Predicting Heat Generation and Temperature in Friction Stir Welding from the Material Properties," *Science and Technology of Welding and Joining,* 12 (2007) 284 – 297.

[16]P. Heurtier, M. J. Jones, C. Desrayaud, J. H. Driver, F. Montheillet, D. Allehaux, "Mechanical and Thermal Modeling of Friction Stir Welding," *Journal of Materials Processing Technology*, 171 (2006) 348 – 357.

[17]T. Sheppard and D. Wright, "Determination of Flow-Stress 1. Constitutive Equation for Aluminum-Alloys at Elevated-Temperature," *Metals Technology*, 6 (1979) pp. 215 – 223.

[18]C. Hamilton, A. Sommers, S. Dymek, "A Thermal Model of Friction Stir Welding Applied to Sc-Modified Al-Zn-Mg-Cu Alloy Extrusions," *International Journal of Machine Tool and Manufacture*, 49 (2009) 230 – 238.

[19]C. Hamilton, S. Dymek and A. Sommers, "A Thermal Model of Friction Stir Welding in Aluminum Alloys," *International Journal of Machine Tool and Manufacture*, 48 (2008) 1120 – 1130.

[20]C. Hamilton, S. Dymek, M. Blicharski, "A Model of Material Flow during Friction Stir Welding," *Materials Characterization*, 59 (2008) 1206 – 1214.

[21]K. Colligan, "Material Flow Behavior during Friction Stir Welding of Aluminum," *Welding Journal*, 78 (1999) 229S – 237S.

[22]A. Guerra, C. Schmidt, J. C. McClure, L. E. Murr, A. C. Nunes, "Flow Patterns during Friction Stir Welding," *Materials Characterization,* 49 (2002) 95 – 101.

[23]A. P. Reynolds, "Visualization of Material Flow in Autogenous Friction Stir Welds," *Science and Technology of Welding and Joining*, 5 (2000) 120 – 124.

[24] C. Hamilton, S. Dymek, O. Senkov, "Characterization of Friction Stir Welded 7042-T6 Extrusions through Differential Scanning Calorimetry," *Science and Technology of Welding and Joining*, 17 (2012) 42 – 48.

Friction Stir Welding and Processing VII
Edited by: Rajiv Mishra, Murray W. Mahoney, Yutaka Sato, Yuri Hovanski, and Ravi Verma
TMS (The Minerals, Metals & Materials Society), 2013

Material Flow and Texture in Friction Extruded Wire

X. Li, W. Tang and A. P. Reynolds

Department of Mechanical Engineering, University of South Carolina
300 Main Street, Columbia, SC 29208, USC

Keywords: Friction Extrusion, Marker Insert Technique, Aluminum Wire, Texture

Abstract

Friction extrusion is a solid state process that can produce high quality, fully consolidated wire from metal chips, powder or billet. In order to broaden the understanding of consolidation and material flow during this process flow visualization experiments were conducted during friction extrusion of AA6061 wire with AA2195 as a marker insert. Variations in material flow with changes in die geometry, rotation rate and extrusion force and location of marker insert are presented and discussed with reference to metallographic cross-sections of resulting wire and remaining billet. The shape of marker material in the wire and hence material flow and consolidation during extrusion is highly dependent on die rotation rate and extrusion force. Grain size and crystallographic texture development and their variation during wire extrusion of AA6061 and AA2195 are also elucidated.

Introduction

In the early 1990's, the friction extrusion technique was invented at The Welding Institute (Cambridge, UK). [1] The friction extrusion die consolidates the material in the billet chamber prior to extrusion by rotating about the center axis and squeezing down with constant Z-force for deformation. The process may be an industrially useful technique for the achievement of several purposes, e.g. recycling of machining waste, consolidation of powder products, manufacturing wire feedstock for additive manufacturing processes and, potentially, as a method of producing nanograin structure in bulk materials and developing novel alloy compositions. [2]

As for other friction based processes, friction extrusion does not require melting material for consolidation. Therefore, compared with conventional recycling processes, it can bring about considerable saving of material, energy and labor when recycling metal chips to produce useful products. With the current situation of high energy costs and the desire to conserve material, the potential of this green and economical process has become increasingly relevant.

In a previous study [3-5], by use of a marker insert technique, the material flow and strain level has been deduced. It was shown that die geometry has an effect on the flow pattern but does not modify the consolidation mechanism fundamentally. [4] In this paper, similarly, AA2195 wire is employed as the marker in an AA 6061 cylinder charge. The variations of the material flow as functions of die geometry, die rotation rate, and extrusion force have been studied by using various experimental parameters. To examine the approach to steady state of the friction extrusion process, a long wire (over 210.0 *mm*) has been extruded and analyzed by inspecting both transverse cross-sections and longitudinal cross-sections. Grain size and its variation during wire extrusion of AA6061 and AA2195 is also elucidated.

Experiment Preparation

In this work, Φ25 mm X 19.2 mm AA 6061 aluminum alloy cylinder is adopted as the extrusion billet charge, and Φ2.54 mm AA 2195 wire is adopted as the marker. Through holes from top to bottom were drilled at 1/3 cylinder cross section radius (1/3r) from center of AA 6061 cylinder and AA 2195 wires were embedded into those holes.

Figure 1. AA6061 Cylinder Samples with AA 2195 markers 1/3 radius position

The friction extrusion devices are composed with the following three parts: dies, chamber, back plate and Process Development System. The dimension of extrusion die is Φ25 mm X 114.3 mm with a Φ2.5 mm through hole at the center, which defines the size of extruded wire. There are two die tip patterns utilized: one has a scrolled surface which is similar to friction stir welding tool shoulder designed for operation at a 0° angle of attack. The scroll die is rotated in a clockwise direction (From top view) so that the scroll pattern can gather plastic material toward the center hole. The other die has a flat smooth surface. This flat die is also rotated in a clockwise direction to facilitate comparison with the scrolled die. All dies were fabricated with H13 too steel. Both die tips are shown in Figure 2.

Figure 2. Friction extrusion scroll die (left) and flat die tips (right)

The inner diameter of stationary 'billet' chamber is 25.4mm, which is little bigger than un-extruded AA 6061 cylinder samples and dies. The chamber has a wider shoulder ring outside with two semicircle breaches for being fixed on the back plate. The chambers were fabricated with O1 tool steel.

A stainless steel plate is adopted as the back plate for supporting the chamber and cylinder samples. In order to preventi slipping between the back plate and cylinder sample, a short bar with 4 mm height is set on the back plate inside 'billet' chamber, this prevents the billet charge from rotating.

For parameter comparison experiments, the USC FSW PDS (University of South Carolina Friction Stir Welding Process Development System) was used to implement the extrusion process. Using this machine, the extrusion process is performed under extrusion force control. Unlike conventional rate controlled extrusion processes, this force control mode can lead to varying die advance speed and extrusion rate. Because of the limited headspace above the die in the FSW PDS, only a small part of the aluminum cylinder would be extruded out to form a wire with 150mm length.

For making longer wires and to study the approach to a steady state condition, a modified milling machine is employed. The milling machine spindle is used to produce die rotation while a hydraulic cylinder mounted on the knee is used to drive the billet chamber charge into the spindle. The milling machine has a hollow drawbar enabling the extrusion of longer wires than is possible on the FSW PDS. Wire lengths of up to 10 feet have been produced.

Friction Wire Extrusion Process

The first step in setting up for extrusion is to shift the die right above the chamber containing the sample and make sure the Z-axis of the machine is coincident with the extrusion axis: hence the Z-force is equivalent to the extrusion force. Move the die down until it contacts with the sample, and then start to spin the die. The heat generated by the friction between the die and aluminum charge increases the temperature immediately. As a result, the aluminum alloy close to the die tip is turned to a plastic state and will flow. With the downward pressure and heat input, the aluminum cylinder expands towards the chamber wall to fill in the small gap (~0.4 mm) between the sample and chamber. When sufficient temperature is reached the wire starts to extrude out. In this friction extrusion processes, the rotation speed, die position and Z-force (extrusion force) were recorded by the FSW system control computer with 10Hz sampling rate. The rotating torque and power were recorded by a torque transducer mounted on the machine spindle (which drives the die rotation): these data were recorded with at least 40Hz data collection rate.

For comparison of parameters, extrusion runs were performed using 3 different rotating speeds combined with 3 different Z forces. Both the scroll die and the flat die were used. See table 1.

Table 1 Adopted friction extrusion parameters (SD: Scroll die; FD: Flat die)

Z-force(*N*) \ Rpm	200	300	400
22250	FD	SD, FD	FD
44500	SD, FD	SD, FD	SD+FD
66750	FD	SD, FD	FD

For long wire extrusion, the constant die rotation rate is 300 *rpm*.

Post-extrusion

Since the geometry of the FSW PDS spindle limits the length of aluminum wire that can be extruded, only the upper part of the cylinder has been extruded out. Meanwhile, pressed and stirred by the spinning die, some aluminum was squeezed up through the gap between the die and chamber. After the friction extrusion, the remnant cylinder became a deformed disk in the chamber. By analyzing the deformation of the disk and tracing the distribution of the marker material, AA 2195, the flow pattern of the material can be determined.

The wire extruded by the FSW PDS was about 150mm long, and it was cut into 6 segments of approximately 25 mm each. Each aluminum wire segment was mounted in epoxy so that transverse cross sections could be observed. The transverse cross-section of every segment was ground, polished and then etched with Keller's Reagent (190ml water, 2ml HF, 3ml HCl, and 5ml HNO_3). The distribution of AA2195 marker in the AA6061 wire cross sections was examined by optical microscopy. The remnant disk was carefully removed from the chamber and necessary machining was applied on top side to form a flat and smooth surface. After that, it was ground and polished. Keller's Reagent was applied on the machined surface. With the etchant chemical attack, AA2195 turned dark while AA6061 remained the same color. Pictures were taken by an optical microscope and by a flat bed scanner. Furthermore, multiple grinding, polishing, etching and picturing procedures were applied to expose successive layers of the disk with each layer being 0.25mm thick. Hence, deformation and flow pattern at different heights were obtained (a form of mechanical tomography). By assembly of all pictures in sequence of height, a three-dimensional image of the marker positions was established. Therefore, the material deformation/flow before entering into the extrusion die hole can be observed.

The extruded wires from the modified milling machine were 2.21*m* and 2.13*m* by scroll die and flat die respectively. The first extruded 7.6*cm* wire was cut off, then, the rest was cut into 7 segments with about 30.5*cm* lengths. The same examination procedures were employed for analyzing FSW machine wire segments. Besides, the grain size of recrystallized AA2195 markers in the extruded wire was measured using the mean linear intercept method.

Results and discussion

Good quality wire was obtained from all of the extrusion parameter sets except 400 rpm and 22250N Z-force. Transverse cross sections (12.7mm to the beginning of wire) of all 1/3r marker wires extruded by flat die are shown in figure 3. All AA2195 showed coaxial multi-spirals pattern. The cross sections at same position of scroll die wire are shown in figure 4. With different rpm and Z-axis force, the number of spiral revolutions and inner radius (the distance from center of cross-section to the inside end of spiral line) are different. Inspection of several transverse cross sections of a single wire indicates that these two values are also varying along the longitudinal direction of the wire. See figure 5.

The effects of process control parameters and length of wire extruded on the number of marker material spiral revolutions and inner radius of the marker are shown in figure 6 and 7, respectively.

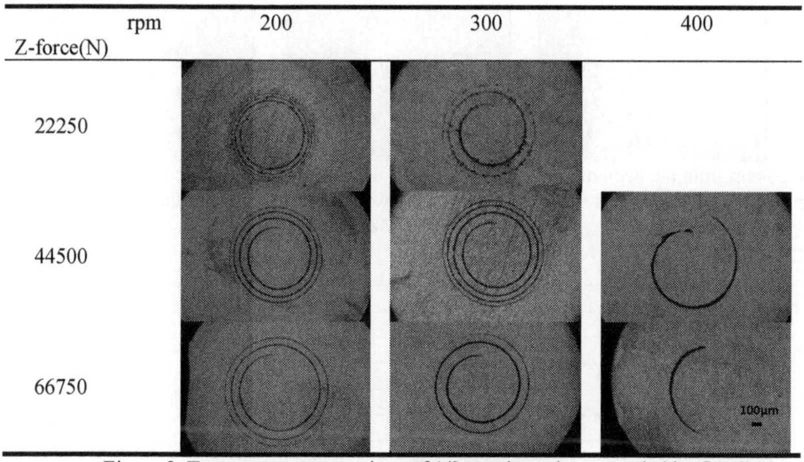

Figure 3. Transverse cross sections of 1/3r marker wires extruded by flat die, 12.7*cm* from the beginning of wire

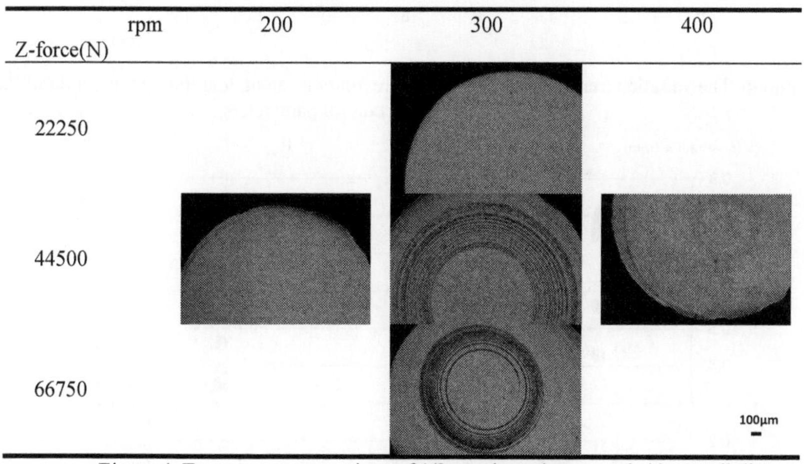

Figure 4. Transvers cross sections of 1/3r marker wires extruded by scroll die, 12.7*cm* from the beginning of wire

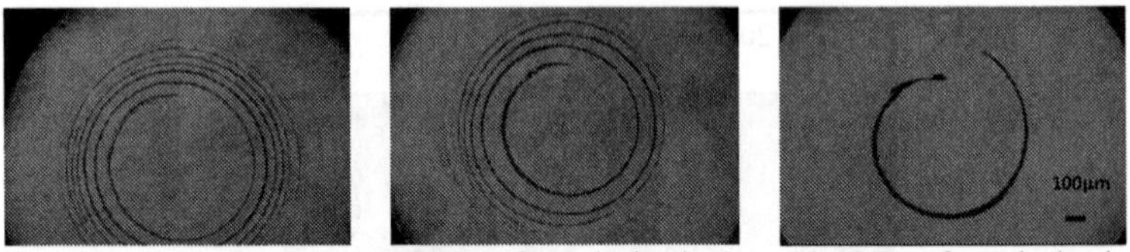

(a) 2.54cm from the beginning (b) 7.62cm from the beginning (c) 12.7cm from the beginning

Figure 5. Spiral changes along the length of the wire extruded using the flat die with 400 *rpm* and 44500*N* Z-force

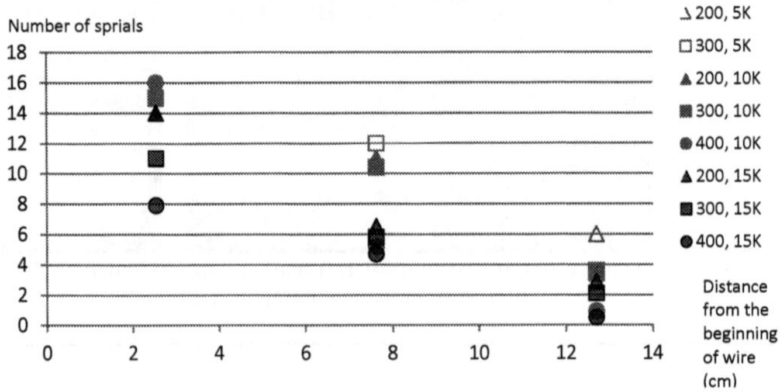

Figure 6. The variation trend of number of spiral revolutions along longitudinal direction of flat die wire with different control parameters

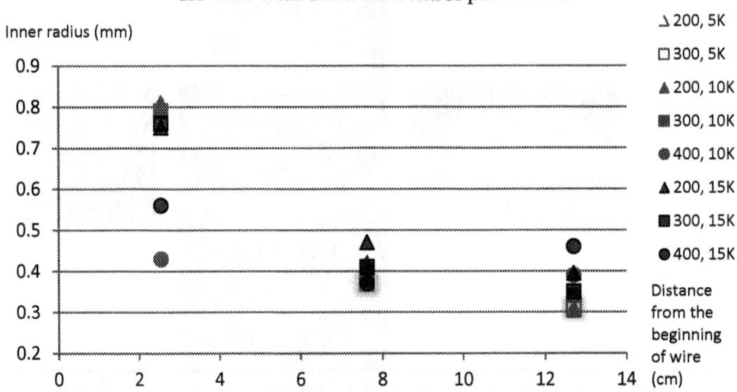

Figure 7. The variation trend of inner radius along longitudinal direction of flat die wire with different control parameters

Several binary pictures of different layers in the remnant disc of a flat die extrusion are presented in figure 8. The outside circle is the edge of the remnant disc; the white shapes inside the disc are 2195 marker. The initial stages of deformation process can be deduced from these images.

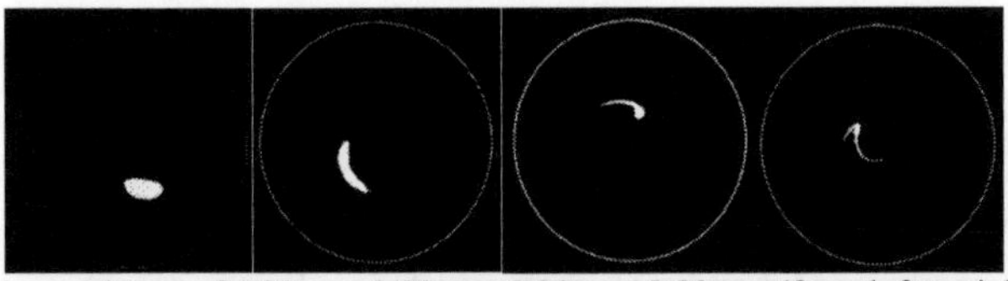

Figure 8. Binary picture of 4.30mm, 3.80mm, 3.30mm, 2.80mm (from left to right) depth of remnant disc

The cross sections of long wire made using the milling machine with the flat die are shown in figure 9. The variation trend of AA 2195 spiral along wire is in accord with flat die wire from FSW PDS machine. Obviously, even such long wire hasn't reached steady state yet.

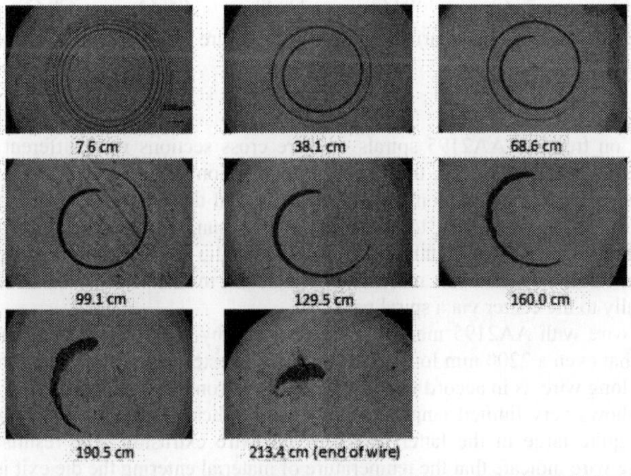

Figure 9. Cross sections of flat die wire made by Supermill machine from 7.6cm to 213.4cm to the beginning of wire (cm)

The grain sizes in the 2195 markers on the cross sections shown in figure 9 were measured. See figure 10. As in FSW, it has been observed that larger grain size in friction extrusion results from higher extrusion temperature. [6]

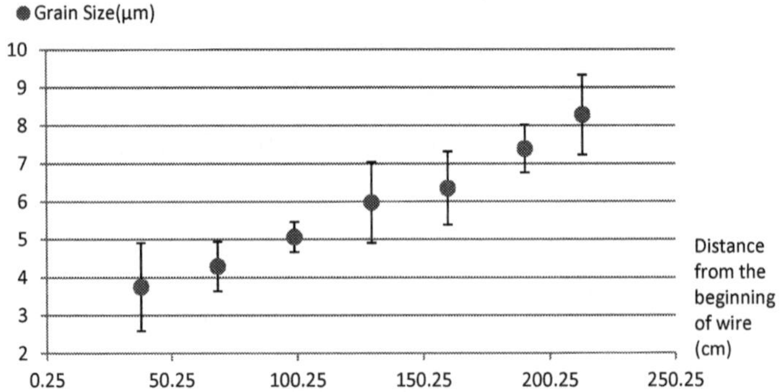

Figure 10. The grain size variation along flat die wire made by Supermill machine

Conclusions

Based on trend of AA2195 spirals on wire cross sections with different positions and series of parameter (figure 3 to 7), the following conclusion can be draw: (1) since rpm and Z-force are constant during the friction extrusion process but the flow pattern is varying along the wire, the ratio of extrusion rate to rotation rate is changing (this parameter is somewhat analogous to the advance per revolution in FSW); (2) flat die and scroll die generate qualitatively similar but quantitatively different material flow; (3) the material at 1/3 radius of the cylinder moves gradually to the center via a spiral path.

Long wire with AA2195 marker has been made by modified milling machine. Results demonstrate that even a 2200 mm long wire has not reach steady state yet. The trend of AA2195 pattern in the long wire is in accord with the short wires from FSW PDS machine. The pattern at end of wire shows very limited tangential movement indicating that the extrusion rate per die revolution is quite large in the latter stages of the wire extrusion. The results of grain size variation along wire indicate that the temperature of material entering the die exit is increasing as the friction extrusion process progresses. This conclusion can also explain the variation trend of AA2195 pattern in different position of wire. When temperature increases, the flow stress of the metal decreases, which leads to better mobility and higher extrusion rate for a given rotation rate and die pressure. For more detailed analysis, orientation imaging microscopy will be adopted to inspect the texture distribution along the wire in future.

Acknowledgements

This work was supported by NASA-EPSCoR grant #520879-USC.

References

1. W.M. Thomas et al., International Patent Application No.PCT/GB92/02203 and GB Patent Application No. 9125978.9, 1991.
2. Gronostajski, J., Matuszak, A., ''The recycling of metals by plastic deformation: an example of recycling of aluminum and its alloy chips'', *Journal of Materials Processing Technology*

(1999) 92–93, 35–41.
3. T.U. Seidel and A.P. Reynolds, ''Visualization of the material flow in AA2195 friction-stir welds using a marker insert technique'', *Metallurgical and materials transactions a-physical metallurgy and materials science* (2001) 2879-2884.
4. X. Li, W. Tang and A.P. Reynolds, "Visualization of Material Flow in Friction Extrusion Processing", *ICAA13: 13th International Conference on Aluminum Alloys.*
5. W. Tang, A.P. Reynolds, ''Production of wire via friction extrusion of aluminum alloy machining chips'', *Journal of Materials Processing Technology* 210 (2010) 2231–2237.
6. Yutaka S. Sato, Mitsunori Urata and Hiroyuki Kokawa, ''Parameters controlling microstructure and hardness during friction-stir welding of precipitation-hardenable aluminum alloy 6063'', *Metallurgical and Materials Transactions A Volume 33, Number 3 (2002), 625-635.*

Friction Stir Welding and Proceesing VII
Edited by: Rajiv Mishra, Murray W. Mahoney, Yutaka Sato, Yuri Hovanski, and Ravi Verma
TMS (The Minerals, Metals & Materials Society), 2013

APPLICATION OF ACOUSTIC EMISSION TECHNIQUE TO MONITOR FRICTION STIR WELDING PROCESS TO PRODUCE DEFECT FREE WELDS

B M Rajaprakash[1], C N Suresha[2], Sarala Upadhya[1]

[1] Department of Mechanical Engineering, University Visvesvaraya College of Engineering, Bangalore University, Bangalore, Karnataka, 560 001,India.
[2] Department of Mechanical Engineering, Jyothy Institute of Technology, Bangalore, Karnataka, 560 062, India.

Keywords: Friction Stir Welding, Acoustic Emission, Online monitoring

Abstract

Friction Stir Welding (FSW) is a solid - state welding technique used for joining of metals and alloys to avoid problems associated with fusion welding. Acoustic Emission (AE) technique had been successfully used to monitor processes like metal cutting, grinding and electron beam welding. An attempt has been made to study the feasibility of application of AE technique to monitor FSW process. Experiments were carried out using aluminium alloy AA 7020-T6 by continuously varying the FSW input process parameters namely tool rotational speed and weld traverse speed during welding. Variation in pattern of AE signals was observed when FSW input parameters were varied. The pattern were found helpful in identifying the defect occurred during welding, as well in developing a model for online monitoring of FSW process to produce quality welds. The developed model has been successfully used to monitor FSW process during welding of aluminum alloy AA6082-T4.

Introduction

FSW was invented at The Welding Institute (TWI) United Kingdom in 1991 as a solid-state joining technique [1]. The basic concept of FSW is remarkably simple. A non-consumable rotating tool with a specially designed pin and shoulder is inserted into the abutting edges of sheets or plates to be joined and subsequently traversed along the joint line as shown in Figure 1 along with FSW terminology.

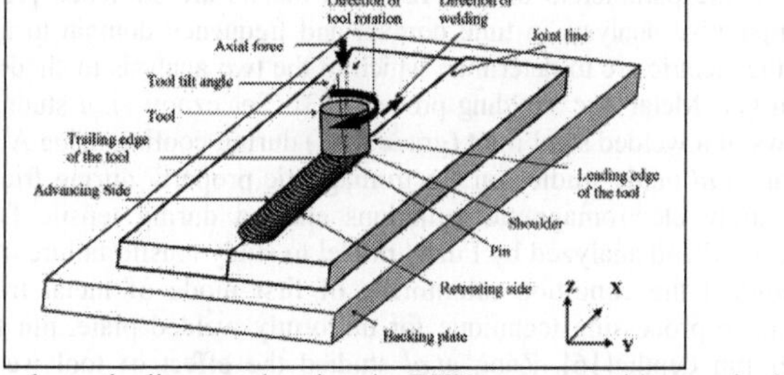

Figure1. A schematic diagram showing the FSW process and its terminology

The advancing side is the side where the velocity vectors of tool rotation and traverse direction are similar and the side where the velocity vectors are opposite is referred as retreating side [2]. The FSW process makes use of: (i) The adiabatic heat generated by localized deformation, and (ii) The friction heat generated between the shoulder of the rotating tool and the surface of the material, to plasticize the material immediately beneath the shoulder. A gradual traverse of the rotating tool along the weld line progressively transfers the material from the leading edge to the trailing edge and this material flow is considered to be a chaotic process. The temperature developed due to friction is found to be about 80% of the melting point of the work piece metal during welding process. The defects associated with solidification of fusion welds, such as porosity, slag inclusion, dendrite structure, etc., are not likely to form in the weld metal produced by the FSW process [3]. Authors in their initial studies made investigation on the role of tool pin profile on tensile strength of welded joints by considering the important FSW parameters like (i) Tool rotational speed, (ii) Weld traverse speed and (iii) Plunge depth, which control the quality of the weld [4].

Acoustic Emission is the class of phenomena whereby transient elastic waves are generated by the rapid release of energy from a localized source or source within a material, or the transient elastic wave generated when a material undergoes deformation [5]. AE testing is different from other non- destructive testing techniques in two regards: (i) It pertains to the origin of the signal. Instead of supplying energy to the object under examination, AE technique simply listens for the energy released by the object. AE tests are often performed on structures while in operation, as this provides adequate loading for propagating defects and triggering AE and (ii) AE testing deals with dynamic processes or changes in a material. This is particularly meaningful because only active features (e.g. crack growth) are highlighted [6].

An advantage of using AE technique as a process monitoring tool is that the frequency range of AE is much higher than that of machine vibrations and ambient acoustic noise. Sources generating AE in different materials are unique due to properties of materials [7]. Lee *et.al* studied the tensile behavior of low Carbon Steel welds by means of AE technique to observe the AE characteristics at base metal, heat affected zone and weld metal of welded joint [8]. Chen *et.al* used A E sensing for in-process tool wear monitoring to correlate the machining process and AE signals [9]. Sokolkin *et.al* studied the use of AE in testing bottoms of welded vertical tanks used for storing oil and oil derivatives to assess the conditions of bottoms, based on the corrosive activity resulting AE [10]. Sharma *et.al* used AE for online monitoring of Electron Beam Welding of TI6AL4V alloy to detect the formation of defects in a material accompanied by dissipation of energy in the form of acoustic waves [11]. V. V. Korchevskii made studies on measurement of the parameters of the AE when metals are stretched [12]. Huanca Cayo *et.al* made the comparative analysis in time domain and frequency domain to the acoustical pressure generated by the electric arc to determine which of the two analysis methods are better to evaluate the stability in Gas Metal Arc Welding process [13]. Ser'eznov *et.al* studied the possibilities of localizing flaws of a welded steel joint (grade CT3) during cooling using AE [14].

Muthukumaran *et.al* made studies on electromagnetic property during friction stir weld failure. During their study electromagnetic radiations emitted during tensile failure of welds were measured, recorded and analyzed by Fuzzy model to study tensile failure of welds [15]. Preetish Sinha *et.al* studied the condition monitoring of first mode of metal transfer in friction stir welding by image processing technique for uniformly welded plate, pin failure while welding and for lesser pin depth [16]. Zang *et.al* studied the effect of tool wear on microstructure, mechanical properties by AE technique during Friction Stir Welding of aluminum alloy AA6061 [17].

It has been observed that AE technique has been successfully adopted in many fields, namely laser welding, health monitoring of aerospace structures, study of tool wear during drilling, turning, milling, etc. Zang *et.al* have used AE technique as an online technique to monitor tool wear in FSW process [17]. However, it was observed that, there is no published information about the use of AE to detect the defects online during FSW process. In this study an attempt has been made to study the feasibility of application of AE technique to identify the occurrence of defects to monitor FSW process, followed by developing a model for online monitoring of FSW process to produce defect free welds.

Experimentation

Experimental details comprises of three stages: (i) Conduction of experiments and sensing of AE signals during welding (ii) Study the behaviour of AE parameters derived from AE signals when FSW parameters are varied to check the applicability of AE technique during FSW and (iii) Development of a model based on threshold values of AE parameters to distinguish between good and bad welds.

The base material used in this study is aluminum alloy AA7020-T6 which is a medium strength precipitation hardenable Al–Zn–Mg alloy, used in aerospace industry for structural applications. These alloys (7000 series) are not recommended for conventional fusion welding. But these alloys can be successfully welded by FSW process. The nominal chemical composition and mechanical properties of the alloy is presented in Table I and Table II respectively. Plates of 5 mm thick were cut to a dimension of 300 mm×75 mm with square edges. Plates were fixed firmly on the FSW machine using suitable clamps. Welding was carried out using conical threaded tool made out of oil hardened steel. The important process parameters considered during the study are tool rotational speed, weld traverse speed and plunge depth. Welds were prepared by varying tool rotational speed and weld traverse speed individually by keeping other parameters constant during welding. This has been done to observe the pattern of AE signals when an input process parameter is varied during welding. Based on experiments conducted earlier by the authors the broad range of process parameters selected to weld is shown in Table III [4].

Table I. Nominal composition (wt %) of Aluminum alloy AA 7020-T6

Zn	Mg	Cu	Cr	Mn	Ti	Si	Al
4.0-5.0	1.0-1.4	0.20	0.1-0.35	Max 0.40	0.08-0.20	Max 0.35	92.10- 93.80

Table II. Mechanical properties of Aluminum alloy AA 7020- T6

Ultimate Tensile strength (N/mm^2)	Yield Strength (N/mm^2)	Elongation %
375	370	5-8

Table III. Process parameters used during welding of AA 7020-T6

Weld No.	Tool Rotational Speed (rpm)	Weld Traverse Speed (mm/min)	Plunge depth (mm)
1.	700-1200	50	4.75
2.	1000	50-100	4.8

AE sensor was mounted on the work piece by using a suitable couplant followed by connecting the sensor fixed to computer through preamplifier and band pass filter. The position of sensor was confirmed by Hsu-Nielsen method as per ASTM E 976-05. Acoustic Emission data was acquired for the selected FSW input process parameters and stored in the computer using AE data acquisition system setup (PCI-2 " 2 channel AE system on card" make: MISTRAS ,Physical Acoustic Corporation, USA) as shown in Figure 2 and processed to determine various AE parameters using AEwin software. The important AE parameters acquired during the studies are RMS, Energy, Amplitude and Counts. Welded joints were subjected to X- ray radiography test to know the position of defects produced during welding. Tensile specimens were prepared as per ASTM - E8 / E8M-08 using Electronic Tensometer (model: PC 2000, make: Kudale Instruments, Pune) by choosing the portion of the welded joint where there were no defects revealed from X –ray radiography image.

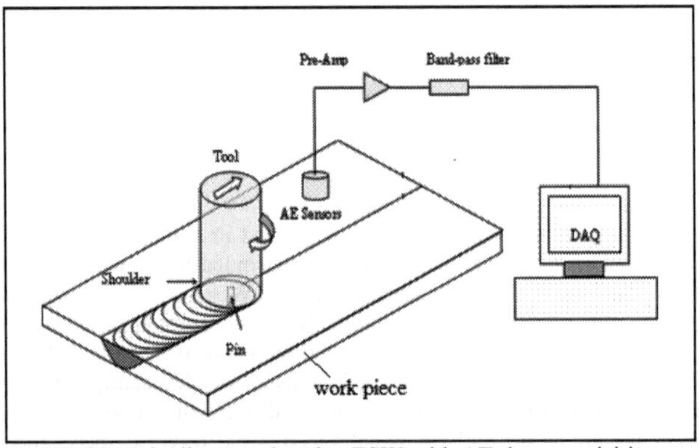

Figure 2. Schematic diagram showing FSW with AE data acquisition system (Courtesy: MISTRAS, Physical Acoustic Corporation, USA)

Results and discussion

The photographs of FSW joints prepared with varying tool rotational speed and traverse speed are shown in Figures 3 and 4 respectively.

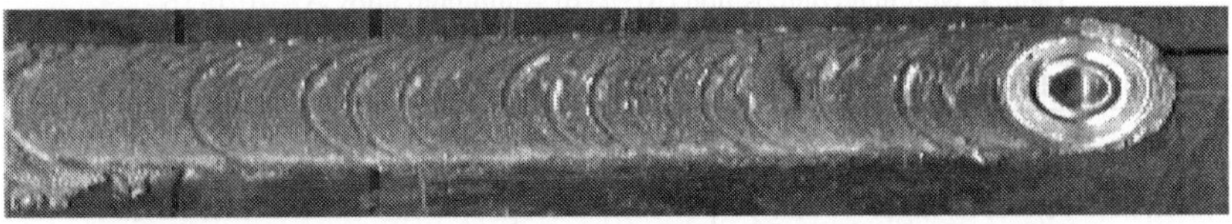

Figure 3.Photograph of FSW joint with varying tool rotational speed (Weld 1)

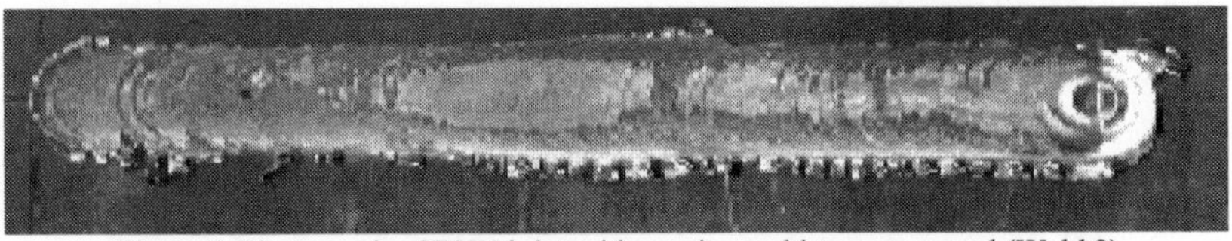

Figure 4. Photograph of FSW joint with varying weld traverse speed (Weld 2)

The screen shots of AE signal pattern of AE parameters acquired during welding for varying tool rotational speed and weld traverse speed are shown in Figures 5 and 6 respectively.

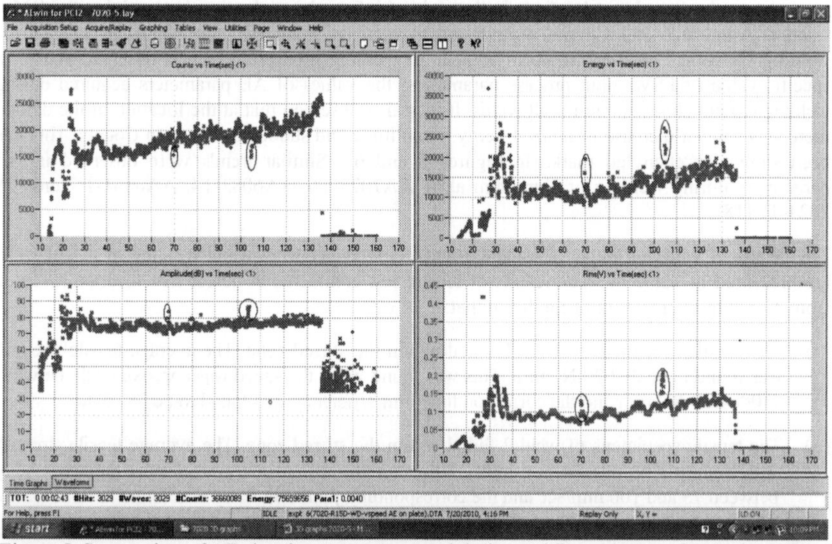

Figure 5. Screen shot of AE signal patterns generated during welding using varying tool rotational speed (weld 1)

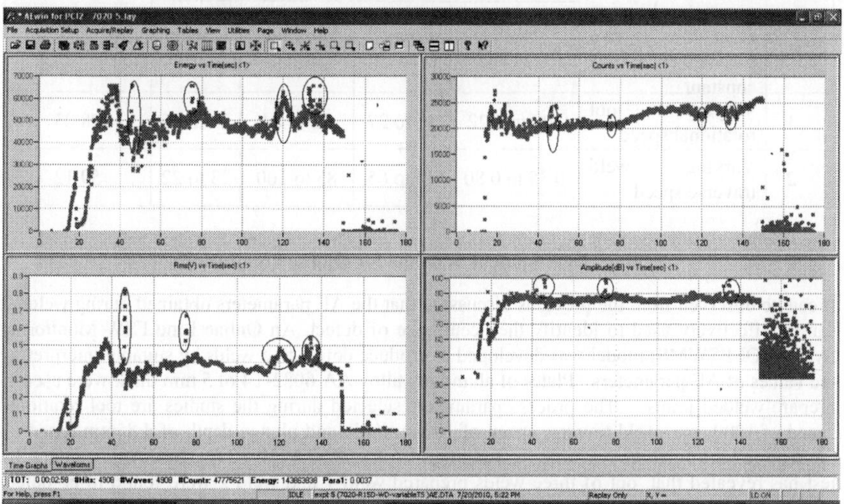

Figure 6. Screen shot of AE signal pattern generated during welding using varying weld traverse speed (weld 2)

In this study the time taken for complete plunge and dwell is about 40 sec. After plunging the tool traverses with the set speed resulting in welding till the preset length of weld is obtained. X-ray radiography test showed defects in the weld at a few places. It was observed that the RMS, Energy, Amplitude and Counts were found to increase marginally during 40 to 135 sec of welding for weld 1 and during 40 to 140 sec of welding for weld 2. It was observed that for a specific value of FSW input process parameters; the values of AE parameters acquired during welding remain almost same for all welds. It was also observed that at the location of the defect, there was sudden increase in RMS, Energy, Amplitude and sudden decrease in Counts. This has been shown as encircled marks in Figures 5 and 6. Similar trends were observed in AE parameters during welding of aluminum alloy AA7075-T6, AA6082-T4, AA6061-T6 and AA 2024-T3 [18].

The details of variation in AE parameters when defects occurred while welding of AA7020-T6 are presented in Table IV. The results of tensile test conducted at defect free location revealed from X- ray radiography image indicate that:

(i) The tensile specimens of weld 1 failed within the stirred zone. The average tensile strength was found to be 292.70 N /mm^2. This was found when the speed was approximately between 1100 and 1200 rpm and the corresponding joint efficiency was found to be 78.05 %.

(ii) The tensile specimens of weld 2 failed within the stirred zone. The average tensile strength was found to be 221.70 N / mm^2. This strength was at a traverse speed approximately between 75 and 100 mm/sec and the corresponding joint efficiency was found to be 59.12 %. The reduction in tensile strength may be due continuous increase in weld traverse speed.

Table IV. Values of AE parameters at location of defect / defects in weld joints for AA 7020-T6

Weld No.	Variation of input parameter keeping other two parameters constant	Sudden change in value of AE parameter				Joint Efficiency (%)
		RMS (v)	Energy (v) x10^3	Amplitude (dB)	Counts x 10^3	
1	Varying tool rotational speed	0.09 to 0.22	16 to 27	72 to 84	20 to 15	78.05
2	Varying weld traverse speed	0.35 to 0.80	50 to 65	85 to 100	23 to 22	59.12

Development of model for Online Monitoring

It can be observed from the previous discussion that the AE parameters obtained during welding can be effectively used to identify the occurrence of defect. An Online Line FSW Monitoring System (OLFSWMS) model was developed to produce defect free welds by suitably interpreting the values of AE parameters. Plates of aluminum alloy AA 6082-T4 of 5 mm thick were used to prepare welded joints. The process parameters selected during the studies are tool rotational speed of 1000 rpm, weld traverse speed of 100 mm/min and plunge depth of 4.85mm based on the reported information available for welding of AA 6082-T4 [19 - 21]. The X ray radiography test has revealed that, out of three welds prepared weld 1 and weld 2 were found to be defect free. It has been observed for weld 3 between 95 sec and 120 sec that there was a sudden increase in RMS, Energy, Amplitude and sudden decrease in Counts resulting in defect. This has been shown as encircled marks in Figure 7. The details of sudden increase in values of RMS, Energy and Amplitude and sudden decrease in Counts during preparation of weld of AA 6082-

T4 (weld 3) are shown in Table V. During online monitoring of FSW process, to produce defect free weld, after noticing the defect, welding was stopped and tool was taken back by 20mm and re- welding was done in the region where the defect had occurred earlier. The screen shot of AE signal pattern acquired during re-welding is shown in Figure 8, wherein there was no sudden variation in RMS, Energy, Amplitude and Counts resulting defect free weld. Tensile specimens were cut in welded region and also in re-welded region. The results of tensile test conducted at defect free location in welded region and re - welded region are based on X- ray radiography test. The efficiency of welded joint at welded region was found to be 79.20 %, while at re - welded region it was 101.90 %. This variation may be attributed to formation of fine grains in the re-welded region.

The pattern of AE signals generated was used to determine threshold values of AE parameters for distinguishing good and bad welds and the same has been presented in Table VI. Based on the threshold values of AE parameters along with action to be taken, OLFSWMS model to distinguish between good and bad welds is shown in Figure 9.

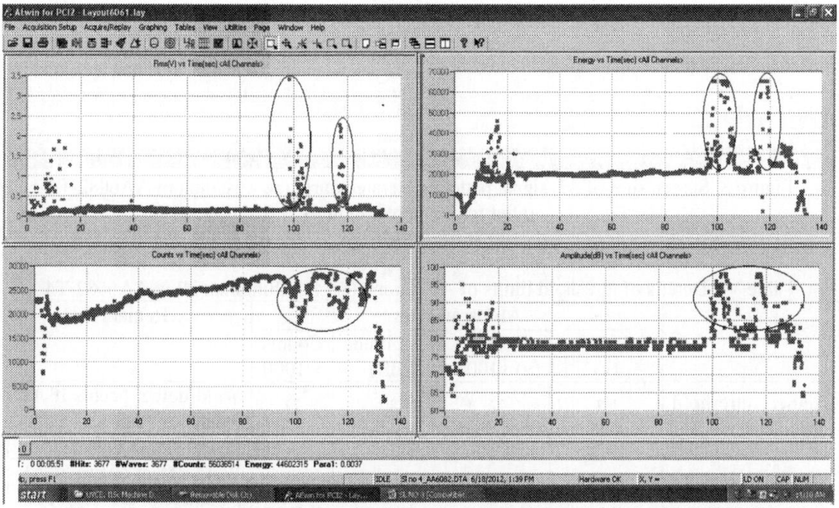

Figure 7. Screen shot of AE signal pattern generated during welding of AA6082-T4 (before retracing /re-welding)

Table V. AE parameters at location of defect / defects in weld joints along with joint efficiency for aluminum alloy AA6082-T4

Weld No.	Sudden change in value of AE parameter				Joint Efficiency (%)	Remarks
	RMS (v)	Energy (v) x1000	Amplitude (dB)	Counts x 1000		
3	0.3 to 3.5	22 to 68	80 to 100	28 to 18	79.20	Before re - welding

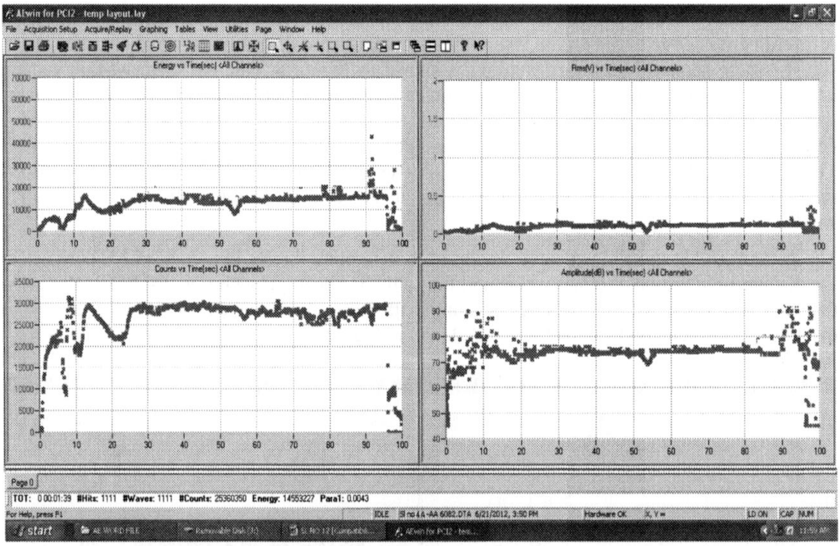

Figure 8. Screen shot of AE signal pattern generated during re - welding of AA6082-T4 (after retracing/re-welding)

Table VI. Details of threshold limits of AE parameters for aluminum alloy AA6082-T4

	AE parameters				Remarks
	RMS (v)	Energy (v) x1000	Amplitude (dB)	Counts x 1000	
Higher threshold limit	0.3	22	80	27	Weld defect occurs if AE parameters are more than higher threshold limit
Lower threshold limit	0.1	12	75	18	Weld defect occurs if AE parameters are less than lower threshold limit

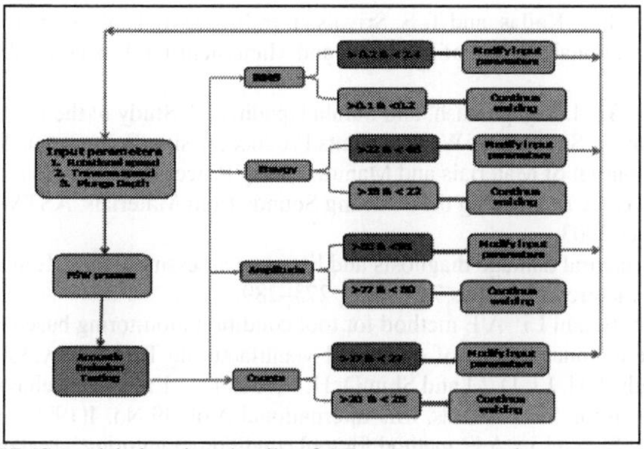

Figure 9. OLFSWMS model showing details of AE parameters and necessary action to be taken during welding of AA6082-T4

Conclusions

It can be concluded from the above studies that:

- The values of AE parameters varied when FSW input parameters were varied, indicating the existence of relationship between them.
- The quality of welded joints can be predicted by studying the pattern of AE signals generated during welding.
- AE technique can be effectively used for online monitoring of FSW process by suitably setting the threshold values of AE parameters to distinguish between good and bad welds.
- The model developed is helpful in identifying the occurrence of good and bad welds along with the suggested action to be taken to produce defect free weld.
- At location of defect in the weld, the process of retracing or re- welding was found useful in producing defect free weld with increased strength of welded joint.
- The outcome of results can be extended to other materials of different thickness during online monitoring of FSW process using AE technique after making repeatability and reliability studies.

Acknowledgement

The authors wish to express sincere thanks to Aeronautical Research & Development Board (ARDB), New Delhi, for financial support rendered through R & D project No: DARO/08/2031473/M/I. Authors are thankful to Prof. Satish V Kailas, Department of Mechanical Engineering, Indian Institute of Science (IISc), Bangalore for extending facilities of FSW setup to carry out this investigation.

References

[1] Rajiv S. Mishra, Murray W. Mahoney-editors. Friction Stir Welding and Processing, ASM International, (2007)1-5.
[2] K. Kumar, Satish V. Kailas. The role of friction stir welding tool on material flow and weld formation, International Journal Materials Science and Engineering A 485 (2008),367–374.

[3] K. Kumar, Satish V. Kailas, and T. S. Srivatsan. Influence of Tool Geometry in Friction Stir Welding, International Journal of Materials and Manufacturing Processes, 23(2008), 188–194.
[4] C. N. Suresha, B. M. Rajaprakash, and Sarala Upadhya. A Study of the Effect of Tool Pin Profiles on Tensile Strength of Welded Joints Produced Using Friction Stir Welding Process, International Journal of Materials and Manufacturing Processes, 26 (2011), 1111–1116.
[5] Mark F. Carlos. A E: Heeding the Warning Sounds from Materials, ASTM Standardization News, October 2003.
[6] Carpinteri. Structural damage diagnosis and life-time assessment by A E monitoring-Engineering Fracture Mechanics 74 (2007), 273–289.
[7] Xiaozhi Chen. Beizhi Li. A E method for tool condition monitoring based on wavelet analysis: International Journal of Advanced Manufacturing Technology,33(2007), 968–976.
[8] Lee .C. S. Huh. J .H. Ll .D. M and Shin D. H. Acoustic Emission Behavior during Tensile Tests of Low Carbon Steel Welds. ISIJ International, Vol. 39 No. 4(1999), 365-370.
[9] Xiaozhi Chen. Beizhi Li. A E method for tool condition monitoring based on wavelet analysis: Int J Adv Manuf Technol, 33(2007) 968–76.
[10] V. Sokolkin, I. Yu. Ievlev, and S. O. Cholakh. Use of AE in Testing Bottoms of Welded Vertical Tanks for Oil and Oil Derivatives, Russian Journal of Nondestructive Testing, Vol. 38, No. 12(2002) 902–908. Translated from Defektoskopiya, No. 12 (2002), 43–51.
[11] Sharma, M.I. Junaidh, K.K. Purushothaman, C.P. Kotwal, J. Paul, Shalini Tripathi, B. Pant and A.S. Sankaranarayanan. Online Monitoring of Electron Beam Welding of TI6AL4V Alloy through A E. Proc. National Seminar on Non-Destructive Evaluation, Dec.7 - 9, 2006.
[12] V. V. Korchevskii. Measurement of the parameters of the A E when metals are stretched, Measurement Techniques, Vol. 49, No. 5(2006), 517-23.
[13] Huanca Cayo, S.C. Absi Alfaro. GMAW process stability evaluation through acoustic emission by time and frequency domain analysis, Journal of Achievements in Materials and Manufacturing Engineering, (2009), Volume 34, Issue 2 : 157-64.
[14] N. Ser'eznov, L. N. Stepanova, E. Yu. Lebedev, S. I. Kabanov, V. N. Chaplygin,S. A. Laznenko, K. V. Kanifadin, and I. S. Ramazanov. Acoustic-Emission Study of the Possibilities of Localizing Flaws of a Welded Joint during Cooling, SSN 1061-8309, Russian Journal of Nondestructive Testing, Vol. 45, No. 5 (2009), 310–316.
[15] Muthukumaran,Kumar Pallav, Vikas Kumar Pandey and Mukherjee. A study on electromagnetic property during friction stir weld failure. Int J Adv Manuf Technol (2008), 36:249-53.
[16] Preetish sihna, Muthukumaran,Sivakumar and Mukherjee. Condition monitoring of first mode of metal transfer in friction stir welding by image processing. Int J Adv Manuf Technol36 (2008),484-489.
[17] W.M. Zeng, Hl W and J zang. Effect of tool wear on microstructure, mechanical properties and A E of friction stir welded 6061 Al alloy. Acta Metallurgica, vol.19, 1 (2006), 9-19.
[18] C N Suresha. Studies on The Performance of the Welded Joints produced by Friction Stir Welding process using A E Technique. (Ph.D. thesis, Bangalore University, Bangalore, 2012),337.
[19] Lars-Erik Svensson and L. Karlsson. First International Symposium on Friction Stir Welding (Session 5), Thousand Oaks, California, USA, 14-16 June 1999.
[20] M. Ericsson, R. Sandstrom and J. Hagstrom. Second International Symposium on Friction Stir Welding (session 8), Gothenburg, Sweden, 26-28 June 2000.
[21] H. Larsson, L. Karlsson, S. Stoltz and E-L Bergqvist. Second International Symposium on Friction Stir Welding (session 4), Gothenburg, Sweden, 26-28 June 2000.

AUTHOR INDEX
Friction Stir Welding and Processing VII

SUBJECT INDEX
Friction Stir Welding and Processing VII

MIX
Papier aus verantwortungsvollen Quellen
Paper from responsible sources
FSC® C105338

If you have any concerns about our products,
you can contact us on
ProductSafety@springernature.com

In case Publisher is established outside the EU,
the EU authorized representative is:
Springer Nature Customer Service Center GmbH
Europaplatz 3, 69115 Heidelberg, Germany

Printed by Libri Plureos GmbH
in Hamburg, Germany